This is the third edition of a successful and well-established text. Thoroughly revised and updated, the book provides a comprehensive introduction to the fundamentals of optics, and to a wide variety of more advanced areas of modern optical science. Several new sections have been added, including discussions of super-resolved imaging, phase-retrieval in optical and X-ray diffraction, phase-conjugate imaging and squeezed-light interferometry.

Throughout, the subject matter is developed by a combination of un-sophisticated mathematics and physical intuition, with particular emphasis being placed on Fourier analysis. This approach is used to describe a range of topics in wave propagation, from Gaussian geometrical optics, through diffraction and imaging systems, to dielectric filters, optical fibres, lasers and quantum optics.

The very broad range of subjects treated, together with the inclusion of many problems and over 300 diagrams and photographs, will make the book of great use to undergraduate and graduate students of physics, and to anyone working in the field of optical science.

Optical Physics

Optical Physics

Third Edition

S. G. Lipson
Technion–Israel Institute of Technology

H. Lipson
University of Manchester Institute of Science and Technology

and

D. S. Tannhauser
Technion–Israel Institute of Technology

CAMBRIDGE
UNIVERSITY PRESS

Published by the Press Syndicate of the University of Cambridge
The Pitt Building, Trumpington Street, Cambridge CB2 1RP
40 West 20th Street, New York, NY 10011-4211, USA
10 Stamford Road, Oakleigh, Melbourne 3166, Australia

First published 1969
Second edition 1981
Third edition 1995

Printed in Great Britain at the University Press, Cambridge

A catalogue record of this book is available from the British Library

Library of Congress cataloguing in publication data

Lipson, S. G. (Stephen G.)
Optical physics / S.G. Lipson, H. Lipson, and D.S. Tannhauser. – 3rd ed.
    p. cm.
Includes bibliographical references and index.
ISBN 0-521-43047-X. – ISBN 0-521-43631-1 (pbk.)
1. Physical optics. I. Lipson, H. (Henry), 1910– .
II. Tannhauser, D. S. (David Stefan), 1927– . III. Title.
QC395.2.L56 1995
535–dc20   94-13138 CIP

ISBN 0 521 43047 X hardback
ISBN 0 521 43631 1 paperback

TAG

Henry S. Lipson, F.R.S., the senior author of this book, died suddenly on 26 April 1991, shortly after the third edition of this book had been suggested by the publishers. Henry's great interest in physical optics, and Fraunhofer diffraction in particular, was born from his research into X-ray crystallography, of which he was one of the pioneers.

He received his B.Sc. in physics from Liverpool University in 1930 and continued his studies with research into crystal structure determination, which led him into close friendship with W. L. Bragg who strongly influenced the rest of his career. Together with A. Beevers, he solved the first complex crystal structure, $CuSO_45H_2O$, which has only minimal symmetry, using Fourier synthesis and much inspired

Henry Lipson, F.R.S., 1910–1991.

guesswork about the phases. This work led to their development of Fourier strips, the computational method which was subsequently used in crystal structure work for about 30 years until electronic computers made it obsolete.

In 1945 he was appointed head of the Physics department at the Manchester College of Technology, later to become Professor of Physics when the College became the University of Manchester Institute of Science and Technology (UMIST). He held this position until he retired in 1977.

The analogy between X-ray and optical diffraction led Henry to devote his efforts from 1948 into devising methods of using optical diffraction to simulate Fourier synthesis, and this way to determine crystal structures. A particular problem which was amenable to this approach was that of superlattices in alloys. It was such work which was, in 1966, the genesis of this book.

Henry is survived by his wife, Jenny, to whom the first two editions were dedicated, two of his three children, and seven grandchildren. This edition of *Optical Physics* is dedicated to his memory.

Contents

Preface to the third edition

We were most encouraged by the publisher's request to revise *Optical Physics* for a third edition. The request involved considerably more work than we had anticipated; on the one hand we were specifically asked to enlarge our coverage of geometrical optics, fibre optics and quantum optics, and on the other hand not to increase the total length, which obviously necessitated rewriting much of the rest of the book! The requests for the two last topics in particular were very welcome. Both fibre optics and quantum optics have taken great strides forward in the last decade, and a basic understanding of them is essential for any student of physics, not only for a specialist. Since we are not mathematicians, we hope that the approach used for these subjects – an analogy with well-known elementary solved problems in quantum mechanics – will appeal to that section of our readership of a similar ilk. We have certainly learnt a lot in preparing both of these topics. We decided to present geometrical optics in a practical form, which we hope makes it attractive to today's students who have an easy familiarity with computers. We limited ourselves mainly to Gaussian optics, which is of most service to the physicist in general.

We have used the previous editions for more than a decade as the basis for two courses. One is an undergraduate course, which has as prerequisite a knowledge of high-school optics and a familiarity with elementary wave theory and quantum mechanics. This course covers most of Chapters 3, 4, 7, 8, 9, 11 and 12. We have also given a graduate course on advanced optics which includes, amongst other material, Chapters 6, 10, 13 and 14.

This new edition also has an expanded and updated bibliography. As far as possible, we have referred to books and reviews, which contain material prepared with pedagogical aims. In addition, we have cited several popularized presentations which are most useful to beginning students. But, particularly in the advanced topics of quantum optics, we have found it unavoidable to refer also to journal articles, through which we hope at least to convey some of the excitement of modern optics research.

It is very sad to record that Henry Lipson participated only in the first meeting of the three authors at which the general outline of the revisions was drawn. He died a few days after that meeting, but of course the imprint of his personal view of optics and the way it should be taught is still very obvious in the presentation. We have thus retained the emphasis on the Fourier transform as a basic tool and on illustrations strongly biased towards crystallographic methods. As far as possible, we have illustrated the theory with photographs of real laboratory results; many of the experiments illustrated in the second edition have been redone and we hope improved as a result.

We have revised and enlarged the problem section which now includes about a hundred problems. Most of them are intended to encourage thinking rather than as a preparation for examinations. We intend to publish a collection of solutions which can be obtained by writing to one of the authors (e-mail PHR22SL @ Technion.TECHNION.AC.IL, PHR07DT @ TX.Technion.AC.IL.)

It is our pleasure to thank the many people who have helped us in preparation of this edition. In particular, we must mention Sam Braunstein, Netta Cohen, Arnon Dar, Michael Elbaum, Baruch Fisher, Kopel Rabinovitch, Erez Ribak, Vassilios Sarafis, John Shakeshaft and Michael Woolfson, who have suggested topics and read and criticized parts of the manuscript. The patience and skill of Mrs Julia Löwnheim in preparing the photographs are gratefully acknowledged, as is the assistance of Michael Dovrat, Gila Etzion, Shimon Kostianovski, Ariel, Carni and Doron Lipson, Carina Reisin and Elizabeth Yudim in preparing the manuscript and figures. We are particularly indebted to Rina Lipson for collating the index. We also acknowledge a grant from the Technion President's fund to cover some of the preparation costs.

S. G. Lipson
D. S. Tannhauser

Preface to the second edition

Since the writing of the first edition the subject of Optics, as studied in universities, has grown greatly both in popularity and scope, and both we and the publishers thought that the time had arrived for a new edition of *Optical Physics*.

In preparing the new edition we have made substantial changes in several directions. First, we have attempted to correct all the mistakes and misconceptions that have been pointed out to us during the nine years the book has been in use. Secondly, we have made one important change in the subject matter: we have absorbed the chapter on Quantum Optics into the rest of the book. During the years, there have appeared many books devoted to laser physics, and it now seems impracticable for a book on physical optics to cover the subject at all satisfactorily in one chapter. However, since some knowledge of the principles of the laser is necessary for the understanding of physical optics today, particularly when coherence is being discussed, we have covered what we feel to be the necessary minimum as parts of Chapters 7 and 8.

In addition to the above changes in the subject matter, we have increased the number of exercises offered to the reader, organized them according to chapter, and provided solutions. We have also included a few suggestions, based on our experience, for student projects illustrating the material in the book.

We are, of course, most grateful to all those who have pointed out to us errors and room for improvement. But in particular we must thank D. S. Tannhauser, I. Senitsky and M. Neugarten who have helped us considerably by reading and criticizing in detail parts of the revised manuscript. We are also extremely grateful to the staff of Cambridge University Press, who have contributed considerably to the elimination of faults and inaccuracies in our manuscript.

S. G. Lipson
H. Lipson
May 1979

Preface to the first edition

There are two sorts of textbooks. On the one hand, there are works of reference to which students can turn for the clarification of some obscure point or for the intimate details of some important experiment. On the other hand, there are explanatory books which deal mainly with principles and which help in the understanding of the first type.

We have tried to produce a textbook of the second sort. It deals essentially with the principles of optics, but wherever possible we have emphasized the relevance of these principles to other branches of physics – hence the rather unusual title. We have omitted descriptions of many of the classical experiments in optics – such as Foucault's determination of the velocity of light – because they are now dealt with excellently in most school textbooks. In addition, we have tried not to duplicate approaches, and since we think that the graphical approach to Fraunhofer interference and diffraction problems is entirely covered by the complex-wave approach, we have not introduced the former.

For these reasons, it will be seen that the book will not serve as an introductory textbook, but we hope that it will be useful to university students at all levels. The earlier chapters are reasonably elementary, and it is hoped that by the time those chapters which involve a knowledge of vector calculus and complex-number theory are reached, the student will have acquired the necessary mathematics.

The use of Fourier series is emphasized; in particular, the Fourier transform – which plays such an important part in so many branches of physics – is treated in considerable detail. In addition, we have given some prominence – both theoretical and experimental – to the operation of convolution, with which we think that every physicist should be conversant.

We would like to thank the considerable number of people who have helped to put this book into shape. Professor C. A. Taylor and

Professor A. B. Pippard had considerable influence upon its final shape – perhaps more than they realize. Dr I. G. Edmunds and Mr T. Ashworth have read through the complete text, and it is thanks to them that the inconsistencies are not more numerous than they are. (We cannot believe that they are zero!) Dr G. L. Squires and Mr T. Blaney have given us some helpful advice about particular parts of the book. Mr F. Kirkman and his assistants – Mr A. Pennington and Mr R. McQuade – have shown exemplary patience in producing some of our more exacting photographic illustrations, and in providing beautifully finished prints for the press. Mr L. Spero gave us considerable help in putting the finishing touches to our manuscript.

And finally we should like to thank the three ladies who produced the final manuscript for the press – Miss M. Allen, Mrs E. Midgley and Mrs K. Beanland. They have shown extreme forbearance in tolerating our last-minute changes, and their ready help has done much to lighten our work.

S. G. L.
H. L.

CHAPTER ONE

History of ideas

1.1 Importance of history

Why should a textbook on physics begin with history? Why not start with what is known now and refrain from all the distractions of out-of-date material? These questions would be justifiable if physics were a complete and finished subject; only the final state would then matter and the process of arrival at this state would be irrelevant. But physics is not such a subject, and optics in particular is very much alive and constantly changing. It is important for the student to understand the past as a guide to the future. To study only the present is equivalent to trying to draw a graph with only one point.

It can also be interesting and sometimes sobering to learn how some of the greatest ideas came about. By studying the past we can sometimes gain some insight – however slight – into the minds and methods of the great physicists. No textbook can, of course, reconstruct completely the workings of these minds, but even to glimpse some of the difficulties that they overcame is worthwhile. What seemed great problems to them may seem trivial to us merely because we now have generations of experience to guide us; or, more likely, we have hidden them by cloaking them with words. For example, to the end of his life Newton found the idea of 'action at a distance' repugnant in spite of the great use that he made of it; we now accept it as natural, but have we come any nearer than Newton to understanding it? By being brought back occasionally to such fundamental problems the physicist is bound to have his wits sharpened; no amount of modern knowledge can produce the same effect. The way to study physics is to ask questions, as the geniuses of the past asked them. The ordinary physics student will find someone to answer them; the good physics student will answer them himself.

The history of optics follows two parallel trails, distinct but highly dependent on one another. The first is concerned with understanding the nature of light; originally the question was whether light consisted of corpuscles obeying Newtonian mechanics, or was a wave motion, and if so in what medium? As the wave nature became clearer, the question of the medium became more urgent, finally to be resolved by the theory of relativity. But the quantum nature of physics re-aroused the wave–particle controversy in a new form, and today many basic questions are still being asked about the interplay between particle and wave representations of light. It also resulted in the invention of the laser, which emits light completely different to that from thermal sources, and which has required the rethinking of many optical concepts.

The second trail follows the applications of optics. Starting with simple refractive imaging devices, well explained by corpuscular con-siderations, the wave theory became more and more relevant as the design of these instruments improved, and it became clear that bounds to their performance existed. But even the wave theory is not quite adequate to deal with the sensitivity of optical instruments, which is eventually limited by quantum theory. A fuller understanding of this is leading us today towards more and more sensitive measurement techniques.

1.2 The nature of light

1.2.1 The first ideas

Some odd ideas about the nature of light were put forward by the ancients, who wanted the sense of sight to be somehow similar to the sense that they knew best – that of touch. But this approach did not make much headway, and it was not until Galileo (1564–1642) introduced the experimental method that progress really began. Galileo was the first scientist effectively to propagate the idea of testing theories by experiment and, as an example, he tried to measure the speed of light. He failed, but even to have thought of the concept was an intellectual triumph.

1.2.2 The basic facts

Let us go back to Galileo's time. What was known about light in the seventeenth century? First of all, it travelled in straight lines. Secondly, it was reflected off smooth surfaces and the laws of reflexion were known. Thirdly, it changed direction when it passed from one medium to another (refraction); the laws for this phenomenon were not so obvious, but they had been established by Snell (1591–1626) and were later confirmed by Descartes (1596–1650). Fourthly, what we now call Fresnel diffraction had been discovered by Grimaldi (1618–63) and by Hooke (1635–1703). Finally, double refraction had been discovered by Bartholinus (1625–98). It was on the basis of these phenomena that a theory of light had to be constructed.

The last two facts were particularly puzzling. Why did shadows reach a limiting sharpness as the size of the source became small, and why did fringes appear on the light side of the shadow of a sharp edge? And why did light passing through a crystal of calcite produce two images while light passing through most other transparent materials produced only one?

1.2.3 The wave–corpuscle controversy

As usual in science when there is inadequate evidence, controversy resulted. Newton (1642–1727) threw his authority behind the theory that light is corpuscular, mainly because his first law of motion said that if no force acts on a particle it will travel in a straight line; he therefore postulated that light corpuscles are not acted upon by ordinary forces such as gravity. Double refraction he explained by some asymmetry in the corpuscles, so that their directions depended upon whether they passed through the crystal forwards or sideways. He envisaged the corpuscles as resembling magnets and the word 'polarization' is still used although this explanation has long been discarded.

Diffraction, however, was difficult. Newton realized its importance and carried out some crucial experiments; he showed that the fringes formed in red light were separated more than those formed in blue light. But when he found that the corpuscular theory could not be made to fit in, he weakly dropped his experiments, saying that he was rather busy! Newton was also puzzled by the fact that light was partly transmitted and partly reflected by a glass surface; how could his corpuscles sometimes go through and sometimes be reflected? He answered this question by propounding the idea of 'fits of reflexion' and 'fits of transmission'; in a train of corpuscles some would go one way and some the other. He even worked out the lengths of these 'fits' (which came close to what we now know as half the wavelength). But the idea was very cumbersome and was not really satisfying.

His contemporary Huygens (1629–95) was a supporter of the wave theory. With it he could account for diffraction and for the behaviour of two sets of waves in a crystal, without explaining *how* the two sets arose. Both he and Newton thought that light waves, if they existed, must be like sound waves, which are longitudinal. It is surprising that two of the greatest minds in science should have had this blind spot. If they had thought of transverse waves, the difficulties of explaining double refraction would have disappeared.

1.2.4 Triumph of wave theory

Newton's authority kept the corpuscular theory going until the end of the eighteenth century, but by then ideas were coming forward that could not be suppressed. In 1801 Young (1773–1829) produced his double-slit fringes (Fig. 1.1) – an experiment so simple to carry out and interpret that the results were incontrovertible. In 1815 Fresnel (1788–1827) worked out the theory of the Grimaldi–Hooke fringes and in 1821 Fraunhofer (1787–1826) produced diffraction patterns in parallel light for which the theory was much simpler. These three men laid the foundation of the wave theory that is still the basis of what is now called physical optics.

The defeat of the corpuscular theory, at least until the days of quantum ideas, came in 1818. In that year, Fresnel wrote a prize essay on the diffraction of light for the French Académie des Sciences on the basis of which Poisson (1781–1840), one of the judges, produced an argument that seemed to invalidate the wave theory by *reductio ad absurdum*. Suppose that a shadow of a perfectly round object

Figure 1.1
Young's fringes.

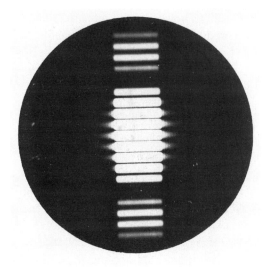

is cast by a point source; at the periphery all the waves will be in phase, and therefore the waves should also be in phase at the centre of the shadow, and there should therefore be a bright spot at this point. Absurd! Then Fresnel and Arago (1786–1853) carried out the experiment and found that there really was a bright spot at the centre (Fig. 1.2). The triumph of the wave theory seemed complete.

1.3 Speed of light

1.3.1 Measurement

The methods that Galileo employed to measure the speed of light were far too crude to be successful. In 1678 Römer (1644–1710) realized that an anomaly in the times of successive eclipses of the moons of Jupiter could be accounted for by a finite speed of light, and deduced that it must be about 3×10^8 ms^{-1}. In 1726 Bradley (1693–1762) made the same deduction from observations of the small ellipses that the stars describe in the heavens; since these ellipses have a period of one year they must be associated with the movement of the Earth.

It was not, however, until 1850 that direct measurements were made, by Fizeau (1819–96) and Foucault (1819–68), confirming the estimates obtained by Römer and Bradley. Knowledge of the exact value was an important confirmation of Maxwell's (1831–79) theory of electromagnetic waves (§1.4.2) which allowed the wave velocity to be calculated from the results of laboratory experiments on static and current electricity. In the hands of Michelson (1852–1931) their

Figure 1.2 The bright spot at the centre of the shadow of a disc.

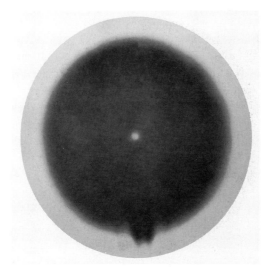

methods achieved a high degree of accuracy – about 0.03%, and subsequently much more accurate determinations have been made.

1.3.2 Refractive index

The idea that refraction occurs because the velocity of light is dependent on the medium dates back to Huygens and Newton. According to the corpuscular theory, the speed of light should be greater in a medium than in air because the corpuscles must be attracted towards the the denser medium to account for the changed direction of the refracted light. According to the wave theory, the waves must travel more slowly in water and 'slew' round to give the new direction. Foucault's method of measurement only required a relatively short path, and the speed of light could therefore be measured directly in media other than air – water, for example. Although the wave theory was by then completely accepted, Foucault's experiment provided additional confirmation. It also provided a method of investigating the effects of motion of the medium on the velocity of light, because it was possible to carry out the measurements when the water was flowing through the apparatus. The results could not be explained on the basis of nineteenth century physics of course, but pre-empted the theory of relativity (§9.4.1).

1.4 Transverse or longitudinal waves?

1.4.1 Polarization

The distinction between transverse and longitudinal waves had been appreciated early in the history of physics; sound waves were found to be longitudinal and water waves were obviously transverse. In the case of light waves, the phenomenon that enabled a decision to be made was that of double refraction in calcite. As we mentioned before, Huygens had pointed out that this property, which is illustrated in Fig. 1.3, must mean that the orientation of the crystal must somehow be related to some direction in the wave, but had failed to appreciate the connexion with transversality of the waves.

1.4.2 Nature of light

These experiments, of course, tell us nothing about the nature of light; they are all concerned with its behaviour. The greatest step towards understanding came from a completely different direction – the theoretical study of magnetism and electricity.

In the first half of the nineteenth century the relationship between magnetism and electricity had been worked out fairly thoroughly, by men such as Oersted (1777–1851), Ampère (1775–1836) and Faraday (1791–1867). In 1864, Maxwell was inspired to combine their results in mathematical form, and in manipulating the equations he found that they could assume the form of a transverse wave equation. The velocity of the wave could be derived from the known magnetic and electric constants. Evaluation of this velocity showed that it was equal to the velocity of light, and thus light was established as an electromagnetic disturbance. This was the beginning of the most brilliant episode in physics, during which different fields and ideas have been brought together and related to one another.

1.5 Quantum theory

1.5.1 The origins

With the marriage of geometrical optics and wave theory (physical optics) it seemed, up to the end of the last century, that no further rules about the behaviour of light were necessary. Nevertheless there remained some basic problems, as the study of the light emitted by hot bodies indicated. Why do such bodies become red-hot at about 600 °C and become whiter as the temperature increases? The great

Figure 1.3 Double refraction in a calcite crystal.

physicists such as Kelvin (1824–1907) were well aware of this problem, but it was not until 1900 that Planck (1858–1947) put forward, very tentatively, an *ad hoc* solution, now known as the quantum theory.

Planck's idea was that wave energy is divided into packets (quanta) whose energy content is proportional to the frequency; the lower frequencies, such as those of red light, are then more easily produced than higher frequencies. The idea was not liked, but gradually scepticism was overcome as more and more experimental evidence in its favour was produced. By about 1920 it was generally accepted, largely on the basis of Einstein's (1879–1955) study of the photoelectric effect (1905) and of Compton's (1892–1962) understanding of energy and momentum conservation in the scattering of X-rays (1923).

1.5.2 Wave–particle duality

But the problem had not really been solved. The equivalence of waves and particles is still difficult to appreciate to those of us who like intuitive physical pictures. The energy of a wave is distributed through space; the energy of a particle is concentrated. Perhaps one of the closest analogies is that of the surf rider carried along and guided by a wave. If he had to pass through a narrow opening he would either miss it and be completely stopped or he would go completely through. The wave guiding him, however, would be curtailed and could no longer guide him with the same accuracy; this is equivalent to diffraction.

1.5.3 Corpuscular waves

As usual in physics one idea leads to another, and in 1924 a new idea occurred to de Broglie (1892–1987), based upon the principle of symmetry. Faraday had used this principle in his discovery of electromagnetism; if electricity produces magnetism, does magnetism produce electricity? De Broglie asked, 'If waves are corpuscles, are corpuscles waves?' By now physicists had learnt not to be sceptical, and within three years his question had been answered. Davisson (1881–1958) and Germer (1896–1971) by ionization methods and G. P. Thomson (1892–1975) by photographic methods, showed that fast-moving electrons could be diffracted by matter similarly to X-rays. Since then other particles such as neutrons, protons and atoms have also been diffracted. Based on these experiments, Schrödinger (1887–1961) in 1928 produced a general wave theory of matter, which appears to be adequate down to atomic dimensions at least.

1.6 Instruments

1.6.1 The telescope

Although single lenses had been known from time immemorial, it was not until the beginning of the seventeenth century that optical instruments as we know them came into being. Lippershey (d. 1619) discovered in 1608, probably accidentally, that two separated lenses could produce a clear enlarged image of a distant object. Galileo seized upon the discovery, made his own telesope, and began to make a series of discoveries – such as Jupiter's moons and Saturn's rings – that completely altered the subject of astronomy. Newton, dissatisfied with the colour defects in the image, invented the reflecting telescope (Fig. 1.4). Since then the telescope has not changed in essence, but it has changed in size and cost.

1.6.2 The microscope

The story of the microscope is quite different. Its origin is uncertain; many people contributed to its early development. New ways of using it are still being found and further fundamental developments are still being made (§12.5).

The microscope originated from the magnifying glass. In the six-teenth and seventeenth centuries considerable ingenuity was exercised in making high-powered lenses; a drop of water or honey could produce wonderful results in the hands of an enthusiast. Hooke (1635–1703) played perhaps the greatest part in developing the compound

Figure 1.4 Newton's reflecting telescope.

microscope, and some of his instruments (Fig. 1.5) already showed signs of future trends in design. One can imagine the delight of such an able experimenter in having the privilege of developing a new instrument and of using it to examine for the first time the world of the very small, depicted in his *Micrographia* (1665).

1.6.3 Resolution limit

In order to put the design of optical instruments on a sound basis, the discipline of geometrical optics was founded, based entirely on the concept of rays of light, which trace straight lines in uniform media and are refracted according to Snell's law at boundaries. Based on these concepts, rules were formulated to improve the performance of lenses and mirrors, in particular by skilful figuring of surfaces and by finding ways in which inevitable aberrrations would cancel one another.

Figure 1.5 Hooke's microscope, from his *Micrographia.*

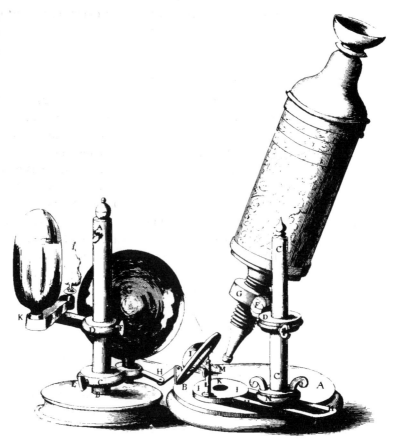

But the view that progress in optical instruments depended only upon the skill of their makers was suddenly brought to an end by Abbe (1840–1905) in 1873. He showed that the geometrical optical theory – useful though it was in developing optical instruments – was incomplete in that it took no account of the wave properties of light. Geometrically, the main condition that is necessary to produce a perfect image is that the rays from any point in the object should be so refracted that they meet together at a point on the image. Abbe showed that this condition is necessarily only approximate; waves spread because of diffraction and so cannot intersect in a point.

He put foward another view of image formation – that an image is formed by two processes of diffraction. As a result, one cannot resolve detail less than about half a wavelength, even with a perfectly corrected instrument. This simple result was greeted by microscopists with disbelief; many of them had already observed detail less than this with good rigidly-mounted instruments. Abbe's theory, however, proves that such detail is erroneous; it is a function of the instrument rather than of the object. Improving lenses further is not the right way to improve microscopes.

1.6.4 Resolving-power challenge

Any fundamental limitation of this sort must not be considered as depressing; it must be regarded as a challenge. Until difficulties are clearly exposed no real progress is possible. Now that it was known where the limitations of optical instruments lay, it was possible to concentrate upon *them* rather than upon lens design. Since resolving power is a function of wavelength, we need to consider new radiations with shorter wavelengths rather than new lens combinations. Ultra-violet light is an obvious choice, but the experimental difficulties are too great to justify a gain of perhaps a factor of two. The radiations that *have* been effective are X-rays and electron waves; these have wavelengths about 10^{-3} or 10^{-4} of those of visible light. They have produced revolutionary results.

1.6.5 X-ray diffraction

X-rays were discovered in 1895, but for seventeen years no one knew whether they were particles or waves. Then, in 1912, a brilliant idea of von Laue (1879–1960) solved the problem; he envisaged the possibility of using a crystal as a (three-dimensional) diffraction grating and the experiment of passing a fine beam of X-rays onto a crystal of copper sulphate (Fig. 1.6) showed definite indications of diffraction, indicating wave-like properties. In addition to answering the question, a new subject – X-ray crystallography – was born.

1.6.6 Electron microscopy

The realization that moving particles also have wave properties (§1.5.3) heralded new imaging possibilities. If such particles are charged they can be deflected electrostatically or magnetically, and so refraction can be simulated and it was found theoretically that suitably-shaped fields could act as lenses so that image formation was possible.

Electrons have been used with great success for this work, and electron microscopes with magnetic (or more rarely electrostatic) 'lenses' are available for producing images with very high magnifications. By using accelerating voltages of the order of 100 kV, wavelengths of about 0.1 Å can be produced, and thus a limit of resolution far better than that with X-rays should be obtainable. In practice, however, electron lenses are quite crude by optical standards and thus only small apertures are possible, which degrade the resolution. The electron microscope developed rapidly in the 1930s, and the limit of resolution soon reached and then passed that of the light microscope. Today, images showing atomic and molecular resolution are available, which have revolutionized fields such as biology (Fig. 1.7).

Figure 1.6 X-ray diffraction patterns produced by a crystal. (*a*) Original results; (*b*) a clearer picture, showing symmetry. (From Ewald, 1962.)

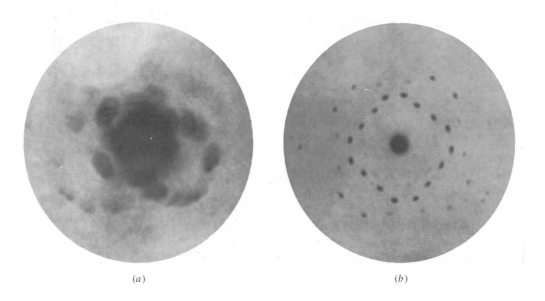

(*a*)   (*b*)

1.7 Recent developments

This history of optics has taken us to about the 1930s. At this stage there was a lull in its progress and some physicists began to say that no more advances could be expected, that it was now possible to regard optics as a closed book, and that it was hardly worth teaching. This attitude was, of course, mistaken; even if optics was finished, it would still form a necessary part of physics teaching for its concepts which permeate the whole of the rest of the subject.

One man largely kept the optics flag flying. This was Zernike (1888–1966), of the Netherlands, who chose to apply his talents to this subject in spite of the blandishments of other branches which appeared to be more exciting. He appreciated the importance of the concept of coherence and showed that there was much more to it than the two extremes of complete coherence and incoherence. He introduced the idea of partial coherence of a single beam of light, and invented the phase-contrast microscope, a device that revitalized biology by eliminating the need for staining and for which he was awarded the Nobel Prize in 1953.

It seemed, however, that perfect coherence – like any other perfection in physics – would remain an unattainable idea, until Townes, of the USA, and Basov and Prokharov, of the USSR, invented the laser, a device for which they were awarded the Nobel Prize in 1964. They managed to arrange that, in effect, all the atoms in a system should

Figure 1.7 Electron microscope image of a virus crystal, magnified 3×10^4, showing resolution of individual molecules (courtesy of R. W. G. Wyckoff).

1.0 μ

vibrate in phase and so produce a completely coherent beam, in which all their radiated energy is concentrated; even with powers as small as 1 mW an extremely intense beam is produced.

With the practical invention of the laser and the theoretical concept of coherence, optics has taken many new directions. Although at any particular moment it seems clear what are the most important ones, we judge it more prudent to leave the writing of the history of this era to the scientists of the next century.

A chapter of this length can do little more than whet the appetite of the student who is interested in the development of optics from the first hesitant concepts to the very considerable body of knowledge that the subject now comprises. Two books which which amplify the matter of this chapter in a very readable manner are those by H. Lipson (1968) and Segré (1984); the historical development of optics in particular is described in the first chapter of Born and Wolf (1980).

CHAPTER TWO

Waves

2.1 Introduction

Optics is the study of wave propagation and its quantum implications. Traditionally, it has centred around visible light waves, but in the modern era the concepts which have developed over the years have been found increasingly useful when applied to many other types of wave, both within and without the electromagnetic spectrum. Wave propagation in a medium is described mathematically in terms of a *wave equation*; this is a differential equation relating the dynamics and statics of small displacements of the medium, and whose solution may be a propagating disturbance. This chapter will be concerned with such equations and their solutions.

The term 'displacements of the medium' is not, of course, restricted to mechanical displacement but can be taken to include any field quantity (continuous function of **r** and *t*) which can be used to measure a departure from equilibrium, and the equilibrium state itself may be nothing more than the vacuum.

Although it is convenient, from an elementary point of view, to study wave equations arising from the mechanical relationships between displacement and velocity, we quickly learn that almost any relationships between derivatives of a field in space and time can replace them. Then the distinction, which is clear in the mechanical sense, between 'static' and 'dynamic' properties may become blurred. For example, in the electromagnetic wave, the variables are electric and magnetic fields. On looking for an equivalence to static and dynamic variables, the electric field – instinctively related to a charge distribution – might possibly be considered as the 'static' variable, while the magnetic field – related to currents, or motion of charges – as the 'dynamic' variable. But what is more important is that derivatives of the two fields are related by Maxwell's equations (5.1–5.4), whose symmetry overshadows any real distinction of this sort.

Once an equation has been set up, to call it a wave equation we require it to have propagating solutions. This means that if we supply initial conditions in the form of a disturbance which is centred around some given position at time zero, then we shall find a disturbance of similar type centred around a new position at a later time. The term we used: 'centred around' is sufficiently loose that it does not require the disturbance to be unchanged, but only refers to the position of its centre of gravity. We simply ask for some definition of a centre which can be applied similarly to the initial and later stages, and which shows propagation. This way we can include many interesting phenomena in our definition, and benefit from the generality. But first we shall consider the simplest case, in which the propagating disturbance is indeed unchanged with time.

2.2 The non-dispersive wave equation in one dimension

2.2.1 Differential equation for a non-dispersive wave

The most important elementary wave equation in one dimension can be derived from the requirement that any solution:

(i)   propagates in either direction ($\pm x$) at a constant velocity v,
(ii) does not change with time, when referred to a centre which is moving at this velocity.

These are restrictive conditions but, just the same, they apply to a very large and diverse group of physical phenomena. The resulting wave equation is called the *non-dispersive wave equation*.

We start with the requirement that a solution $f(x, t)$ of the equation must be unchanged if we move the origin a distance $x = \pm vt$ in time t (Fig. 2.1). This gives two equations:

$$f(x, t) = f(x - vt, 0) \tag{2.1}$$

$$f(x, t) = f(x + vt, 0) \tag{2.2}$$

where f is any function which can be differentiated.† The argument

$$(x \pm vt) \equiv \phi_{\pm} \qquad (2.3)$$

is called the *phase* of the wave.

Differentiating (2.1) by x and t respectively, we have

$$\frac{\partial f}{\partial x} = \frac{\mathrm{d}f}{\mathrm{d}\phi_-}; \qquad \frac{\partial f}{\partial t} = -v\frac{\mathrm{d}f}{\mathrm{d}\phi_-} \qquad (2.4)$$

and for (2.2)

$$\frac{\partial f}{\partial x} = \frac{\mathrm{d}f}{\mathrm{d}\phi_+}; \qquad \frac{\partial f}{\partial t} = v\frac{\mathrm{d}f}{\mathrm{d}\phi_+}. \qquad (2.5)$$

Equations (2.4) and (2.5) can be reconciled to a single equation by a second similar differentiation followed by eliminating $\mathrm{d}^2f/\mathrm{d}\phi^2$ between the pairs; either equation gives

$$\frac{\partial^2 f}{\partial x^2} = \frac{\mathrm{d}^2 f}{\mathrm{d}\phi^2}; \qquad \frac{\partial^2 f}{\partial t^2} = v^2\frac{\mathrm{d}^2 f}{\mathrm{d}\phi^2}$$

whence

$$\frac{\partial^2 f}{\partial x^2} = \frac{1}{v^2}\frac{\partial^2 f}{\partial t^2} \qquad (2.6)$$

of which (2.1) and (2.2) are the most general solutions. Equation (2.6) is known as the *non-dispersive wave equation*, because the solutions imply that the form of the wave does not alter it as it progresses.

Although (2.6) has a general solution (2.1) and (2.2), there is a particular solution to it which is more important because it satisfies a large class of equations known as *wave equations*. This solution is

† This is not intended to an accurate mathematically-acceptable statement, but simply states our basic requirements. Of course, from the physical point of view there will always be restrictions on the magnitude, size of derivative etc. of the disturbance which leaves no permanent trace on the particular medium, to a certain degree of accuracy.

Figure 2.1 An arbitrary disturbance moves at a constant velocity.

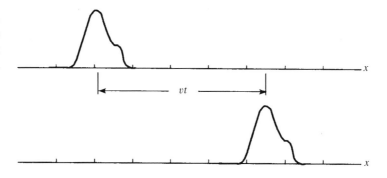

a *simple-harmonic wave* of amplitude a, which we shall write in its complex exponential form:

$$f(x,t) = a \exp\left[2\pi i \left(\frac{x}{\lambda} - vt\right)\right]$$

where v is the frequency in cycles per unit time and λ is the wavelength. A more tidy expression can be written in terms of

the *spatial frequency* or *wave-number* $\quad k \;=\; 2\pi/\lambda,$

the *angular frequency* $\quad \omega \;=\; 2\pi v.$

The latter is just the frequency expressed in units of radians per second, and we shall generally refer to it simply as 'frequency'. These give

$$f(x,t) = a \exp[i(kx - \omega t)]. \tag{2.7}$$

It is easy to verify that this function satisfies (2.6), and that the velocity is given by

$$v = \omega/k ; \tag{2.8}$$

this is called the *phase velocity* or *wave velocity*.

2.2.2 Harmonic waves and their superposition

One particular value of using simple harmonic waves is that, as we shall see in Chapter 3, any other waveform can be built up out of these by superposition. Now *if the wave equation is linear in f,* the propagation of a number of simple-harmonic waves superposed can easily be studied by considering the propagation of each of the components separately, and then recombining. In the case of the non-dispersive wave equation this is easy.

Consider an elementary wave with wavenumber k, for which $\omega = kv$:

$$f_k(x,t) = a_k \exp[i(kx - \omega t)] = a_k \exp(i\phi). \tag{2.9}$$

Take an initial ($t = 0$) superposition of such waves

$$g(x,0) = \sum_j f_{k_j}(x,0) = \sum_j a_{k_j} \exp(ik_j x). \tag{2.10}$$

At time t, each of the elementary waves has evolved as in (2.9) so:

$$g(x,t) \;=\; \sum_j a_{k_j} \exp[i(k_j x - \omega_j t)] \tag{2.11}$$

$$=\; \sum_j a_{k_j} \exp[ik_j(x - vt)] \tag{2.12}$$

$$=\; g(x - vt, 0). \tag{2.13}$$

In words, the initial function $g(x, 0)$ has propagated with no change at velocity v; (2.13) is equivalent to (2.1). It is important to realize that this simple result arose because of the substitution of kv for ω in (2.11). If different frequencies travel at different velocities (a *dispersive* wave) our conclusions will be modified (§2.8).

2.2.3 Example of a non-dispersive wave

To illustrate the non-dispersive one-dimensional wave equation we shall consider a compressional wave in a continuous medium, a fluid. If the fluid has compressibility K and density ρ, the equilibrium-state equation is Hooke's law:

$$P = K \frac{\partial \eta}{\partial x}, \tag{2.14}$$

where P is the local pressure, i. e. the stress, and η the local displacement from equilibrium. The differential $\partial \eta / \partial x$ is thus the strain. A dynamic equation relates the deviation from the equilibrium state (uniform and constant P) to the local acceleration:

$$\rho \frac{\partial^2 \eta}{\partial t^2} = \frac{\partial P}{\partial x}. \tag{2.15}$$

Equations (2.14) and (2.15) lead to a wave equation

$$\frac{\partial^2 \eta}{\partial x^2} = \frac{\rho}{K} \frac{\partial^2 \eta}{\partial t^2}. \tag{2.16}$$

Thus the waves are non-dispersive, with wave velocity

$$v = \left(\frac{K}{\rho} \right)^{\frac{1}{2}}. \tag{2.17}$$

The wave equation (2.16) is valid provided that the stress–strain relationship (2.14) remains linear, i.e. for small stress. It would not describe shock waves, for example, for which the stress exceeds the elastic limit. Another example of a non-dispersive wave equation is that derived by Maxwell for electromagnetic waves which will be discussed in depth in Chapter 5.

2.3 Dispersive waves

2.3.1 Equation for a dispersive wave in a linear medium

In general, wave equations are not restricted to second derivatives in x and t. Provided that the equation remains linear in f, derivatives of other orders can occur; in all such cases, a solution of the form

$f = a\exp[i(kx - \omega t)]$ is found. For such a wave one can replace $\partial f/\partial t$ by $-i\omega f$ and $\partial f/\partial x$ by ikf, so that if the wave equation can be written

$$p\left(\frac{\partial}{\partial x}, \frac{\partial}{\partial t}\right) f = 0, \tag{2.18}$$

where p is a polynomial function of $\partial/\partial x$ and $\partial/\partial t$, which operates on f, the result will be an equation

$$p(ik, -i\omega) = 0 \tag{2.19}$$

which is called the *dispersion equation*.

For example, we shall first return to the non-dispersive equation (2.6) and see it in this light. We had

$$\frac{\partial^2 f}{\partial x^2} = \frac{1}{v^2}\frac{\partial^2 f}{\partial t^2}, \tag{2.20}$$

which can be written

$$\left[\left(\frac{\partial}{\partial x}\right)^2 - \frac{1}{v^2}\left(\frac{\partial}{\partial t}\right)^2\right] f = 0. \tag{2.21}$$

Thus, from (2.18) and (2.19)

$$(ik)^2 - \frac{1}{v^2}(-i\omega)^2 = 0 = \frac{\omega^2}{v^2} - k^2, \tag{2.22}$$

implying

$$\omega/k = \pm v. \tag{2.23}$$

2.3.2 Example of a dispersive wave equation: Schrödinger's equation

A dispersive wave equation which rivals Maxwell's electromagnetic wave equation in its importance is Schrödinger's wave equation for a non-relativistic particle of mass m, moving in a potential field V, which we quote here in its one-dimensional form:

$$i\hbar\frac{\partial\psi}{\partial t} = \frac{-\hbar^2}{2m}\cdot\frac{\partial^2\psi}{\partial x^2} + V(x)\psi. \tag{2.24}$$

Here $|\psi|^2\delta x$ is the probability of finding the particle in the region between x and $x + \delta x$, and ψ is called the *probability amplitude*. Using (2.18) and (2.19) we can immediately write down the dispersion equation as:

$$i\hbar(-i\omega) = \frac{-\hbar^2}{2m}\cdot(ik)^2 + V(x), \tag{2.25}$$

$$\hbar\omega = \frac{(\hbar k)^2}{2m} + V(x). \tag{2.26}$$

We identify $\hbar\omega$ as the total energy E of the particle (Planck: $\hbar\omega = h\nu$) and $\hbar k$ as its momentum, $p = mv$ (de Broglie: $\hbar k = h/\lambda$). Thus, (2.26) becomes:

$$E = \frac{p^2}{2m} + V(x) = \frac{1}{2}mv^2 + V(x) \tag{2.27}$$

or total energy = kinetic energy + potential energy. In this case the dispersion equation expresses the Newtonian mechanics, while the wave equation is the quantum mechanical equivalent.

2.4 Complex wavenumber, frequency and velocity

Solution of a dispersion equation such as (2.26) may give rise to complex values for k or ω, and it is important to give a physical interpretation to such cases. The velocity ω/k may also be complex as a result.

2.4.1 Complex wavenumber: attenuated waves

Suppose that the frequency ω is real, but the dispersion relation then leads us to a complex value of $k \equiv k_1 + ik_2$. We then have

$$\begin{aligned} f &= a\exp[\mathrm{i}(k_1 + ik_2)x - \mathrm{i}\omega t] \\ &= a\exp(-k_2 x)\exp[\mathrm{i}(k_1 x - \omega t)]. \end{aligned} \tag{2.28}$$

This describes a propagating wave, with velocity $v = \omega/k_1$, attenuated progressively by the factor $\exp(-k_2 x)$. Thus its amplitude decays by a factor e^{-1} in every *characteristic decay distance* of length k_2^{-1} (Fig. 2.2).

Figure 2.2 An attenuated harmonic wave (*a*) as a function of x at $t = 0$, (*b*) as a function of t at positions $x_1 < x_2 < x_3$.

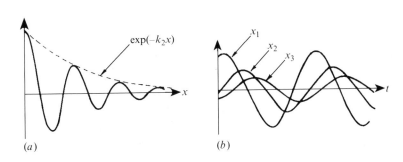

2.4.2 Imaginary velocity: evanescent waves

Sometimes the wavenumber turns out to be purely imaginary ($k_1 = 0$). Now, in (2.28) the wave clearly has no harmonic space-dependence at all; it is a purely exponential function of x, but still oscillates in time with frequency ω (Fig. 2.3). It is then called an *evanescent wave*.

2.4.3 The diffusion equation

The diffusion equation can be considered as a wave equation, although it has many commonly-known solutions which are not wave-like. We shall assume here that it applies to heat diffusion, in deference to Fourier's application of wave theory to this subject which laid the mathematical foundation for much of this book. The wave equation in one dimension arises from the heat conduction equation along, for example, a bar with negligible heat losses from its surfaces,

$$q = -\kappa\, \partial\theta/\partial x, \tag{2.29}$$

which relates heat flux q per unit area to temperature θ in a medium with thermal conductivity κ and specific heat s per unit volume. Conservation of heat requires

$$s\, \partial\theta/\partial t = -\partial q/\partial x \tag{2.30}$$

whence

$$\frac{\partial\theta}{\partial t} = D\frac{\partial^2\theta}{\partial x^2}, \tag{2.31}$$

where $D = \kappa/s$. The dispersion relation (2.19) is thus

$$i\omega = \mp Dk^2 \tag{2.32}$$

giving, for real ω,

$$k = \left(\frac{\omega}{2D}\right)^{\frac{1}{2}}(1 + i). \tag{2.33}$$

$\frac{1}{2}\left(1+i\right)^2 = i$

Figure 2.3 An evanescent wave (a) as a function of x at $t = 0$, (b) as a function of t at positions $x_1 < x_2 < x_3$.

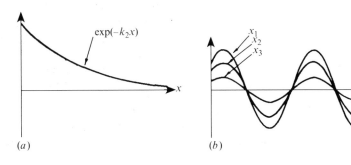

$\exp(-k_2 x)$

(a)   (b)

If one end of the bar is subjected to alternating heating and cooling so that its temperature rise above ambient θ can be written as

$$\theta(0, t) = \theta_0 \exp(-i\omega t) \qquad (2.34)$$

the wave is propagated along the bar in the form

$$\theta(x, t) = \theta_0 \exp\left\{i\left[\left(\frac{\omega}{2D}\right)^{\frac{1}{2}}(1 + i)x - \omega t\right]\right\}$$

$$= \theta_0 \exp\left[-\left(\frac{\omega}{2D}\right)^{\frac{1}{2}} x\right] \exp\left\{i\left[\left(\frac{\omega}{2D}\right)^{\frac{1}{2}} x - \omega t\right]\right\}. \qquad (2.35)$$

This wave is attenuated along the bar with characteristic decay distance $(2D/\omega)^{\frac{1}{2}}$. The propagated disturbance is still a wave, however; the phase of oscillation progresses regularly with x.

Now suppose that the same bar has an initial temperature distribution

$$\theta = \theta_0 \exp(ikx) \quad (k \text{ real}) \qquad (2.36)$$

impressed upon it at time $t = 0$, and the temperature distribution is left to its own devices. From (2.35), we write the subsequent temperature distribution

$$\begin{aligned} \theta(x, t) &= \theta_0 \exp[i(kx - \omega t)] \\ &= \theta_0 \exp[-(Dk^2 t)] \exp(ikx). \end{aligned} \qquad (2.37)$$

There is no oscillatory time-dependence; the spatial dependence $\exp(ikx)$ remains unchanged, but its amplitude decays to zero with time-constant $(Dk^2)^{-1}$. This is a wave evanescent in time. Thus the heat-diffusion equation illustrates both types of behaviour; it supports a wave attenuated in distance for a real frequency, or evanescent in time for a real wavelength.

2.5 Group velocity

The non-dispersive wave equation (2.6) has the property that disturbances of all frequencies travel with the same velocity. The result, according to §2.2.2, is that a wave of any form propagates undistorted. On the other hand, in many media we find that:

– waves of different frequencies propagate with different velocities.
– the form of the wave becomes distorted as it progresses.

Let us suppose that we have a *wave-group*, which is a wave of given ω_0 and k_0 whose amplitude is modulated so that it is limited to a restricted region of space at time $t = 0$ (Fig. 2.4).

It is clear that all the energy associated with the wave is concentrated in the region where its amplitude is non-zero. Such a wave can be built up by superposition of many component waves whose frequencies and wavenumbers are approximately ω_0 and k_0. Methods for calculating their amplitudes will be discussed in Chapter 4; we do not need them explicitly here. At a given time, the maximum value of the wave-group envelope occurs at the point where all the component waves have the same phase, and thus reinforce one another. We shall now show that this point travels at the *group velocity*, a well-defined but different velocity from that of the individual waves themselves; this is also the velocity at which energy is transported by the wave.

If the maximum of the envelope corresponds to the point at which the phases of the components are equal, then

$$\frac{d\phi}{dk} = \frac{d}{dk}(kx - \omega t) = 0 \tag{2.38}$$

at that point. The velocity at which the maximum moves is then given by

$$v_g = \frac{x}{t} = \frac{d\omega}{dk} \tag{2.39}$$

which is the basic expression for the group velocity. It can be reformulated in several ways in terms of $\lambda = 2\pi/k$, $\omega/k = v$ and $v = \omega/2\pi$, such as

$$v_g = v - \lambda \, dv/d\lambda. \tag{2.40}$$

Figure 2.4
A wave-group.

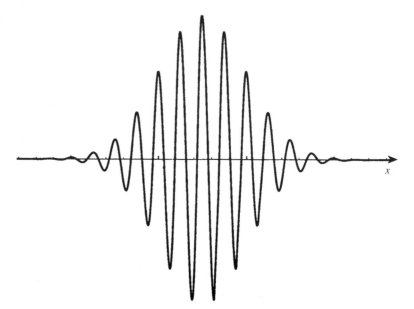

In general, of course, $d\omega/dk$ is not a constant. Since the wave-group consists of components around k_0, the value of $d\omega/dk$ evaluated at k_0 is implied. To a first approximation the envelope of the wave-group propagates at this velocity with little change if the wave-group is not too short; since the phase velocity v is not in general equal to v_g, the individual waves can be seen to move relative to the envelope in either a forward or backward direction. Such a behaviour can easily be observed in water waves, since these waves have quite strong dispersion.

The waveform is usually distorted with propagation. In §2.8 we shall analyze an illustrative example of this. Very strong dispersion makes the subject quite complicated and can lead to apparently paradoxical situations in which v_g appears to be greater than the velocity of light (Brillouin, 1960).

2.6 Waves in three dimensions

The reader might consider it a trivial excercise to repeat in three dimensions the analysis in §2.2.1 in which we derived the non-dispersive wave equation in one dimension; but it is not! The reason is that, even in a non-dispersive medium, a three-dimensional wave in general changes its profile as it propagates. Consider, for example, the spherical acoustic waves emanating from a source in air (which is a good non-dispersive acoustic medium at audible frequencies). As we go further from the source the sound intensity weakens like r^{-2}, so the amplitude of the disturbance is obviously changing with distance.

There is, however, one important type of wave which can propagate in three dimensions without change. This is the *plane wave*.

2.6.1 Plane waves

A plane wave which propagates at velocity v in direction $\hat{\mathbf{n}}$ has the general form equivalent to (2.1):

$$f(\mathbf{r}, t) = f(\mathbf{r} \cdot \hat{\mathbf{n}} - vt, 0). \qquad (2.41)$$

As in (2.3), the phase ϕ is then

$$\phi = \mathbf{r} \cdot \hat{\mathbf{n}} - vt. \qquad (2.42)$$

This is a constant on any plane satisfying $\mathbf{r} \cdot \hat{\mathbf{n}} - vt = \text{const.}$ Such a plane is called a *wavefront* and is a plane of constant phase normal to the direction of propagation $\hat{\mathbf{n}}$ (Fig. 2.5).

2.6.2 Wave equation

On this basis we can derive the wave equation. We have (2.41)

$$f(\mathbf{r}, t) = f(\mathbf{r} \cdot \hat{\mathbf{n}} - vt, 0) = f(\phi). \tag{2.43}$$

For this function, the time and spatial derivatives are:

$$\partial/\partial t = -v \, d/d\phi, \tag{2.44}$$

$$\nabla \equiv \left(\frac{\partial}{\partial x}, \frac{\partial}{\partial y}, \frac{\partial}{\partial z} \right) = \left(\frac{\partial \phi}{\partial x}, \frac{\partial \phi}{\partial y}, \frac{\partial \phi}{\partial z} \right) \frac{d}{d\phi} = \hat{\mathbf{n}} \frac{d}{d\phi} \tag{2.45}$$

where (2.42) has been used to calculate $\partial \phi / \partial x$ etc. Thus, from (2.43)

$$\frac{\partial^2 f}{\partial t^2} = v^2 \frac{d^2 f}{d\phi^2} \tag{2.46}$$

and from (2.45)

$$\nabla \cdot (\nabla f) = (\hat{\mathbf{n}} \cdot \hat{\mathbf{n}}) \frac{d^2 f}{d\phi^2} = \frac{d^2 f}{d\phi^2}, \tag{2.47}$$

whence

$$\nabla \cdot \nabla f \equiv \nabla^2 f = \frac{\partial^2 f}{\partial x^2} + \frac{\partial^2 f}{\partial y^2} + \frac{\partial^2 f}{\partial z^2} = \frac{1}{v^2} \frac{\partial^2 f}{\partial t^2}. \tag{2.48}$$

So far, f has been considered as a scalar, but the same analysis can be repeated for each component of a vector field, giving

$$\nabla^2 \mathbf{f} = \frac{1}{v^2} \frac{\partial^2 \mathbf{f}}{\partial t^2}. \tag{2.49}$$

This is the three-dimensional non-dispersive wave equation.

Figure 2.5 A plane wave with wave-vector **k**.

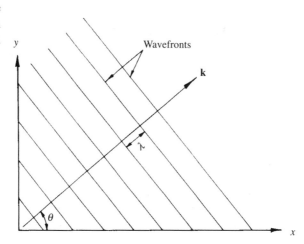

Following (2.7), a harmonic plane wave can be constructed by replacing **f** by **a** exp[i(**k** · **r** − ωt)]. The phase is

$$\phi = \mathbf{k} \cdot \mathbf{r} - \omega t = k\left(\frac{\mathbf{k}}{k} \cdot \mathbf{r} - \frac{\omega}{k}t\right) = k\left(\frac{\mathbf{k}}{k} \cdot \mathbf{r} - vt\right) \tag{2.50}$$

from which it follows, using (2.42), that $\hat{\mathbf{n}} = \mathbf{k}/k$ which is the unit vector in the direction **k**. The wavefront is therefore *normal to the direction of propagation* **k**. The magnitude of **k** is $k = 2\pi/\lambda$, as in (2.7). Because of this relationship, λ is not a vector, even though it has both magnitude and direction, because its components do not combine in the prescribed manner.

Dispersion equations in three dimensions are derived in a similar manner to those in one dimension, by using the substitutions $\nabla = i\mathbf{k}$, $\partial/\partial t = -i\omega$. This will give, in general, a vector equation of the form $p(i\mathbf{k}, -i\omega) = 0$ (cf (2.19)).

2.6.3 Spherical and cylindrical waves

Other possible waves in three dimensions are the *spherical wave*, for which the wavefronts are spheres,

$$\mathbf{f}(\mathbf{r}, t) = \mathbf{A}(\mathbf{r})\exp[i(kr - \omega t)], \tag{2.51}$$

and the *cylindrical wave*, for which the wavefronts are cylinders,

$$\mathbf{f}(\mathbf{r}, t) = \mathbf{A}(\mathbf{r})\exp\{i[k(x^2 + y^2)^{\frac{1}{2}} - \omega t]\}. \tag{2.52}$$

Problems concerning polarization arise with these waveforms, and in this book their use will be limited to the scalar-wave theory (Chapter 7).

2.7 Waves in inhomogeneous media

Propagation of a simple harmonic wave in a homogeneous medium is a relatively simple matter, but when the medium is inhomogeneous, problems arise which defy analytical treatment except in the simplest cases. Two important principles, developed by Huygens and Fermat, go a long way to simplifying the physics of such situations.

2.7.1 Huygens's construction

As we have seen, Huygens was a staunch advocate of the wave theory and introduced several ideas which have stood the test of time. One of these was the wavefront which we have already described as a surface of constant phase (§2.6.1). Huygens considered the wave more as a transient phenomenon, which was emitted from a point at a certain instant. Then the wavefront is defined as the surface which the wave disturbance has reached at a given time. In principle, a wavefront can have almost any shape, but only plane, spherical or occasionally ellipsoidal wavefronts have any analytical importance.

Huygens pointed out that if one knows the wavefront at a certain moment, the wavefront at a later time can be deduced by considering each point on the first one as the source of a new disturbance. The new disturbance is a spherical wave. At the later time, the spherical waves will have grown to a certain radius, and the new wavefront is the envelope of all the new disturbances (Fig. 2.6(a)).

The mathematical embodiment of this idea is the Kirchhoff diffraction integral, which will be discussed in Chapter 7. The new disturbances are known as *Huygens's wavelets*. On the basis of this principle, one can easily see that spherical, plane and cylindrical waves will retain their shapes, but other shapes can become distorted.

The construction can be applied to anisotropic materials (Chapter 6) by realizing that, if the wave velocity is a function of direction, the wavelets are ellipsoids and not spheres. An important corollary then follows when we consider the progress of a wavefront of limited spatial extent. The direction of energy flow (**Π**, the *ray direction*), is given by joining the origin of each wavelet to the point at which it touches the

Figure 2.6 Huygens's construction: (a) in an isotropic medium, (b) in an anisotropic medium.

(a)

(b)

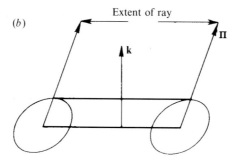

envelope, and this direction does not always coincide with that of the wave-vector **k** (Fig. 2.6(*b*)).

It is sometimes argued that Huygens's principle is deficient in that it predicts a backward wave as well as a forward one. This is not so. If we wish to find out how a wavefront A progresses during a time t, we can divide t into two parts, say t_1 and t_2. The wavefront A will develop into B after t_1 and B into C after t_2. If we take backward waves into account the wavefront C is accompanied by C' and C'' which lie at positions depending on the exact values of t_1 and t_2. All such waves will interfere destructively and only the forward wave, which depends on $t = t_1 + t_2$ alone, remains (Fig. 2.7). In the Kirchhoff treatment of Huygens's construction, which we discuss in §7.2, this problem is solved in a natural manner. Huygens's construction is most useful in getting a physical picture of wave propagation under conditions where an exact calculation defeats us; an illustration follows in §2.9.

2.7.2 Fermat's principle

Suppose that a light wave is emitted from a point source in an inhomogeneous medium, and can 'choose' between several possible routes to an observer. Which one will it take? Fermat originally stated that it will choose that which takes the minimum time, thus illustrating Nature's concern with economy! The law of rectilinear propagation in a homogeneous medium is an obvious result, and the laws of reflexion and refraction can also be deduced. In using this principle it is convenient to define the *optical path* from A to B as

$$\overline{AB} = \int_A^B \mu(s)ds \qquad (2.53)$$

where $\mu(s)$ is the *refractive index*, $c/v(s)$, at the point a distance s along the route. The time taken for light to travel from A to B is then $\overline{AB}/c$.

Fermat's principle can only be understood properly in terms of interference, but the concepts needed are elementary. Supposing that light waves propagate from A to B by *all possible* routes AB_j, unrestricted initially by rules of geometrical or physical optics (those rules will

Figure 2.7 The backward wave problem in Huygens's construction.

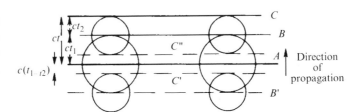

emerge). The various optical paths $\overline{AB}_j$ will vary by amounts greatly in excess of one wavelength, and so the waves arriving at B will have a large range of phases and will tend to interfere destructively. But if there is a shortest route AB_0, and the optical path varies *smoothly* through it, then a considerable number of neighbouring routes close to AB_0 will have optical paths differing from $\overline{AB}_0$ by second-order amounts only, $< \lambda$, and will therefore interfere constructively. Waves along and close to this shortest route will thus dominate (Fig. 2.8) and AB_0 will be the route along which the light is seen to travel. The same argument also shows that a possible route is that when $\overline{AB}_j$ is *maximum*. Sometimes profligate Nature takes the *longest* path!

Fermat's principle also applies to imaging. If all the routes from A to B have the *same* optical path, the waves traveling by all of them will interfere constructively at B. Then no particular route for the light is chosen. A and B are then *conjugate points* and each is the image of the other.

We conclude with an example, relevant to the next chapter, to show the derivation of Snell's law of refraction from Fermat's principle. There is a plane interface I between regions of refractive indices μ_1 and μ_2 (Fig. 2.9). Two neighbouring routes from A to B pass through C_1 and C_2, where $C_1 C_2 = d$. Then the *difference* in optical path between the two is (when $d \ll AC_1, BC_1$)

$$\mu_1 X C_2 - \mu_2 Y C_1 = \mu_1 d \sin \hat{\imath} - \mu_2 d \sin \hat{r}. \qquad (2.54)$$

Figure 2.8 Fermat's principle. The ticks along the rays indicate units of one wavelength.

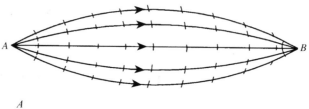

Figure 2.9 Snell's law proved by Fermat's principle.

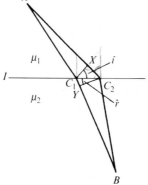

This is zero (indicating minimum or maximum optical path $\overline{AB}$) when

$$\mu_1 \sin\hat{\imath} = \mu_2 \sin\hat{r}.$$

Thus Fermat's principle leads to *Snell's law of refraction*.

2.8 Advanced topic: propagation and distortion of a wave group in a dispersive medium

In §2.2.2 we considered the propagation of the superposition of a number of elementary waves of the form

$$F(x,t) \equiv \sum_j f_{k_j}(x,t) = \sum_j a_{k_j} \exp[\mathrm{i}(k_j x - \omega_j t)]. \qquad (2.55)$$

Let us replace the summation by an integral

$$F(x,t) = \int_{-\infty}^{\infty} f(k,x,t)\mathrm{d}k = \int_{-\infty}^{\infty} a(k)\exp[\mathrm{i}(kx - \omega t)]\mathrm{d}k \qquad (2.56)$$

where $\omega(k)$ is defined. Now we shall take the particular case of a *Gaussian wave-group* (§4.4.6, §4.6.5) in which $a(k)$ is the Gaussian function

$$a(k) = \left(\frac{2\pi}{\sigma}\right)^{-\frac{1}{2}} \exp\left[\frac{-(k-k_0)^2\sigma^2}{2}\right], \qquad (2.57)$$

centred on $k = k_0$ and having variance σ^{-1}.

The integral (2.56) can be performed for $t = 0$ (§4.4.6) and the result (Fig. 2.10(a)) is:

$$F(x,0) = \exp(-x^2/2\sigma^2) \exp(\mathrm{i}k_0 x). \qquad (2.58)$$

The envelope $\exp(-x^2/2\sigma^2)$ of this wave-group peaks at $x = 0$ and has half-peak-width (§4.4.6) $w = 2.36\sigma$.

Now let us consider the propagation of this wave-group in three media:

- a non dispersive medium, for which $\omega = vk$;
- a linearly dispersive medium, for which $\omega = v_g k + \alpha$ (the phase velocity is not constant in this example);
- a quadratic dispersive medium, for which $\omega = v_g k + \alpha + \beta(k - k_0)^2$.

In the first case, as we saw in §2.2.2, we write $x' = x - vt$ and immediately find $F(x,t) = F(x',0) = F(x - vt, 0)$: the wave-group propagates unchanged at velocity v.

In the second case, substitution for ω in (2.56) gives us

$$F(x,t) = \int_{-\infty}^{\infty} a(k)\exp\{\mathrm{i}[kx - t(v_g k + \alpha)]\}\mathrm{d}k. \qquad (2.59)$$

Substituting x' for $x - v_g t$ gives, as above,

$$F(x,t) = e^{-i\alpha t} F(x',0) = e^{-i\alpha t} F(x - v_g t, 0). \qquad (2.60)$$

Thus the wave-group is propagated with unchanged envelope at velocity v_g ($= d\omega/dk$, the group velocity), but the phase of the wave changes with propagation; this is the process of 'individual waves moving through the envelope' mentioned in §2.5.

Neither of the above examples invoked explicitly the form of $a(k)$, and the results would be true for any form of wave-group. This is because $d\omega/dk$ is not a function of k. But in the third example, the way in which the wave envelope behaves depends on its initial form. Equation (2.55) becomes

$$
\begin{aligned}
F(x,t) &= (2\pi/\sigma)^{-\frac{1}{2}} \int_{-\infty}^{\infty} \exp\left[\frac{-(k_0 - k)^2 \sigma^2}{2}\right] \\
&\quad \times \exp\{i[kx - (v_g k + \alpha + \beta(k + k_0)^2)t]\}dk \\
&= (2\pi/\sigma)^{-\frac{1}{2}} e^{-i\alpha t} \int_{-\infty}^{\infty} \exp\left[-(k - k_0)^2 (\frac{\sigma^2}{2} + i\beta t)\right] \exp(ikx')dk.
\end{aligned}
$$
$$(2.61)$$

Formidable as it may look, the integral can be evaluated just like any other Gaussian integral (§4.4.6)† and gives

$$F(x',t) = (1 + 4\beta^2 \sigma^{-4} t^2)^{-\frac{1}{4}} \exp\left[\frac{-x'^2}{2(\sigma^2 + 4\beta^2 t^2 \sigma^{-2})}\right] e^{i\psi(t)} \qquad (2.62)$$

The form is not complicated. The envelope is centred on the point $x' = (x - v_g t) = 0$, so that once again we see the peak propagating at the group velocity. But the half-width and amplitude of the envelope have changed. The half-peak-width is now $2.36(\sigma^2 + 4\beta^2 t^2 \sigma^{-2})^{\frac{1}{2}}$ which is always greater than the initial $w = 2.36\sigma$, and continues to grow with time. In concert with this, the amplitude falls so that total energy is conserved. Notice that the narrower the original pulse (smaller σ_e, or larger σ), the faster the pulse broadens; for a given time of travel, there is a particular initial pulse-width which gives the narrowest final pulse. The phase factor $\psi(t)$ is complicated and is best calculated numerically. It contains the factor αt which occurred in the linear dispersive media, but also extra factors; it shows evidence of *chirping*, a variation of local k with distance. A numerically calculated example‡ is shown in Fig. 2.10. Study of the propagation of pulses in dispersive media is very important in designing optical communication systems (§§10.2–10.3) and this calculation is only a modest beginning.

† This is the reason that we chose the Gaussian wave group as an example!
‡ We are grateful to Yoav Oreg for providing this example.

2.9 Advanced topic: gravitational lenses

An interesting recent application of Huygens's and Fermat's principle to astrophysics is the explanation of multiple stellar images by gravitational lensing. The idea that a massive body can deflect a light wave is not new; Newton thought of it as a result of the mass of his assumed light corpuscles, but today it is considered as arising from the distortion of space near a massive body as predicted by general relativity. The effect and Einstein's prediction of its size were first confirmed in 1919 by observations of the apparent deflexion of star positions near to the direction of the Sun at the time of a complete eclipse. Similar radio observations have been made since then.

Recently, several high-resolution radio and infra-red images of very distant quasars have been observed to have several components, all with the same spectra (indicating a common source) and with measurable time delays between non-periodic features (such as bursts). These have been interpreted as resulting from optical effects induced by massive galaxies situated near the line of sight.

It is easiest to understand the effect by replacing the distortion of space-time by an effective refractive index μ_g which is related to the gravitational potential Φ and the velocity of light c by

$$\mu_g = 1 - 2\Phi/c^2 . \tag{2.63}$$

Since masses always attract, Φ is negative and thus $\mu_g > 1$.

First let us calculate the deflexion of the light from a star near the Sun. For a point (or spherical) mass M, $\Phi = -MG/r$ where G is the gravitational constant. Now the deflexion is very small in all cases, (<arcseconds) so that to a good approximation the optical path from a quasar, Q, to us, O, is a straight line whose closest approach to the

Figure 2.10 (*a*) A Gaussian wave-group at $t = 0$; (*b*) computed form of the wave-group after propagating for a distance through a medium with a quadratic dispersion equation.

Figure 2.11 Geometry of the source Q, gravitational lensing mass M and observer O.

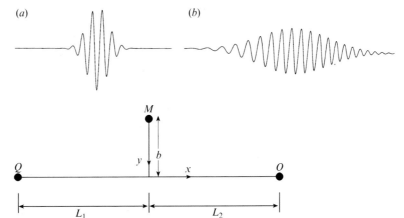

mass is a distance b. This is usually called, in deference to nuclear physics, the *impact parameter* (Fig. 2.11). The optical path length is

$$\bar{l} = \overline{QO} = \int_Q^O \mu_g(\mathbf{r})d\mathbf{r} = \int_{-L_1}^{L_2} \left(1 + \frac{2MG}{c^2(x^2+b^2)^{\frac{1}{2}}}\right) dx$$

$$= L_1 + L_2 + \frac{2MG}{c^2}(\ln 4 + \ln L_2 + \ln L_1 - 2\ln b)\text{(2.64)}$$

where we and the quasar are at distances L_2 and L_1, respectively, from M. This is shown in Fig. 2.12(a). Clearly, the optical path becomes longer the closer the path gets to the central mass, and so, in a sense, the mass acts like a converging lens, with a profile like that shown in Fig. 2.12(b). The angle of deflexion of the light, α, can be found from Huygens's principle. The local wavefront at the observer is the surface of constant $\bar{l}$ (equal time of propagation QO and QO'), and this is at angle $\alpha = d\bar{l}/dy$ for small angles. Thus

$$\alpha = \frac{d\bar{l}}{dy} = \frac{L_1}{L_2+L_1} \cdot \frac{d\bar{l}}{db} = \frac{4MG}{c^2b} \cdot \frac{L_1}{L_2+L_1} \simeq \frac{4MG}{c^2b}, \qquad (2.65)$$

Figure 2.12
Gravitational
lensing: (a) optical
path $\bar{l}$ as a function
of b for a point mass;
(b) profile of a plastic
lens used to simulate
gravitational lensing
in the laboratory.

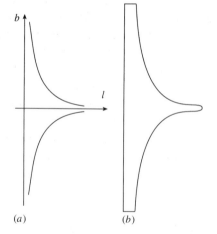

Figure 2.13
(a) Wavefronts
leaving the
gravitationally-
modified region;
(b) development of
the wavefront W
into W_1, W_2 etc, by
Huygens's
construction,
showing the
formation of three
distinct images, I_1, I_2
and I_3 instead of the
expected image I.

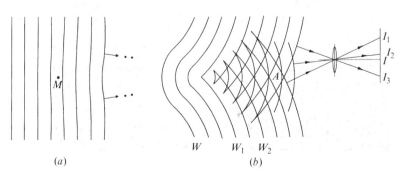

Figure 2.14 Image formed after light from a point source passes through a gravitational lens: (a) imaged through the lens of Fig. 2.12(b) on axis; (b) as (a), but off-axis (a fifth point, near the centre, exists but is too weak to be seen in the photograph); (c) the observed structure of the source Q2237+0305, obtained in the near infra-red by the JPL wide-field telescope, shows five distinct images with the same red-shift, 1.695. The scale-bar shows 2 arcsec. Photograph courtesy of NASA. A false colour representation of this image appears in Paczyński and Wambsganss (1993).

the approximation being valid when $L_2 \ll L_1$ (solar≪stellar distance). This was the deflexion observed in the 1919 experiment.

Now let us get a qualitative picture of the lensing which occurs when the Earth, a distant quasar and a massive lensing galaxy at intermediate distance lie very close to a straight line, so that b is extremely small. Astronomers assure us that there is a reasonable chance for such an occurrence. Assume that the point mass has finite size. This destroys the singularity of (2.64) at $b = 0$, and has allowed us to round off the centre of the lens in (Fig. 2.12(b)). We use this to construct the wavefront leaving the lensing mass region (Fig. 2.13(a)). Now we invoke Huygens' construction to continue the propagation of the wavefront. We find a situation as in Fig. 2.13(b). The propagation we see in this figure is just the Huygens's picture of the development of the wavefront which emerged from the lensing-mass region, even in the complete absence of any further distortion. At sufficient distance from the mass a region occurs in which the wavefront is multivalued. The margins of this region are a *caustic*, which is a pair of lines of cusps. Clearly an observer within the caustic observes three different wavefronts simultaneously, each with a different shape, and therefore sees three separate images. Notice also that the three wavefronts passing through A are not at the same optical distance from the source – hence the differing time delays observed. The above argument has been developed in two dimensions. Had the observer, the lensing galaxy and the source been in a straight line, it is clear from the resultant axial symmetry that the image observed would be a central point surrounded by a ring of light. If the lens is not spherically symmetrical, or the system not coaxial, a three-dimensional argument, similar to the one above, gives five point images. Fig. 2.14 shows

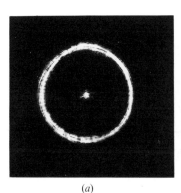

(a)

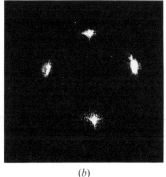

(b)

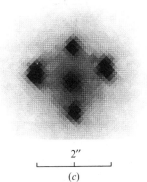

2″

(c)

an observed image which is thought to be gravitationally distorted by a galaxy close to the line of sight, compared with laboratory photographs of a point source as seen through a plastic lens with a profile like Fig. 2.12(*b*).

CHAPTER THREE

Geometrical optics

3.1 Introduction

If this book were to follow historical order, the present chapter should have preceded the previous one, since lenses and mirrors were known and studied long before wave theory was understood. However, once we have grasped the elements of wave theory, it is much easier to appreciate the strengths and limitations of geometrical optics, so logically it is really more appropriate to put this chapter here. Essentially, geometrical optics, which considers light waves as rays which propagate along straight lines in uniform media and are related by Snell's law (§§2.7.2, 5.4.2) at interfaces, has a similar relationship to wave optics as classical mechanics does to quantum mechanics. For geometrical optics to be strictly true, it is important that the sizes of the elements we are dealing with be large compared with the wavelength λ. This means that we can neglect diffraction, which otherwise prevents the exact simultaneous specification of the positions and directions of rays on which geometrical optics is based. From the practical point of view, geometrical optics answers most questions about optical instruments extremely well and in a much simpler way than wave theory could do; it fails only in that it can not define the limits of performance such as resolving power, and does not work well for very small devices such as optical fibres. These will be dealt with by wave theory in Chapters 10 and 12.

The plan of the chapter is first to treat the classical ray theory of thin lens systems in the *paraxial approximation*. This approximation assumes that for a given system an *optical axis*, z, can be defined and that all rays are almost parallel to it; we follow with an elegant and useful reformulation of paraxial geometrical optics which uses matrices and linear algebra. Except for one or two examples, we shall not deal in any depth with non-paraxial optics, which complicate matters by introducing aberrations. Good references for a deeper study of geometrical optics are Kingslake (1978, 1983) and Welford (1986).

The main elements from which optical systems are built are:

- thin lenses, for example spectacles or magnifying glasses, which converge or diverge bundles of light rays;
- compound lenses which, in addition to the above, are designed to correct various aberrations; examples are achromatic doublets and microscope objectives;
- plane mirrors or prisms which change the direction of the optical path and often serve to invert an image as in binocular field glasses;
- spherical or parabolic mirrors which replace lenses in large telescopes or in instruments working with wavelengths absorbed by common optical materials.

We shall limit the treatment in this chapter to lens systems only. The reason is that curved mirrors behave quite similarly to lenses but introduce minus signs into the equations, which add little to the physics but help to confuse the reader. The treatment of mirrors can be found in the more specialized texts.

Most lenses have spherical surfaces for the purely practical reason that such surfaces are easy to grind and polish. Spherical surfaces have no special optical significance which distinguishes them from paraboloidal, ellipsoidal or other analytical surfaces. All give rise to aberrations, and all are perfect solutions for some special cases which are too rare to be worth mentioning here. An exception is the aplanatic system (§3.9) in which a spherical surface turns out to be a perfect solution which can be applied to many important problems such as the design of microscope objectives.

3.2 The philosophy of optical design

Before discussing details and calculations, we shall briefly describe the steps which are taken when designing a lens-based imaging system. This in fact includes almost all applications of geometrical optics.†

(*a*) Decide in general how to solve your problem.

(*b*) Draw a ray diagram for object and image on the axis, using paraxial optics (§ 3.3).

---

† An important exception which springs to mind is the design of radiation collection systems, which are discussed by Welford and Winston (1989).

(*c*) Draw a similar paraxial diagram for an off-axis object. This will clarify the magnification of the system and emphasize any problem caused by light rays missing the finite apertures of the optical elements. Such a problem is called *vignetting* (see §3.4.2).

We emphasize the importance of stages (*b*) and (*c*), which will usually fulfil the requirements of the average physicist!

(*d*) Solve the problem in detail using matrix optics (§3.5) or a computer-based optical design program to find the best location of the lenses.
(*e*) Consider effects of large angle rays and aberrations.
(*f*) Consider possible use of aspherical lenses etc.

Stages (*e*) and (*f*) involve technical aspects which are outside the scope of this book.

3.3 Classical optics in the Gaussian approximation

The small angle or paraxial approximation is often known as *Gaussian optics*. In real life, rays propagating through lenses do not usually make small angles with the optical axis; two typical situations are shown in Fig. 3.1 where we have traced the rays through each lens by using Snell's law. We see that a bundle of rays entering parallel to z does *not* meet at a point on the axis. This is an example of *spherical aberration*, which will be further discussed in §3.8.3. However, in most of this chapter we shall try to avoid the problems created by large angles, and we shall therefore assume that the angles θ of all rays with respect to z is small enough that $\theta \approx \sin\theta \approx \tan\theta$. This defines the scope of Gaussian optics, which in practice does an excellent job even under conditions where the approximation is invalid! We also need to linearize Snell's law of refraction (§2.7.2), $\mu_1 \sin\hat{\imath} = \mu_2 \sin\hat{r}$, to the form $\mu_1\hat{\imath} = \mu_2\hat{r}$. This requires $\hat{\imath}$ and $\hat{r}$ to be small too, so that all refracting surfaces must be almost normal to z. It follows that their radii must be large compared to the distance of rays from the optical axis.

3.3.1 Sign conventions

It is helpful to have a consistent convention for the use of positive and negative quantities in optics. In this book, we use a simple Cartesian convention, as shown in Fig. 3.2, together with the rules that surfaces concave to the right have a positive radius and that angles of rays ascending to the right are taken as positive. It is also assumed that light rays go from left to right as long as we limit ourselves to lens systems. Other conventions – based upon whether real or virtual images or objects are involved – lead to difficulties. Although there are no such direct applications in this book, anyone who wishes to extend his knowledge of lens aberrations, for example, will have to use a Cartesian system and the above convention for spherical surfaces.

When drawing ray diagrams the y axis has to be scaled by a large factor, otherwise small angles are invisible. Therefore, spherical surfaces will appear in the diagrams to be planar, whatever their radii of curvature.

3.3.2 The imaging equation for a single thin lens in air

We treat first a single refracting spherical surface of radius R_1 located at $z = 0$ as in Fig. 3.2 with $\mu = 1$ on the left, and consider a ray originating from a point object O located on the axis at $z = u$. In accordance with the sign convention, distances to the left of the vertex V are taken as negative. A ray from O passes through the surface at $y = y_1$. It is refracted from angle $\hat{i}$ to angle $\hat{r}$ and consequently seems to have originated from a virtual image I located at $z = v$. We then have:

$$\hat{i} - \phi = -y_1/u, \quad \hat{r} - \phi = -y_1/v, \quad \phi = y_1/R_1 , \tag{3.1}$$

and it follows that

$$\frac{\hat{i}}{\hat{r}} = \mu = \frac{y_1(\frac{1}{R_1} - \frac{1}{u})}{y_1(\frac{1}{R_1} - \frac{1}{v})} , \tag{3.2}$$

which simplifies to

$$-\frac{\mu}{v} + \frac{1}{u} = \frac{1}{R_1}(1 - \mu) . \tag{3.3}$$

The μ in the first term of (3.3) indicates that v refers to a region of refractive index μ.

Figure 3.1 Spherical aberration of simple lenses: (a) biconvex; (b) plano-convex. The two lenses have the same paraxial focal length.

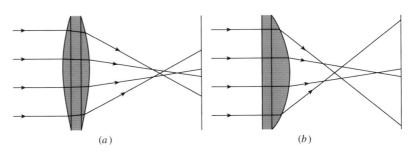

(a) (b)

The position $z = v'$ of the image generated by a thin lens in air can now be derived by using (3.3) for a second surface with radius R_2 and replacing the object position by the v just calculated (Fig. 3.3). This replacement implies that vertices of the two surfaces coincide geometrically, i.e. that the lens is indeed thin compared with v and u. The roles of μ and 1 are interchanged and we get

$$-\frac{1}{v'} + \frac{\mu}{v} = \frac{1}{R_2}(\mu - 1) . \tag{3.4}$$

We now find by substitution of μ/v from (3.3) that

$$-\frac{1}{v'} + \frac{1}{u} + \frac{1}{R_1}(\mu - 1) = \frac{1}{R_2}(\mu - 1) . \tag{3.5}$$

Therefore we obtain the well-known formula:

$$-\frac{1}{u} + \frac{1}{v'} = \frac{1}{f} , \tag{3.6}$$

where

$$\frac{1}{f} = (\mu - 1)(\frac{1}{R_1} - \frac{1}{R_2}) . \tag{3.7}$$

The object and image distances u and v' are called *conjugates*. When $u \to \infty$, $v' \to f$. This means that all rays entering the lens parallel to the axis cut the axis at the focus F (Fig. 3.4). The quantity f is called the *focal length* and its reciprocal $1/f$ the *power* of the lens. When f is measured in metres, the unit of power is the *dioptre*. A

Figure 3.2 Deviation of a ray by a single refracting surface. In the Gaussian approximation, $y_1 \ll R_1$ so that $VC \ll R_1$ and is taken as zero in the theory. According to the sign convention, in this diagram u and v are negative, R_1 and angles $\hat{i}$ and $\hat{r}$ are positive.

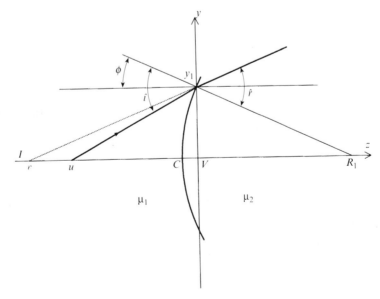

lens with $f > 0$ is called a *converging* lens and one with $f < 0$ is *diverging*. Only the former can produce a real image of a physical object.

The *focal plane*, $\mathscr{F}$, is defined as the plane through F perpendicular to the optical axis. All rays entering the lens at a given small angle α to the axis will converge on the same point in $\mathscr{F}$. One can see this by reference to Fig. 3.5 which shows a general ray a entering at angle α

Figure 3.3 Image formation by a thin lens in air.

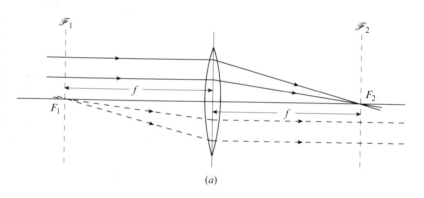

Figure 3.4 Focal planes and focal lengths of a thin lenses in air: (a) converging; (b) diverging. The paths of various rays are shown.

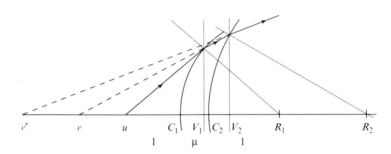

(a)

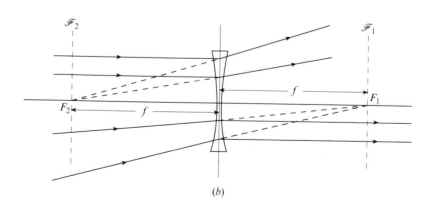

(b)

after intersecting the x axis at some u. It cuts $\mathscr{F}$ at P, with height d, and v is the conjugate to u. We have

$$y = -u\alpha , \qquad d = \beta(f - v) = \frac{y}{v}(v - f) \qquad (3.8)$$

and, using (3.6), it follows that $d = f\alpha$ independent of u. The easiest way to find P is to use the undeviated ray b at α through the centre of the lens.

3.4 Ray tracing through simple systems

Paraxial ray tracing is an important tool in optical design, and involves following through the optical system the routes of several paraxial rays leaving an off-axis point object. There are three types of rays which can be traced simply through a thin lens (see Figs. 3.4 and 3.5):

(1) all rays through a focal point on one side continue parallel to the axis on the other side and *vice versa*;
(2) any ray passing through the centre of the lens continues in a straight line;
(3) all rays of a parallel bundle on one side of a lens go through a single point in the focal plane on the other side, and *vice versa*; the point can be found by drawing the one ray of the bundle which passes through the centre.

Using these rays, we can generally get a good picture of the optical properties of a complete system. We shall treat two examples, the magnifying glass and the astronomical telescope. In the second example we shall also introduce the concepts of vignetting and stops.

Figure 3.5 All rays at a given angle to the axis converge to a single point in the focal plane.

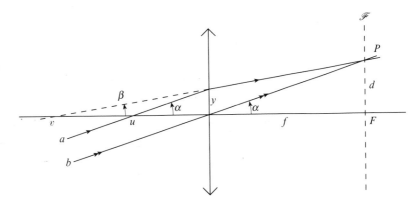

3.4.1 The magnifying glass

The magnifying glass – or its more elaborate form, the eyepiece – is the simplest optical instrument and its function should be clearly understood. Its main purpose is to allow us to bring an object to a position well within the *near point* (the closest distance at which the eye can produce a sharply focused image, about 25 cm for young people with normal eyesight), while creating a virtual image at or beyond the near point, often at infinity. This is illustrated in Fig. 3.6.

For a magnifying glass, or any optical instrument which forms a virtual image, the *magnifying power* is a more useful quantity than the linear magnification (the ratio of image to object size). It can be defined in two equivalent ways:

(*a*) the ratio between the angle subtended at the eye by the image and the angle which would be subtended at the eye by the object if it were situated at the near point, which is at a distance D from the eye;

(*b*) the ratio between the linear dimensions of the retinal image produced with the instrument and those of the largest clear retinal image which can be produced without it, i.e. when the object is at the near point.

Normally the magnifying lens is close to the eye. The magnifying power is then (Fig. 3.6)

$$M = \frac{y'}{y} = \frac{D}{d} . \tag{3.9}$$

Since $1/d - 1/D = 1/f$, we find that

$$M = 1 + \frac{D}{f} ; \tag{3.10}$$

usually $D \gg f$ so that the approximation $M = D/f$ can be used.

The single lens is now used only for informal purposes. For precise work eyepieces are generally used; an example is described in §3.4.3.

Figure 3.6 Ray diagram for a magnifying glass. The object height is y and that of the image y'. In practice, $f \ll D$ and the object distance would be nearly equal to f.

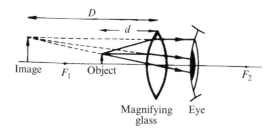

3.4.2 The astronomical telescope, and some remarks about stops

A telescope converts a bundle of parallel rays making an angle α with the optical axis to a second parallel bundle with angle β. The ratio β/α is called the *angular magnification*; it equals the ratio between the retinal image sizes obtained with and without the telescope. Fig. 3.7(a) shows a simple telescope based on two lenses, L_1 (objective) and L_2 (eyepiece) with focal lengths f_1 and f_2; the distance between the lenses is $f_1 + f_2$, so that an object at infinity produces a real image in the common focal plane.

One's first instinct is to place the eye at E_0 immediately behind L_2, but we shall see that this may unnecessarily limit the field of view. Let us try to analyze the light throughput by means of the paraxial ray diagram. A bundle of parallel rays (*aaa*) from a distant point on the optical axis enters the objective L_1 parallel to the axis and leaves through the eyepiece L_2, entering the iris of the observer's eye. Now consider a bundle of parallel rays (*bbb*) from a second distant point, not on the axis. We assume, for the moment, that the aperture of L_2 is very large. The rays enter L_1 at an angle to the axis, and may miss the iris E_0 because of its limited size. This is called *vignetting* (§3.2). However, it should be clear from the diagram that if we move the eye further back from L_2, so that the iris is now at E, rays from the

Figure 3.7 Ray tracing through a telescope with angular magnification 3. (a) Simple astronomical telescope; (b) with the addition of a field lens. The exit pupil is at E in both drawings.

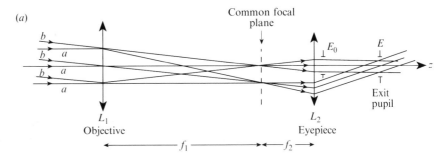

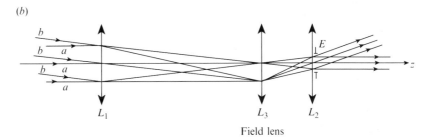

oblique bundle (*bbb*) will also enter it. From the figure, one can see that E is in the plane of the image of L_1 formed by L_2; this image is called the *exit pupil*, which will be defined formally below. Positioning the eye in the plane of the exit pupil allows it to receive light from off-axis points and thus maximizes the field of view.

Now we have assumed so far that the aperture of L_2 was very large, and did not limit the light throughput. If L_2 is finite in size, vignetting of the oblique bundle might occur there. To avoid that problem, we need to add another lens L_3, called a *field lens* (Fig. 3.7(*b*)). This is placed in the common focal plane, where it has no effect on the intermediate image, and is designed to create an image of L_1 on L_2. Then, it is easy to see that an oblique bundle of rays entering L_1 must leave through the centre of L_2. The exit pupil has now been moved to coincide with L_2 and so in this case the best place for the eye is indeed close behind L_2. Vignetting can now occur at the field lens, but since this is in the intermediate image plane its aperture simply forms a sharp edge to the image, which limits the angular field of view. This is called a *field stop*, and an actual ring aperture is usually placed there for aesthetic reasons.

In the end, certain practical considerations win, and it is usual to put the exit pupil somewhat behind L_2, for the observer's comfort, and to place the field lens not quite in the intermediate image plane, so that dust on it will not spoil the image and possibly to allow for cross-hairs or a reticle scale to be placed in the field stop. Both of these compromises require that the lenses be slightly larger than the absolute minimum requirement.

It is useful, in discussing optical instruments, to have some additional definitions at our disposal. The aperture which limits the amount of light passing through an optical instrument from a point on its axis is called the *aperture stop*; in this case it is L_1 (if L_2 is slightly larger than needed), but can in principle be any aperture or lens in the instrument. A complex optical system is usually designed so that the most expensive item, or the one most problematic to design, is the aperture stop, so that it can be fully used.† The image of the aperture stop in the optical elements following it is then the exit pupil. Clearly, the eye or a camera lens should be placed at this position. If the aperture stop is not the first element in the system, then the image of the aperture stop in the lenses *preceding* it is called the *entrance*

---

† Two examples: a large-aperture telescope mirror or lens will always be made the aperture stop because of its cost. In optical systems involving mechanical scanners, the scanner is usually the aperture stop so as to minimize its size and hence its moment of inertia.

pupil. The theory of stops has many other applications; for example, a camera lens always includes an adjustable aperture stop, which controls the light intensity in the film plane, and whose position within the compound lens is calculated to give a uniformly bright image at off-axis points.

We should also point out that ray tracing allows the designer to calculate the optimum sizes for his components. For example in the case of the telescope, since the size of the pupil of the observer's eye is given anatomically, this in turn determines the sizes of L_1 and L_2 which can be utilized for a given magnification.

3.4.3 Compound eyepieces

The field lens and the eyepiece are often combined in a single unit called a *compound eyepiece*, or simply *eyepiece* for short. This has several advantages, amongst them the possibility of using the two lenses to correct aberrations, and the inclusion of a field stop. An example of this is the Ramsden eyepiece, shown in Fig. 3.8.

3.4.4 The microscope

The essential principle of the microscope is that an objective lens with very short focal length (often a few mm) is used to form a highly magnified real image of the object. The object is therefore just outside the focal plane of the objective.† An eyepiece is then used to magnify this image further. A ray diagram is shown in Fig. 3.9. After the objective, the ray diagram is identical to that of a telescope, and the same considerations about field lens and exit pupil apply. Because of its intricate design and the need to use large angles of refraction, the

---

† In order that the object be placed *exactly* in the focal plane of the objective, so that each object point will create a parallel bundle of rays, an additional weak lens (tube lens) is often added to produce this magnified image. Many microscopes include further components such as beam-splitters, polarizers etc. between the objective and the tube lens. These components do not affect the principle, but their design is simpler if they operate in parallel light.

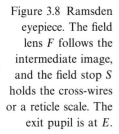

Figure 3.8 Ramsden eyepiece. The field lens F follows the intermediate image, and the field stop S holds the cross-wires or a reticle scale. The exit pupil is at E.

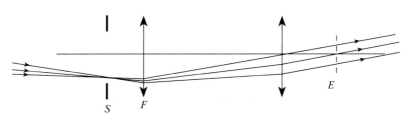

design centres around the objective which contains the aperture stop. It will be seen that the final image is virtual and inverted.

The objects typically observed through a microscope contain detail as small as, or smaller than, the wavelength of light, and so geometrical optics only give us a very general description of the imaging process. A full picture of the capabilities and limitations of microscope imaging only emerges with the use of wave optics. This will be described in Chapter 12.

3.5 The matrix formalism of the Gaussian optics of axially symmetric refractive systems

It is very cumbersome to extend the algebraic type of analysis we saw in §3.3.2 to more complicated systems. A much more convenient method, which utilizes the fact that equations (3.2)–(3.10) are linear as a result of the approximation (3.1), uses matrices to make the calculation of the optical properties of even the most complicated systems quite straightforward, and is also particularly convenient for numerical computations (Problem 3.10).

The propagation of a ray through an axially symmetric system of lenses consists of successive refractions and translations. As mentioned earlier, the direction of propagation of a ray through the system will be taken from left to right. We shall treat only rays which lie in a plane containing the optical axis z; we shall ignore what are called *skew*

Figure 3.9 Ray diagram for microscope showing position of exit pupil.

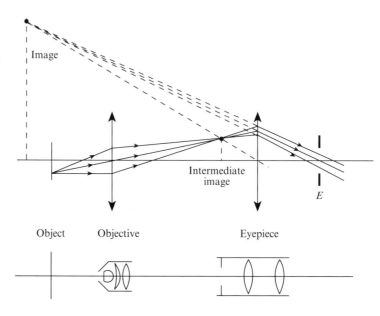

rays, which add no new information in the paraxial approximation.†
Since the system has rotational symmetry around z, a ray at $z = z_0$
is specified if we know its algebraic distance y from the axis and its
inclination $dy/dz = \theta$; it is therefore sufficient to follow the rays in
one (y, z) plane.

3.5.1 The translation and refraction matrices

Let us consider first a ray which propagates in a straight line in a
uniform medium of index μ (Fig. 3.10). It has height y_1 and inclination
θ_1 at $z = z_1$, and y_2 and $\theta_2 = \theta_1$ at $z = z_1 + t$. Then

$$y_2 = y_1 + t\theta_1 , \qquad (3.11)$$

$$\theta_2 = \theta_1 . \qquad (3.12)$$

These equations can be described by the matrix equation between
vectors $(y, \mu\theta)$:‡

$$\begin{pmatrix} y_2 \\ \mu\theta_2 \end{pmatrix} = \begin{pmatrix} 1 & t/\mu \\ 0 & 1 \end{pmatrix} \begin{pmatrix} y_1 \\ \mu\theta_1 \end{pmatrix} = \mathsf{T} \begin{pmatrix} y_1 \\ \mu\theta_1 \end{pmatrix} \qquad (3.13)$$

which defines T, the translation matrix from z_1 to $z_1 + t$.

A second matrix describing the refraction as a ray passes through
a surface with radius R from a medium with index μ_1 to a medium
with index μ_2 is derived as follows (see Fig. 3.11). Snell's law $\mu_1 \sin \hat{\imath} = \mu_2 \sin \hat{r}$ gives

$$\mu_1 \sin(\phi + \theta_1) = \mu_2 \sin(\phi + \theta_2) , \qquad (3.14)$$

† The projection of a skew ray on any plane containing the optical axis is itself a valid
ray in the paraxial approximation. This is not true at large angles.
‡ We use the product $\mu\theta$ rather than θ alone for later convenience.

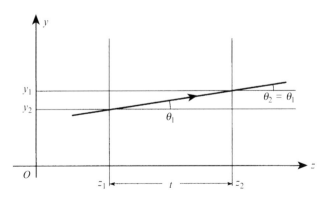

Figure 3.10
Representing the
translation matrix.

which becomes, for small angles,

$$\mu_1\phi + \mu_1\theta_1 = \mu_2\phi + \mu_2\theta_2 \ . \tag{3.15}$$

With $\phi = y_1/R$ we get

$$\mu_2\theta_2 = \mu_1\theta_1 - (\mu_2 - \mu_1)y_1/R \ . \tag{3.16}$$

Note that $\phi > 0$. Since z_1 and z_2 coincide we have $y_1 = y_2$ and so we can define a refraction matrix R by the following equation:

$$\begin{pmatrix} y_2 \\ \mu_2\theta_2 \end{pmatrix} = \begin{pmatrix} 1 & 0 \\ \frac{\mu_1-\mu_2}{R} & 1 \end{pmatrix} \begin{pmatrix} y_1 \\ \mu_1\theta_1 \end{pmatrix} = R \begin{pmatrix} y_1 \\ \mu_1\theta_1 \end{pmatrix}. \tag{3.17}$$

A general matrix M_{21} which connects a ray at z_1 with its continuation at z_2 performs the operation

$$\begin{pmatrix} y_2 \\ \mu_2\theta_2 \end{pmatrix} = M_{21} \begin{pmatrix} y_1 \\ \mu_1\theta_1 \end{pmatrix}, \tag{3.18}$$

where M_{21} is a product of T and R matrices. Since $\det\{R\} = \det\{T\} = 1$, $\det\{M_{21}\} = 1$. We used the combination $\mu\theta$, and not θ alone, to make these determinants unity.

3.5.2 Matrix representation of a simple lens

As we saw in §3.3.2 a simple lens consists of an optically transparent medium, with refractive index μ, bounded by two spherical surfaces (Fig. 3.12). The line joining the centres of the two spheres defines the optical axis z, and the system is symmetric about it. Initially we shall assume that the medium outside the lens has unit refractive index. The vertices of the lens (the points at which the surfaces cut the axis)

Figure 3.11 Representing the refraction matrix. In the spirit of Gaussian optics, V and C coincide.

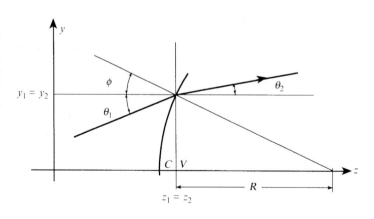

are z_1 and z_2 where $t = z_2 - z_1$. The matrix M_{21} between $z = z_1$ and $z = z_2$ is derived as follows:

$$
\begin{pmatrix} y_2 \\ \theta_2 \end{pmatrix} = \begin{pmatrix} 1 & 0 \\ \frac{\mu - 1}{R_2} & 1 \end{pmatrix} \begin{pmatrix} 1 & t/\mu \\ 0 & 1 \end{pmatrix} \begin{pmatrix} 1 & 0 \\ \frac{1-\mu}{R_1} & 1 \end{pmatrix} \begin{pmatrix} y_1 \\ \theta_1 \end{pmatrix}
$$

$$
= M_{21} \begin{pmatrix} y_1 \\ \theta_1 \end{pmatrix}. \tag{3.19}
$$

where

$$
M_{21} = \begin{pmatrix} 1 + \frac{t(1-\mu)}{\mu R_1} & t/\mu \\ (\mu - 1)(\frac{1}{R_2} - \frac{1}{R_1}) - \frac{t(1-\mu)^2}{R_1 R_2 \mu} & 1 + \frac{(\mu-1)t}{R_2 \mu} \end{pmatrix}. \tag{3.20}
$$

For a thin lens we assume that t is small enough for the second term in the lower left element to be negligible; since $\mu - 1$ is of the order of 1, this means that $t \ll |R_1 - R_2|$. Then, putting $t = 0$, we have

$$
M_{21} = \begin{pmatrix} 1 & 0 \\ (\mu - 1)\left(\frac{1}{R_2} - \frac{1}{R_1}\right) & 1 \end{pmatrix} = \begin{pmatrix} 1 & 0 \\ -\frac{1}{f} & 1 \end{pmatrix} \tag{3.21}
$$

where the focal length is the same as that defined in (3.7). We shall emphasize that the matrix M_{21} summarizes what we already expect from a thin lens in air.

(1) An incident ray parallel to the axis at height y_1, i.e. $(y_1, 0)$, leaves the lens at the same height, but is tilted downwards, if f is positive, to cut the axis at distance f, and is represented by $(y_1, -y_1/f)$.

(2) If $R_1 = R_2$, we have a spherical shell. We might expect this to have infinite focal length, but t can not be neglected (since $R_1 - R_2 = 0$) and the full expression for f from (3.20) must be used (Problem 3.7).

(3) If $R_1^{-1} > R_2^{-1}$, as in Fig. 3.12, and $\mu > 1$, the lens is converging. This relationship means that the lens is thickest in the centre.

If the lens is surrounded by media of refractive indices μ_1 on the left and μ_2 on the right, it is easy to repeat the calculation and show that, in the case where t is negligible,

$$
\begin{pmatrix} y_2 \\ \mu_2 \theta_2 \end{pmatrix} = M_{21} \begin{pmatrix} y_1 \\ \mu_1 \theta_1 \end{pmatrix} \tag{3.22}
$$

$$
\text{where} \quad M_{21} = \begin{pmatrix} 1 & 0 \\ \frac{\mu - \mu_2}{R_2} + \frac{\mu_1 - \mu}{R_1} & 1 \end{pmatrix}. \tag{3.23}
$$

We shall return to this situation in §3.7.2.

3.5.3 Object and image space

A system of lenses is limited by its left and right vertices, V_1 and V_2. It is useful to define the *object space* as a space with origin at V_1 and the *image space* with origin at V_2. To the left of V_1 we can put a real object, and we can project a real image onto a screen if it is to the right of V_2. Both spaces also have 'virtual' parts. For instance, a virtual image can be formed to the left of V_2 by a magnifying glass, but one cannot put a screen at the location of this image; similarly, a virtual object can be produced to the right of V_1 by some preceding optics.

3.6 Image formation

The formation of images is the most common task of an optical system, and we shall now see how it is described by the matrices. Consider a general system, extending from z_1 to z_2 and described by the matrix $\mathsf{M}_{21} = \begin{pmatrix} A & B \\ C & D \end{pmatrix}$. This matrix performs the operation

$$\begin{pmatrix} y_2 \\ \mu_2\theta_2 \end{pmatrix} = \begin{pmatrix} A & B \\ C & D \end{pmatrix} \begin{pmatrix} y_1 \\ \mu_1\theta_1 \end{pmatrix}. \tag{3.24}$$

If this system forms at z_2 an image of an object at z_1, then the (x, y) planes at z_1 and at z_2 are called *conjugate planes*. Imaging means that y_2 must be independent of $\mu_1\theta_1$; in other words, all rays leaving a point (y_1, z_1) *in any direction* θ must arrive at *the same point* (y_2, z_2). Since the point of arrival is independent of the angle θ_1, B must be 0.

Figure 3.12 Ray diagram for image formation by a thin lens in air with all quantities positive. In the spirit of the Gaussian approximation, V_1 and C_1 coincide, as do V_2 and C_2.

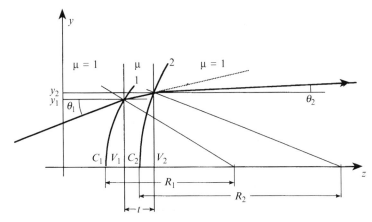

It also follows, since the determinant of M_{21} is unity, that $AD = 1$. The *linear magnification* produced by the system is

$$m = \frac{y_2}{y_1} = A \ . \tag{3.25}$$

A ray originating at $(0, z_1)$ with angle θ_1 will pass through $(0, z_2)$ with angle θ_2. The ratio between the ray angles is the *angular magnification*:

$$\frac{\theta_2}{\theta_1} = D\frac{\mu_1}{\mu_2} = \frac{1}{m}\frac{\mu_1}{\mu_2} \ . \tag{3.26}$$

3.6.1 Imaging by a thin lens in air

Now let us return to the thin lens (§3.5.2). We put an object at $z_1 = u$ (where u is negative), a thin lens at $z = 0$ and find an image at $z_2 = v$. Using (3.21), we write the matrix of the complete system of Fig. 3.12:

$$\begin{pmatrix} A & B \\ C & D \end{pmatrix} = \begin{pmatrix} 1 & v \\ 0 & 1 \end{pmatrix} \begin{pmatrix} 1 & 0 \\ -1/f & 1 \end{pmatrix} \begin{pmatrix} 1 & -u \\ 0 & 1 \end{pmatrix}$$

$$= \begin{pmatrix} 1 - v/f & -u + v + vu/f \\ -1/f & 1 + u/f \end{pmatrix}. \tag{3.27}$$

Since the system is image-forming,

$$B = -u + v + vu/f = 0 \ , \tag{3.28}$$

$$\text{or} \quad -\frac{1}{u} + \frac{1}{v} = \frac{1}{f} \ . \tag{3.29}$$

We have therefore recovered (3.6). The linear magnification is $m = 1 - v/f = v/u$ while for a ray with $y_1 = 0$ the angular magnification is $1 + u/f = 1/m = u/v$.

Another way of expressing the imaging comes from the fact that if $B = 0$, then $AD = 1$. Therefore we have

$$(1 - v/f)(1 + u/f) = 1 \ , \tag{3.30}$$

$$\text{or} \quad (f - v)(u + f) = f^2 \ , \tag{3.31}$$

which is called *Newton's equation*. Remember that u is negative. This equation is very useful; we shall see that it applies to any lens, not just a thin lens, and its usefulness derives from the fact that it does not refer to the vertices of the lens, but only involves the image point, object point and foci. We emphasize that (3.29) and (3.31) are not independent, but each can be derived from the other.

3.6.2 Telescopic or afocal systems

If $C = 0$, θ_2 does not depend on y_1 and a bundle of parallel rays entering the system will emerge as a bundle of parallel rays, but at a different angle. A system with this property is called *telescopic* or *afocal*. From an object at infinity it creates an image at infinity. Two common telescopic systems are the simple astronomical telescope (§ 3.4.2) and the Galilean telescope (which has a diverging eyepiece and therefore does not invert the image).

3.7 The cardinal points and planes

Let us consider an imaging lens system in air which is represented, between its vertices V_1 at z_1 and V_2 at z_2, by the general matrix $M_{21} = \begin{pmatrix} a & b \\ c & d \end{pmatrix}$, which replaces $\begin{pmatrix} 1 & 0 \\ -1/f & 1 \end{pmatrix}$ of the thin lens in air from §3.6.1. Instead of (3.27) we then have

$$\begin{pmatrix} A & B \\ C & D \end{pmatrix} = \begin{pmatrix} 1 & v \\ 0 & 1 \end{pmatrix} \begin{pmatrix} a & b \\ c & d \end{pmatrix} \begin{pmatrix} 1 & -u \\ 0 & 1 \end{pmatrix}$$

$$= \begin{pmatrix} a + vc & b - au + v(d - cu) \\ c & d - cu \end{pmatrix} \qquad (3.32)$$

where we recall that u and v are measured from V_1 and V_2 respectively in the positive z direction. Once again, the imaging condition is represented by $B = 0$, which leads to

$$b - au + vd - vcu = 0 . \qquad (3.33)$$

If $B = 0$, it follows again that $AD = 1$, which gives

$$(a + vc)(d - cu) = 1 . \qquad (3.34)$$

The clue to simplification is then given by comparing (3.34) with Newton's equation (3.31) for the thin lens. Clearly the two are similar when we write (3.34)

$$(fa - v)(fd + u) = f^2, \qquad (3.35)$$

where we define $-1/c$ as the focal length f (as it was for the thin lens). By putting $v = \infty$ and $u = -\infty$ respectively, the *focal points* F_1 and F_2 are then found to be at $z_1 + d/c$ and $z_2 - a/c$. Next we write (3.35) in the form

$$\{f - [v - (a - 1)f]\}\{f + [u - (1 - d)f]\} = f^2 , \qquad (3.36)$$

and use the definitions

$$u_p = (d-1)/c, \tag{3.37}$$

$$v_p = (1-a)/c, \tag{3.38}$$

$$f = -1/c. \tag{3.39}$$

to write this as

$$[f - (v - v_p)][f + (u - u_p)] = f^2. \tag{3.40}$$

This is the same as Newton's equation provided we measure the object and image distances from the *principal points* H_1 at $z = z_1 + u_p$ and H_2 at $z = z_2 + v_p$ respectively. (3.36) can then be written

$$-\frac{1}{u - u_p} + \frac{1}{v - v_p} = \frac{1}{f}. \tag{3.41}$$

It is easy to show from the above equations that the linear magnification is now $m = A = (v - v_p)/(u - u_p)$ and the angular magnification, as usual, is $1/m$.

The *principal planes* $\mathscr{H}_1$ and $\mathscr{H}_2$, normal to z through H_1 and H_2, are defined in many texts as conjugate planes with unit magnification. On substituting $u = u_p$, $v = v_p$ into (3.32) we find immediately:

$$\begin{pmatrix} A & B \\ C & D \end{pmatrix} = \begin{pmatrix} 1 & 0 \\ C & 1 \end{pmatrix}; \tag{3.42}$$

and see that the principal planes are indeed conjugate ($B = 0$) with unit linear and angular magnifications ($A = 1$). We remind the reader that positive unit magnification means that an *upright* image of the same size as the the object is formed. There is a situation which might be confused with this where $u - u_p = -2f$, $v - v_p = 2f$, but this has magnification -1.

The four points H_1, H_2, F_1 and F_2 are four of six *cardinal points* which represent the lens system matrix for ray-tracing purposes. The other two (nodal points N_1 and N_2, which coincide with H_1 and H_2 when the object and image spaces have the same refractive index) will be discussed later (§3.7.2). We summarize the positions of the principal and focal points in terms of the matrix elements of the system, emphasizing that it is immersed in a medium of unit refractive index.

– Principal points: H_1 at $z = (d-1)/c + z_1$, H_2 at $z = (1-a)/c + z_2$.
– Focal points: F_1 at $z = d/c + z_1$, F_2 at $z = -a/c + z_2$.

Clearly, $F_1 H_1 = H_2 F_2 = -1/c = f$, so that each focal point is a distance f from the related principal point.

3.7.1 Geometrical meaning of the focal and principal points

If a bundle of rays parallel to z enters the lens system, we have $u = -\infty$, whence $v - v_p = f$ so that the bundle is focused to F_2. An oblique incident bundle at angle α focuses to $y = \alpha f$ in the *focal plane*, $\mathscr{F}_2$, which is normal to z through F_2. Similarly, any ray passing through F_1 leaves the system parallel to the axis and the focal plane $\mathscr{F}_1$ is normal to z through F_1. These are just like in a thin lens.

For a thin lens $\mathscr{H}_1$ and $\mathscr{H}_2$ are both in the plane of the lens but in a general lens system they may be somewhere else (see example in §3.7.3). They generally do not coincide. Since F_1 is at a distance f to the left of $\mathscr{H}_1$, $\mathscr{H}_1$ can be interpreted as the position of the thin lens of focal length f which would focus light from a point source at F_1 to a parallel beam travelling to the right; similarly, $\mathscr{H}_2$ is the plane of the same thin lens when it is required to focus a parallel beam incident from the left to F_2.

For the purposes of ray tracing through a lens system in air we use the cardinal points as follows (Fig. 3.13(*a*)). Any incident ray (1) passing through F_1 leaves the system parallel to the axis at the height it reaches $\mathscr{H}_1$, as if there were a thin lens in that plane. Likewise, a ray (2) incident from the left and parallel to the axis goes through F_2 as if it had been refracted by a lens in $\mathscr{H}_2$. Any incident ray (3) through the point H_1 exits through H_2 since $\mathscr{H}_1$ and $\mathscr{H}_2$ are conjugate; moreover, the incident and exiting rays are parallel because the angular magnification between the principal planes of a system in air is unity. Thus any ray through the system can be traced by finding its intersection with $\mathscr{H}_1$, and continuing it from $\mathscr{H}_2$ at the same height (unit magnification), using an auxiliary parallel ray through a focus to help find the exit direction.

A convenient way to visualize ray tracing through a complete system in air is carried out by the following steps, shown schematically in Fig. 3.13(*b*). Given the data on the system,

(1) find the cardinal points, F_1, F_2, H_1 and H_2, and mark them along the z axis on a piece of paper, together with V_1 and V_2;
(2) fold the paper so that the planes $\mathscr{H}_1$ and $\mathscr{H}_2$ coincide, and the z-axis remains continuous (this needs two parallel folds, one along $\mathscr{H}_1$ and the other along the mid-plane between $\mathscr{H}_1$ and $\mathscr{H}_2$);
(3) trace rays as if the (coincidental) principal planes were a thin lens (§3.4);
(4) unfold the paper. The rays that are drawn on it represent their paths outside $V_1 V_2$. Within $V_1 V_2$, further information is necessary to complete them (see §3.7.3 for an example).

3.7.2 Lens systems immersed in media: nodal points

Although many optical systems have $\mu = 1$ in both object and image spaces, this is not a requirement. Indeed, the eye lens has vitreous fluid ($\mu = 1.336$) in the image region. The most general system will have $\mu = \mu_1$ in its object space and $\mu = \mu_2$ in its image space. The thin lens of §3.5.2 then has a matrix (3.23):

$$M_{21} = \begin{pmatrix} 1 & 0 \\ \frac{\mu - \mu_2}{R_2} + \frac{\mu_1 - \mu}{R_1} & 1 \end{pmatrix}, \tag{3.43}$$

where μ is the refractive index of the lens material. Its focal length is f, where

$$-\frac{1}{f} = \frac{\mu - \mu_2}{R_2} + \frac{\mu_1 - \mu}{R_1}. \tag{3.44}$$

Figure 3.13
(*a*) Tracing rays through a general optical system in air, using the principal and focal points.
(*b*) The paper-folding method.

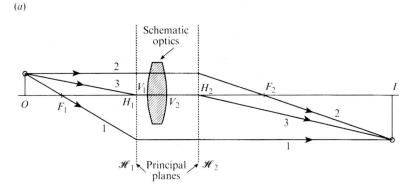

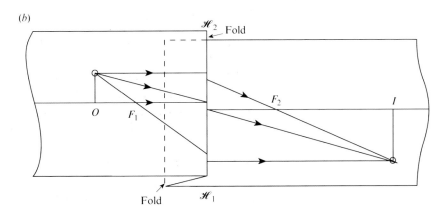

Replacing u by u/μ_1 and v by v/μ_2 in the matrices of (3.27) gives immediately

$$-\frac{\mu_1}{u} + \frac{\mu_2}{v} = \frac{1}{f} \, . \tag{3.45}$$

The focal lengths are then $\mu_1 f$ and $\mu_2 f$ on the left and right. For the general system described by (3.32) we use the same replacement and the following results can be derived straightforwardly. Newton's law (3.31) becomes

$$\left(\frac{-a\mu_2}{c} - v\right)\left(\frac{-d\mu_1}{c} + u\right) = \frac{\mu_1\mu_2}{c^2} = \mu_2\mu_1 f^2 \, . \tag{3.46}$$

Once again, we have principal planes $\mathcal{H}_1$ and $\mathcal{H}_2$ in the positions of thin lenses equivalent to the system as we had in §3.7.1. Now $\mathcal{H}_1$ is at $z = z_1 + \mu_1(d-1)/c$ and $\mathcal{H}_2$ is at $z = z_2 + \mu_2(1-a)/c$. As before, the planes $\mathcal{H}_1$ and $\mathcal{H}_2$ are conjugate with unit linear magnification. But the angular magnification between the principal planes is now, from (3.26),

$$D\mu_1/\mu_2 = \mu_1/\mu_2 \tag{3.47}$$

which is not unity. In order to complete the ray tracing by the method described in §3.7.1, we need to locate a pair of conjugate axial points N_1 and N_2 at $z_1 + u_N$ and $z_2 + v_N$ (the *nodal points*) related by unit *angular* magnification. This requires $D = 1/A = \mu_2/\mu_1$, whence

$$A = \frac{\mu_1}{\mu_2} = a + \frac{v_N c}{\mu_2} \qquad \Rightarrow v_N = \frac{\mu_1 - \mu_2 a}{c} \tag{3.48}$$

$$D = \frac{\mu_2}{\mu_1} = d - \frac{u_N c}{\mu_1} \qquad \Rightarrow u_N = \frac{\mu_1 d - \mu_2}{c} \, . \tag{3.49}$$

Simple subtraction gives $H_1 N_1 = H_2 N_2 = (\mu_1 - \mu_2)/c$. We leave the reader to devise a paper-folding method of ray tracing (§3.7.1) when $\mu_1 \neq \mu_2$. It involves two separate folding procedures.

3.7.3 Examples: meniscus lenses and the telephoto system

A simple experiment will show that the principal planes of a meniscus lens, an asymmetrical lens having the centres of curvature of both surfaces on the same side of it, do not coincide with its plane. You can see this by determining the two focal points of a strong, thick, positive spectacle lens (reading glasses) by imaging a bright distant object. It is easy to see that they are not the same distance from the lens. Since the object and image spaces are both in air, this is an indication that the principal planes are displaced to one side.

A more dramatic demonstration of the function of the cardinal planes is provided by a telephoto system. This can be defined as a system where the focal length is considerably larger than the distance V_1F_2. It is widely used in cameras to achieve a relatively short lens system with a long focal length, thus creating a highly magnified image on the film. An example is shown in Fig. 3.14 where the details are given.

It essentially consists of a pair of thin lenses, the weaker one being positive and the stronger one negative. They are separated by somewhat more than the algebraic sum of their focal lengths, so that the combination acts as a weak converging lens. Calculation of the positions of H_1 and H_2 shows both of them to be situated on the same side of the lens. Each point on a distant object is a source of parallel rays and so it will be imaged on $\mathscr{F}_2$. For light incident from the object space, the system behaves like a thin lens situated at H_2 with a focal length H_2F_2 (which is 160 cm in the example), while the physical length of the hardware is V_1F_2 which is only 69 cm.

Essentially the same telephoto system but with a different distance $L_1L_2 = 12.1$ cm can be used in a diffractometer for classroom demonstration of Fraunhofer diffraction patterns (see Appendix 2). Here each group of parallel rays belongs to one order of diffraction from the mask. The telephoto system enables one to project the diffraction pattern with a size determined by f (which is 16 m in the example) on a screen at distance V_2F_2 (only 6.4 m) from the optics. This gives a real image of the diffraction pattern 2.5 times larger than could be obtained with a simple lens in the same position.

Figure 3.14 A telephoto system made of two lenses L_1 and L_2 having focal lengths 20 cm and -8 cm respectively, separated by 13 cm. (*a*) Cardinal points, (*b*) ray trace for an axial object at $u = -\infty$. The equivalent thin lens at $\mathscr{H}_2$ would refract the rays as shown by the broken line.

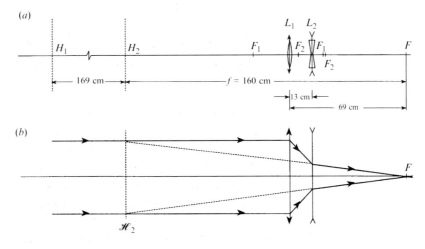

3.7.4 Experimental determination of cardinal points for a system in air

The principal foci F_1 and F_2 of a converging lens system can first be found by locating the image of a distant object, or by finding the position of an object that produces an image in the same place when a plane mirror is placed behind the lens system. We can then find a pair of conjugate positions. If the distances of these points from F_1 and F_2 are p and p', we can then use Newton's equation (3.31) in the form $pp' = f^2$ to find the focal length. Having found the principal foci and the focal length, we have now located the principal planes.

3.8 Aberrations

The Gaussian approximation does not, of course, apply to real lenses and mirrors. Unfortunately, the moment we abandon the approximation that $\sin \hat{\imath} \simeq \tan \hat{\imath} \simeq \hat{\imath}$, the subject of geometrical optics loses its elegant simplicity and becomes very technical. However, since these technicalities are often of great practical importance, we shall skim very superficially over a few examples of lens aberrations and their correction. Much more complete discussions are given by Kingslake (1978) and Welford (1986).

3.8.1 The monochromatic aberrations

No image-forming instrument can produce a perfect image of a finite-sized object, and the best that can be done in practice is to ensure that the aberrations which would be most disturbing in a particular experiment are made as small as possible, sometimes at the expense of making others larger.

The classification of aberrations now usually accepted, which makes their practical importance rather clear, was first introduced by von Seidel in about 1860, and therefore predates the diffraction theory of imaging (Chapter 12). Ideally, a perfect lens imaging a luminous point object would create an exactly spherical wavefront converging on a point in the image plane. The position of this point in the plane would be related linearly to that of the point object in its plane by the magnification m. Some of the named aberrations – *spherical aberration, coma, astigmatism* – describe the blurring of the image point. Others describe the deviation of the best image point from its expected position; *curvature of field* tells us how far in front or behind the image plane we shall find it and *distortion* tells us how big are the deviations from a uniform magnification. All these aberrations are functions of the object point position (x, y, z) and also depend on the lens parameters, including in particular the aperture size which was not relevant to paraxial calculations.

From the point of view of wave theory, the exiting wavefront simply does not coincide with the sphere converging on the 'right' point, as defined by Gaussian optics. If we calculate the deviation of the real wavefront from the expected sphere in the exit pupil, we derive a function Δ which describes the aberrations. The Seidel classification breaks this function down into a linear superposition of radial and angular functions, each one of which corresponds to one of the named aberrations mentioned above (and there are higher-order ones which we did not mention, too). Any one of them which does not introduce a deviation greater than about $\lambda/2$ is negligible, and will not significantly affect imaging.

Two examples are as follows. If the coordinates in the exit pupil are (ρ, θ) and the object is on the axis, *spherical aberration* corresponds to the function

$$\Delta(\rho, \theta) = A\rho^4. \tag{3.50}$$

If the object is at lateral distance x from the axis, there is an additional term proportional to x called *coma*:

$$\Delta(\rho, \theta) = Bx\rho^3 \cos\theta . \tag{3.51}$$

This approach can profitably be developed further. We consider Δ as a phase error $k_0\Delta$, where $k_0 = 2\pi/\lambda$, and then calculate the form of the distorted image of a point as the Fraunhofer diffraction pattern (§8.2.5) of the 'phase object' $f(\rho, \theta) = \exp[-ik_0\Delta(\rho, \theta)]$. The above two examples are shown in Fig. 3.15.

Figure 3.15 The intensity distribution in images of points distorted by (a) spherical aberration and (b) coma.

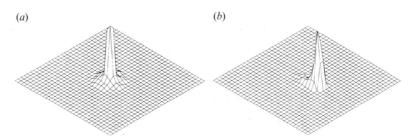

(a)   (b)

3.8.2 Chromatic aberration

In addition to the monochromatic aberrations, a simple lens system has cardinal points whose positions depend on refractive indices, which are a function of λ. Mirror systems are of course free of such defects. The refractive index $\mu(\lambda)$ of a transparent medium is always a decreasing function of λ, but varies from material to material (§13.4.2).

For a simple thin lens, the power is given by (3.7): $f^{-1} = (\mu - 1)(R_1^{-1} - R_2^{-1})$. A combination of two or more lenses made from different glasses can then be designed, by a suitable choice of radii of curvature R for each component, to have equal focal lengths at two or more specified wavelengths. The most common implementation has two components with radii of curvature R_1, R_2 and S_1, S_2 respectively; it is called an *achromatic doublet* or *achromat*. The refractive index variation $\mu(\lambda)$ of each glass is specified by its *dispersive power*, defined for visible optics systems by

$$\omega = \frac{\mu_b - \mu_r}{\mu_y - 1} , \qquad (3.52)$$

where μ_b, μ_y and μ_r are the indices for blue, yellow and red light respectively (usually $\lambda = 4861\,\text{Å}, 5876\,\text{Å}$ and $6563\,\text{Å}$; for other spectral regions different wavelengths would be appropriate). The *Abbe number* $V = \omega^{-1}$ is often listed in glass tables. It is a simple calculation to show that the focal lengths for blue and red light are equal if

$$(\mu_{bF} - \mu_{rF})(R_1^{-1} - R_2^{-1}) + (\mu_{bC} - \mu_{rC})(S_1^{-1} - S_2^{-1}) = 0 , \qquad (3.53)$$

where the two types of glass are indicated by the second suffices F and C.† In terms of the individual focal lengths of the two lenses for yellow light, f_F and f_C, this can be expressed

$$\frac{\omega_F}{f_F} + \frac{\omega_C}{f_C} = 0, \qquad (3.54)$$

and the combined power is of course

$$f^{-1} = f_F^{-1} + f_C^{-1}. \qquad (3.55)$$

A cemented doublet has a common interface, $R_2 = S_1$, and so the two equations (3.54) and (3.55) determine three radii of curvature. One degree of freedom therefore remains; we shall see in §3.8.3 how this can be used to correct another aberration. An uncemented doublet, $R_2 \neq S_1$, has two free parameters.

---

† The letters F and C stand for Flint and Crown glass, which are the commonest glasses used for chromatic correction, but of course could just as well represent other materials.

3.8.3 Correction of spherical aberration

Ray diagrams showing spherical aberration were shown in Fig. 3.1. Comparison of (*a*) and (*b*) suggests that some form of compensation for the defect might be achieved by *bending* the lens by adding a constant to each of R_1^{-1} and R_2^{-1}; this will not change the focal length which is determined only by their difference. However, it turns out that bending can not eliminate the spherical aberration of a single lens completely if the object is at infinity. It *can* do so if the object is closer than the focal length, as will be seen in §3.9 for the aplanatic system (which is a particular form of bent meniscus lens). When the object is at infinity it turns out, for $\mu = 1.6$ for example, that the best one can do to reduce spherical aberration is to use a ratio $R_2/R_1 = -12$. The result is close to a plano-convex lens, with the flatter side facing the image (*not* like Fig. 3.1(*b*)!). This tends to divide the refraction more-or-less equally between the two surfaces, which is a good rule of thumb to follow if aplanatic conditions are inappropriate.

When an achromatic doublet is used, the extra degree of freedom (§3.8.2) can be employed to correct spherical aberration. In this case, good correction can be achieved by bending the lens even for an object at infinity, and most refractive telescope objectives are designed this way. It is also common to use cemented achromats even for laboratory experiments using monochromatic light so as to take advantage of their correction for spherical aberration.

3.8.4 Coma and other aberrations

The *Abbe sine rule*, which will be proved in §12.2.2 by diffraction methods,† states that if a ray leaves a point object at angle θ_1 and converges on the image at angle θ_2 such that

$$\frac{\sin \theta_1}{\sin \theta_2} = \text{constant}, \tag{3.56}$$

both spherical aberration and coma are absent. The constant is, of course, the angular magnification, which can be seen by making θ very small, when the paraxial equations will apply. The aplanatic system (§3.9) satisfies this condition, but it is easy to see that a thin lens does not, since the ratio between the tangents of the angles is constant.

Distortion, however, is smallest in lens systems which are symmetrical about their central plane. If the magnification is -1, this can be seen (Problem 3.11) as resulting from the reversibility of light rays,

---

† It can be proved by geometrical optics also, but one proof seems enough!

but it is found to be approximately true at other magnifications too. The problems facing the lens designer now become clear if one contemplates, for example, correcting distortion (indicating a symmetrical system), spherical aberration and coma (requiring an asymmetrical bent lens) simultaneously. The solution has to be sought with a larger number of component lenses.

In general, the severity of all aberrations is a strong function of the aperture of the lens system. This is often expressed in terms of a dimensionless number, the *f-number* of the lens, defined as the ratio of focal length to aperture stop diameter (*small f*-number means *large* aperture). The brightness of the image, which determines the exposure needed in photography, is proportional to $(f\text{-number})^{-2}$.

3.9 Advanced topic: the aplanatic objective

A system which has no spherical aberrations in spite of large-angle rays is the aplanatic sphere. Despite its being a particular application of Snell's law with no fundamental significance, it is widely used in optical design, particularly in microscope objectives when a limit of resolution near the theoretical maximum is required.

The object is placed at A, for which $u = -R - R/\mu$, measured from V, the right vertex of the sphere. We shall show that in this case an image is formed at A', $v = -(R + R\mu)$ (see Fig. 3.16) and that this relation holds for all angles.

We apply the geometrical sine law to the triangle ACP and see immediately that $\sin \hat{\imath} / \sin \alpha = 1/\mu$. By Snell's law, $\sin \hat{\imath} / \sin \hat{r} = 1/\mu$ and so $\hat{r} = \alpha$. The triangles ACP and PCA' are therefore similar and it follows that

$$A'C/R = R/(R/\mu) , \tag{3.57}$$

i.e.

$$A'V = A'C + R = R\mu + R . \tag{3.58}$$

Therefore rays diverging from a point distant R/μ from the centre of a sphere of radius R will, after refraction, appear to be diverging from a point distant μR from the centre. Since no approximations are involved, the result is correct for all angles. For example, if $\mu = 1.50$ a beam with a semi-angle of $64°$ ($\sin 64° = 0.90$) will emerge as a beam with a semi-angle of $37°$ ($\sin 37° = 0.60$).

Since the imaging is perfect for all angles, including small ones, the formal optical properties of the aplanatic sphere can be handled by matrix optics. The optical system consists of one refracting surface and is described by the matrix

$$\begin{pmatrix} 1 & 0 \\ \frac{1-\mu}{R} & 1 \end{pmatrix}$$

(3.59)

(note that the radius of the surface is $-R$!). The principal planes pass through the vertex. The focal lengths are $f_1 = R\mu/(\mu - 1)$ and $f_2 = R/(\mu - 1)$. Since $u = -(R + R/\mu)$ we confirm from (3.45) that $v = -(R + R\mu)$. The virtual image formed is μ^2 larger than the object.

At first sight it might appear that the aplanatic properties are rather useless, since the object is immersed in the sphere,† but they are in fact widely used in two ways.

First, the sphere can be cut by a section passing through its internal aplanatic point; the specimen is placed near this point and immersed in a liquid (cedarwood oil is used) of the same refractive index as the glass (Fig. 3.17(a)). This has the additional advantage that the wavelength in the medium is smaller than in air, which improves the resolution. The system is known as *oil-immersion* and is used almost universally for microscopes of the highest resolution.

The second way in which the principle can be used involves putting the object at the centre of curvature of the first concave face of a lens and making this same point the inner aplanatic point of the second surface. All the deviation then occurs at the second surface and the image is formed at the outer aplanatic point (Fig. 3.17(b)). It is easy to show that the magnification of such a lens is μ. In this case we have corrected spherical aberrations by bending the lens (§3.8.3).

The complete microscope objective illustrated in Fig. 3.17(c) uses both applications of the aplanatic principle in two successive stages. The semi-angle of the emergent beam which started at 64° is then

† The magnified virtual image of a small goldfish, swimming in a spherical globe of water at distance R/μ from the centre, would be perfect.

Figure 3.16
Aplanatic points of a
sphere of radius R,
drawn for $\mu = 1.50$.
Triangles ACP and
PCA' are similar.

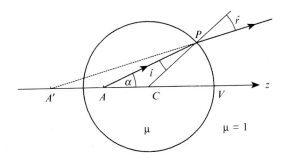

reduced to 24°. The virtual image is then re-imaged to infinity by an additional relatively weak converging lens.

The freedom from coma of the aplanatic sphere can be seen from the fact that all points at distance R/μ from the centre of the sphere

Figure 3.17 Application of the aplanatic points: (a) imaging a point immersed in oil; (b) imaging an external point; (c) a microscope objective using both of the above applications. O is the object and I, I_1 and I_2 images.

(a)

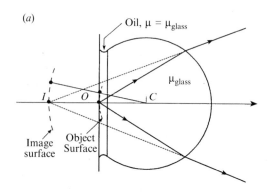

(b)

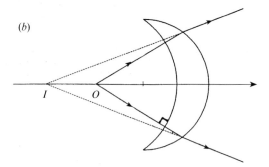

(c)

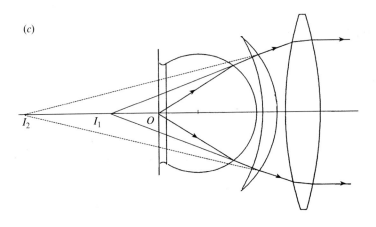

are aplanatic points. Thus, if we ignore the curvature of the surface on which these points lie, we deduce that all points on a plane object will form a plane image that is free from spherical aberration and coma.

3.10 Advanced topic: the spherical Fabry–Perot resonator

Most laser resonators are constructed from two spherical mirrors, usually concave towards one another, so that the light is 'trapped' between them. The idea is that light travelling at a small angle to the axis will not diverge to larger and larger angles after multiple reflexions but will stay within the resonator (Fig. 3.18(a)).

Although we have intentionally avoided discussing spherical mirrors in this chapter, this particular problem deserves mention, because of its importance in lasers (§14.5.1). We can convert it to an equivalent lens system, and investigate that by using matrices (in fact, most mirror systems are best dealt with this way). A spherical mirror of radius R has focal length $f = R/2$.† So when the light is reflected backwards and forwards between the two mirrors of radii R_1 and R_2, separated by L (positive values of R mean concave sides facing one another),

† This leads us to the conclusion, for example, that an object at $u = -R = -2f$ is imaged at $v = 2f = R$; taking into account reversal of the direction of the light, the object and image coincide, and the linear magnification is $v/u = -1$.

Figure 3.18 (a) Tracing a ray through a stable spherical Fabry–Perot resonator; (b) the equivalent infinite periodic set of thin lenses.

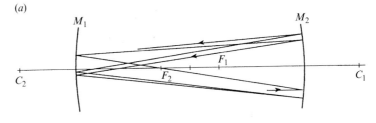

(a)

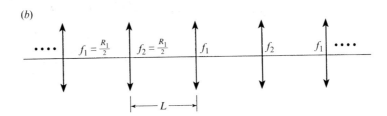

(b)

this is equivalent to a repeated pair of lenses, with focal lengths $R_1/2$ and $R_2/2$ as in Fig. 3.18(*b*). This is a periodic system, one period of which is represented by the matrix M_P, where

$$\mathsf{M}_P = \begin{pmatrix} 1 & L \\ 0 & 1 \end{pmatrix} \begin{pmatrix} 1 & 0 \\ -\frac{2}{R_2} & 1 \end{pmatrix} \begin{pmatrix} 1 & L \\ 0 & 1 \end{pmatrix} \begin{pmatrix} 1 & 0 \\ -\frac{2}{R_1} & 1 \end{pmatrix}. \tag{3.60}$$

Multiplying the matrices gives

$$\mathsf{M}_P = \begin{pmatrix} 1 - \frac{2L}{R_2} - \frac{4L}{R_1} + \frac{4L^2}{R_1 R_2} & 2L - \frac{2L^2}{R_2} \\ -2\left(\frac{1}{R_1} + \frac{1}{R_2}\right) + \frac{4L}{R_1 R_2} & 1 - \frac{2L}{R_2} \end{pmatrix}. \tag{3.61}$$

When light passes through N periods of the system, equivalent to being reflected back and forth N times in the mirror system, we have the matrix M_P^N. To see its convergence properties, it is easiest to diagonalize it. This means, essentially, 'rotating' the vector (h, θ) to a new vector $(ah + b\theta, -bh + a\theta)$, where $a^2 + b^2 = 1$, for which the matrix is diagonal. The technique for doing this is described in any text on linear algebra, and consists of solving the secular equation

$$\det\{\mathsf{M}_P - \lambda \mathsf{I}\} = 0 \tag{3.62}$$

for its two solutions, λ_1 and λ_2 . The diagonal matrix is $\mathsf{M}_D \equiv \begin{pmatrix} \lambda_1 & 0 \\ 0 & \lambda_2 \end{pmatrix}$. Since $\det\{\mathsf{M}_P\} = 1$, (3.62) is easily shown to give:

$$\lambda^2 - \left[4\left(1 - \frac{L}{R_1}\right)\left(1 - \frac{L}{R_2}\right) - 2\right]\lambda + 1 = 0 \tag{3.63}$$

whence λ_1 and λ_2 can be found. Before writing down the solutions, we shall look at their significance. The determinant of M_D is unity, and so $\lambda_1 \lambda_2 = 1$. The possible solutions of the quadratic equation (3.63) can be divided into two groups.

(i) Real solutions, λ_1 and λ_1^{-1}, for which we shall show that the rays diverge, and the resonator is unstable. We define λ_1 to be the larger solution, and exclude $\lambda_1 = \lambda_2 = 1$.

(ii) Complex solutions, which are of the form $\lambda_1 = e^{i\alpha}$, $\lambda_2 = e^{-i\alpha}$, including the solution $\alpha = 0$ ($\lambda_1 = \lambda_2 = 1$). For these values the rays do not diverge and the resonator is stable.

Consider case (i). The matrix M_D^N is then

$$\mathsf{M}_D^N = \begin{pmatrix} \lambda_1^N & 0 \\ 0 & \lambda_1^{-N} \end{pmatrix}. \tag{3.64}$$

After a large enough number N of passes, λ_1^{-N} will be small enough to be negligible, and we can write

$$
\begin{pmatrix} ah_N + b\theta_N \\ -bh_N + a\theta_N \end{pmatrix} \simeq \begin{pmatrix} \lambda_1^N & 0 \\ 0 & 0 \end{pmatrix} \begin{pmatrix} ah_1 + b\theta_1 \\ -bh_1 + a\theta_1 \end{pmatrix} , \qquad (3.65)
$$

where h_N and θ_N are the height and angle after N passes. The solution to these equations is clearly that h_N and θ_N are proportional to λ_1^N, and therefore diverge as N increases. The rays therefore get farther and farther from the axis.

For case (ii), (3.64) becomes

$$
\mathsf{M}_D^N = \begin{pmatrix} e^{Ni\alpha} & 0 \\ 0 & e^{-Ni\alpha} \end{pmatrix} \qquad (3.66)
$$

and the solution to (3.65) is periodic, with period $2\pi/\alpha$ cycles. This means that h and θ just oscillate about the axis, with finite amplitude.

The condition for stability is therefore for the solutions of (3.63) to be unity or complex i.e.:

$$
-1 \leq 2\left(1 - \frac{L}{R_1}\right)\left(1 - \frac{L}{R_2}\right) - 1 \leq 1 , \qquad (3.67)
$$

or, equivalently,
$$
0 \leq \left(1 - \frac{L}{R_1}\right)\left(1 - \frac{L}{R_2}\right) \leq 1 . \qquad (3.68)
$$

Examples of stable and unstable resonators are shown in Fig. 3.19.

Figure 3.19 Stable and unstable resonators.

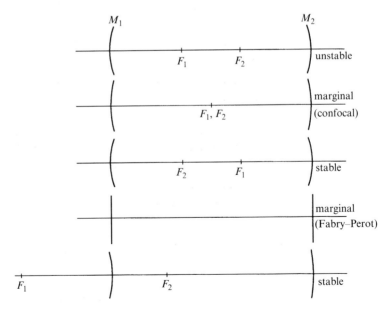

The most commonly-used stable resonator for gas lasers is called the *confocal resonator*, and is marginally stable, with $R_1 = R_2 = L$, i.e. $\lambda = 1$; the two mirrors then have a common focus in between them. A pair of parallel plane mirrors ($R_1 = R_2 = \infty$) is also marginally stable and is found in solid-state lasers.

We should point out that although from the point of view of geometrical optics it is possible to choose the apertures of the two mirrors of a stable resonator such that no rays ever leak out of it, when diffraction is taken into account there will always be some losses, and in the design of a laser these diffraction losses, as well as those arising from imperfect reflexion by the mirrors, have to be offset by the amplification of the active medium (§14.4) before the laser oscillates. On the other hand, if the medium amplifies strongly enough, even mildly unstable resonators can be tolerated.

CHAPTER FOUR

Fourier theory

4.1 Introduction

J. B. J. Fourier (1763–1830) was one of the French scientists of the time of Napoleon who raised French science to extraordinary heights. He was essentially an applied mathematician, and the work by which his name is now known was his contribution to the theory of heat transmission. He was faced with the problem of solving the one-dimensional heat-diffusion equation (2.31) for the development of the temperature distribution $\theta(x, t)$ in a body

$$\frac{\partial \theta}{\partial t} = D \frac{\partial^2 \theta}{\partial x^2} \tag{4.1}$$

for which he knew some initial conditions – the temperature as a function $\theta(x, 0)$ of position x at $t = 0$. This equation, which is a wave equation of the general type dealt with in §2.3.1, has analytic solutions if θ is a sinusoidal function of x,

$$\theta(x, 0) = \theta_0 \sin \, kx \, , \tag{4.2}$$

the solution then being an exponential with characteristic time related to k:

$$\theta(x, t) = \theta_0 \exp(-Dk^2 t) \sin \, kx. \tag{4.3}$$

As a sinusoidal initial distribution is a very artificial example, Fourier devised a method of expressing any periodic function, or any non-periodic function restricted to a body of regular shape, as the sum of a series of sinusoidal terms of various wavelengths for each of which the equation (4.1) could be solved individually. As the equation is homogeneous in θ, the solutions can then be added. This principle of expressing an arbitrary function as the sum of a set of sinusoidal terms is called *Fourier theory* and has found applications far beyond the boundaries of heat-transmission theory. Since optics is concerned with light waves and their interactions with obstacles which represent regions in which the wave equations must be solved under different conditions from their surroundings, Fourier theory is obviously a basic tool which we shall use repeatedly. The intention of this chapter is to make the theory familiar to readers, and to derive some of the more important results and ideas for later use. A more complete discussion can be found in many texts, for example Walker (1988).

4.2 Analysis of periodic functions

4.2.1 Fourier's theorem

Fourier's theorem states that any periodic function $f(x)$ can be expressed as the sum of a series of sinusoidal functions which have wavelengths which are integral fractions of the wavelength λ of $f(x)$. To make this statement complete, zero is counted as an integer, giving a constant leading term to the series:

$$f(x) = \frac{1}{2}C_0 + C_1 \cos\left(\frac{2\pi x}{\lambda} + \alpha_1\right) + C_2 \cos\left(\frac{2\pi x}{\lambda/2} + \alpha_2\right) + \dots$$

$$+ C_n \cos\left(\frac{2\pi x}{\lambda/n} + \alpha_n\right) + \dots \tag{4.4}$$

The ns are called the *orders* of the terms, which are harmonics. The following argument demonstrates the theorem as reasonable. If we cut off the series after the first term, choice of C_0 allows the equation to be satisfied at a discrete number of points – at least two per wavelength. If we add a second term the number of points of agreement will increase; as we continue adding terms the number of intersections between the synthetic function and the original can be made to increase without limit (Fig. 4.1). This does not prove that the functions *must* be identical when the number of terms becomes infinite; there are examples which do not converge to the required function, but the regions of error must become vanishingly small.

This reasoning would, of course, apply to basic functions other than sine waves. The sine curve, however, being the solution of all wave equations, is of particular importance in physics, and hence gives Fourier's theorem its fundamental significance.

4.2.2 Fourier coefficients

Each term in the series (4.4) has an *amplitude* C_n and a *phase angle* α_n. The latter quantity provides the degree of freedom necessary for relative displacements of the terms of the series along the x-axis. The determination of these quantities for each term of the series is called *Fourier analysis*.

It is, however, not always convenient to specify an amplitude and phase; we can express each term in the form:

$$C_n \cos(nk_0 x + \alpha_n) = A_n \cos nk_0 x + B_n \sin nk_0 x , \qquad (4.5)$$

where $A_n = C_n \cos \alpha_n$ and $B_n = -C_n \sin \alpha_n$, and $k_0 = 2\pi/\lambda$. The series (4.4) is then written as

$$f(x) = \frac{1}{2}A_0 + \sum_1^\infty A_n \cos nk_0 x + \sum_1^\infty B_n \sin nk_0 x . \qquad (4.6)$$

The process of Fourier analysis consists of evaluating the pairs (A_n, B_n) for each value of n.

4.2.3 Complex Fourier coefficients

The real functions $\cos\theta$ and $\sin\theta$ can be regarded as real and imaginary parts of the complex exponential $\exp(i\theta)$. Algebraically, there are many advantages in using the complex exponential, and in this book we shall use it almost without exception. We can write (4.6) in the form

$$f(x) = \tfrac{1}{2}A_0 + \sum F_n \exp(ink_0 x) , \qquad (4.7)$$

Figure 4.1
Intersections between
a square wave and its
series terminated
after (*a*) the first and
(*b*) the third term.

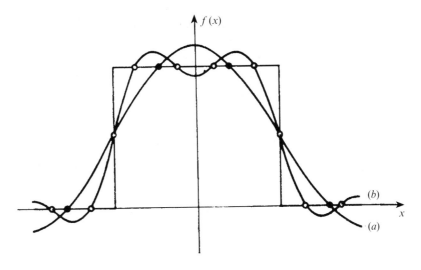

where the range of summation is as yet unspecified. Now let us equate (4.7) and (4.6) for a real $f(x)$. We then have:

$$\sum_{\infty} F_n[\cos(nk_0x) + i\sin(nk_0x)]$$
$$= \sum_1 [A_n\cos(nk_0x) + B_n\sin(nk_0x)] . \qquad (4.8)$$

If we assume that the ranges of the summation are identical and then equate equivalent cosine and sine terms independently, we get:

$$F_n = A_n; \quad iF_n = B_n . \qquad (4.9)$$

This leads to $iA_n = B_n$, which can not be true since A_n and B_n are both real! We have to carry out the complex summation in (4.7) from $n = \infty$ to $+\infty$ in order to solve the problem. There are then two independent complex coefficients, F_n and F_{-n}, corresponding to the pair A_n, B_n, and we then have, on comparing terms in (4.8):

$$F_n + F_{-n} = A_n; \quad i(F_n - F_{-n}) = B_n , \qquad (4.10)$$

whence

$$F_n = \tfrac{1}{2}(A_n - iB_n) = \tfrac{1}{2}C_n \exp(i\alpha_n) , \qquad (4.11)$$
$$F_{-n} = \tfrac{1}{2}(A_n + iB_n) = \tfrac{1}{2}C_n \exp(-i\alpha_n) . \qquad (4.12)$$

The Fourier series is therefore written in complex notation as:

$$f(x) = \sum_{-\infty}^{\infty} F_n \exp(ink_0x) , \qquad (4.13)$$

where $F_0 = \tfrac{1}{2}A_0$. So far, the function $f(x)$, and hence A_n and B_n, have been assumed to be real. In then follows from (4.11) and (4.12) that F_n and F_{-n} are complex conjugates:

$$F_n = F_{-n}^* . \qquad (4.14)$$

In general, however, a complex function $f(x)$ can be represented by complex A_n and B_n which bear no such relationship.

4.3 Fourier analysis

For most functions of importance in physics, Fourier analysis can be carried out analytically by a process that depends on an obvious property of a sinusoidal function – that its integral over a complete number of wavelengths is zero. Consequently, the integral of the product of two sinusoidal functions with integrally-related wavelengths over a complete number of cycles of both functions is also zero with one exception: when the two wavelengths are equal and the two sine functions are not in quadrature, the integral is non-zero. Therefore, if we integrate the product of $f(x)$ (wavelength λ) with a sine function of wavelength λ/m, the result will be zero for all the Fourier components of $f(x)$ except the mth, which has wavelength λ/m, and the value of the integral will then give the amplitude of the coefficient F_m.

To express this mathematically let us find the mth Fourier coefficient by multiplying the function $f(x)$ by $\exp(-imk_0 x)$ and integrating over a complete wavelength λ. It is convenient to replace x by the angular variable $\theta = k_0 x$ and then to take the integral I_m over the range $-\pi \leq \theta \leq \pi$, which is one wavelength. Then

$$
\begin{aligned}
I_m &= \int_{-\pi}^{\pi} f(\theta)\ \exp(-im\theta)\mathrm{d}\theta \\
&= \int_{-\pi}^{\pi} \sum_{-\infty}^{\infty} F_n\ \exp(in\theta)\exp(-im\theta)\mathrm{d}\theta .
\end{aligned}
\tag{4.15}
$$

Every term in the summation is sinusoidal, with wavelength $\lambda/|m-n|$, with the exception of the one for which $n = m$. The sinusoidal terms, being integrated over $|m-n|$ wavelengths, do not contribute; so that

$$
I_m = \int_{-\pi}^{\pi} F_m \mathrm{d}\theta = 2\pi F_m .
\tag{4.16}
$$

Thus we have a general expression for the mth Fourier coefficient:

$$
F_m = \frac{1}{2\pi} \int_{-\pi}^{\pi} f(\theta)\ \exp(-im\theta)\mathrm{d}\theta .
\tag{4.17}
$$

Note that it includes the zero term, the mean value of $f(\theta)$:

$$
F_0 = \frac{1}{2\pi} \int_{-\pi}^{\pi} f(\theta)\mathrm{d}\theta .
\tag{4.18}
$$

4.3.1 Even and odd functions

A function is said to be *even* or *symmetric* if $f(\theta) = f(-\theta)$, and *odd* or *antisymmetric* if $f(\theta) = -f(-\theta)$ (see Fig. 4.2). Let us return for a moment to the formulation (4.6) of the Fourier series in terms of the sine and cosine functions. Now a periodic even function must be expressed as a sum of cosine functions only, since the sine terms make contributions of opposite sign at $+\theta$ and $-\theta$. Thus $B_n = 0$ and it follows from (4.11) and (4.12) that

$$
\text{even function:} \quad F_n = F_{-n} .
\tag{4.19}
$$

If, in addition, the function is real, so that (4.14) is true, we find

$$\text{real even function:} \quad F_n = F^*_{-n} = F_{-n} , \quad (4.20)$$

implying that F_n is real.

Similarly, for an odd function, we must have coefficients $A_n = 0$ and

$$\text{odd function:} \quad F_n = -F_{-n} , \quad (4.21)$$

$$\text{real odd function:} \quad F_n = F^*_{-n} = -F_{-n} , \quad (4.22)$$

implying that F_n is purely imaginary in the latter case. We see that in all these cases the symmetry of $f(x)$ is also present in $F(k)$.

4.3.2 The square wave

We shall illustrate the above analysis with a simple example, *the square wave*. This has value 1 over half its period ($-\pi/2$ to $\pi/2$) and -1 over the other half ($\pi/2$ to $3\pi/2$) (Fig. 4.2(a)). The function as defined above is *real and even*; a_n is therefore real. If possible, it is often worthwhile choosing the position of the origin to make a function even, as the mathematics is usually simpler; if we had chosen to make the function equal to 1 from $-\pi$ to 0 and -1 for 0 to π it would have been odd and its coefficients all imaginary (Fig. 4.2(b)). This effect – the altering of the phase of all coefficients together by a shift of origin – is often important (§4.4.4); the form of the function determines

Figure 4.2 A square wave (a) as an even function, (b) as an odd function.

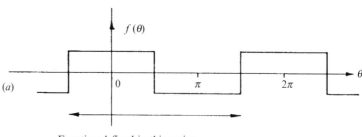

(a)

Function defined in this region.

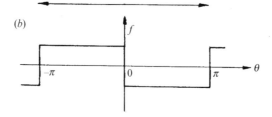

(b)

the relative phases of the coefficients only. For the even function, Fig. 4.2(*a*),

$$f(\theta) = 1 \quad \in (-\pi/2 \le \theta \le \pi/2); \quad f(\theta) = -1 \quad \in (\pi/2 \le \theta \le 3\pi/2),$$
(4.23)

$$
\begin{aligned}
F_n &= \frac{1}{2\pi} \int_{-\pi}^{\pi} f(\theta) \ \exp(-in\theta) d\theta \\
&= \frac{1}{2\pi} \int_{-\pi/2}^{\pi/2} \exp(-in\theta) d\theta - \frac{1}{2\pi} \int_{\pi/2}^{3\pi/2} \exp(-in\theta) d\theta \\
&= \frac{1}{n\pi} \sin \frac{n\pi}{2} [1 - \exp(-in\pi)] \ .
\end{aligned}
$$
(4.24)

Thus we have, evaluating F_0 from (4.18),

$$F_0 = 0, \quad F_{\pm 1} = \frac{2}{\pi}, \quad F_{\pm 2} = 0, \quad F_{\pm 3} = -\frac{2}{3\pi},$$

$$F_{\pm 4} = 0, \quad F_{\pm 5} = \frac{2}{5\pi} \cdots$$

4.3.3 Reciprocal space in one dimension

We can think of the Fourier coefficients F_n as a function $F(n)$ of n. As $F(n)$ is non-zero only for integral values of n, the function can be considered as being defined for non-integral values but as having zero value there; the positive half of the function $F(n)$ which represents the series for a square wave can therefore be drawn as in Fig. 4.3. Given this drawing, we could simply reconstruct the original square wave by summing the series it represents, except that it gives no information about the wavelength λ of the original wave. This defect can be simply remedied. Written in terms of x, the expression for F_n is

$$F_n = \frac{1}{\lambda} \int_{\text{one wavelength}} f(x) \ \exp(-ink_0 x) \, dx \ .$$
(4.25)

Figure 4.3 Positive half of the functions $F(n)$ and $F(k)$ for a square wave.

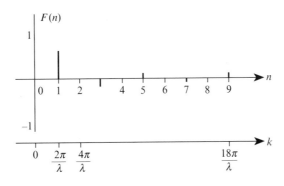

Information about the wavelength λ is included in (4.25) where $k_0 \equiv 2\pi/\lambda$ and the variable $k = nk_0$ is used rather than n; this corresponds to a harmonic of wavelength λ/n. This is shown in Fig. 4.3. The function (4.25) becomes

$$F(k) = \frac{1}{\lambda} \int_{\text{one wavelength}} f(x) \, \exp(-\mathrm{i}kx) \, \mathrm{d}x \,. \qquad (4.26)$$

It is useful now to compare the functions $F(k)$ as λ changes. In Fig. 4.4 this comparison is carried out, the scales of k and x being the same in (a), (b), (c). Clearly the scale of $F(k)$ is inversely proportional to that of $f(x)$. For this reason (k proportional to $1/\lambda$) the space whose coordinates are measured by k is called *reciprocal space*; real space has coordinates measured by x and reciprocal space by x^{-1}. So far, of course, we have discussed a purely one-dimensional space; the extension to two and three dimensions is simple, and will be discussed in Chapter 8.

4.3.4 Analysis of a general function

In general, the integration involved in (4.17) is much more difficult than the example quoted. In many cases, even when $f(x)$ is a simple analytic function, the integral cannot be evaluated analytically and ap-

Figure 4.4 Square waves of different scales and their Fourier coefficients $F(k)$. The waves are assumed to continue from $-\infty$ to $+\infty$.

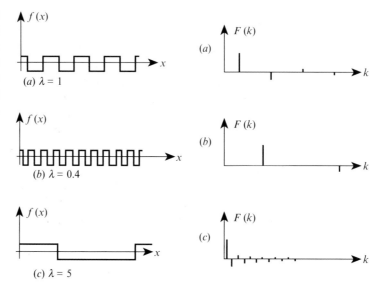

proximate or numerical methods must be used. The process, however, is the same in principle; the integral

$$\frac{1}{2\pi} \int_{-\pi}^{\pi} f(\theta) \, \exp(-in\theta) \mathrm{d}\theta \tag{4.27}$$

has to be evaluated for a series of values of n and, because of the large number of multiplications involved, techniques for reducing the work to a minimum are very valuable. A very efficient computer technique, called the *Fast Fourier Transform,* or *FFT*, is now generally used (Brigham, 1988).

4.4 Non-periodic functions

Although crystals, which have sets of atoms repeating accurately in three dimensions, are almost ideally periodic, matter on the macroscopic scale is usually not so. Natural objects sometimes simulate periodicity in their growth, but this is never precise and most objects which we have to deal with optically (i.e. on a scale greater than the wavelength of light) are completely non-periodic. Since this book is concerned with light and real objects we may therefore ask why Fourier methods are of any importance, since they apply to periodic functions only. The answer is that the theory has an extension, not visualized by Fourier himself, to non-periodic functions. The extension is based upon the concept of the Fourier transform.

4.4.1 Fourier transform

We have seen in §4.2.1 that a periodic function can be analyzed into harmonics of wavelengths ∞, λ, $\lambda/2$, $\lambda/3,\ldots$, and we have shown by Fig. 4.4 how the form of the function $F(k)$ depends on the scale of λ. When our interest turns to non-periodic functions we can proceed as follows. Construct a wave of wavelength λ in which each unit consists of some non-periodic function (Fig. 4.5). We can always make λ so large that an insignificant amount of the function lies outside the one-wavelength unit. Now allow λ to increase without limit, so that the repeats of the non-periodic function separate further and further. What happens to the function $F(k)$? The spikes approach one another as λ increases, but one finds that the envelope of the tips of the spikes remains invariant; it is determined only by the unit, the original non-periodic function. In the limit of $\lambda \to \infty$ the spikes are infinitely close to one another, and the function $F(k)$ has just become the envelope. This envelope is called the *Fourier transform* of the non-periodic function. The limiting process is illustrated in Fig. 4.5.

Admittedly, this suggests that the Fourier series for a non-periodic function is a set of spikes at discrete but infinitesimally-spaced frequencies rather than a continuous function. The argument does not show that in the limit $\lambda \to \infty$ the function becomes continuous, although physically the difference may seem rather unimportant. From the mathematical point of view it is better to work in reverse. We now *define* the Fourier transform of a function $f(x)$ as

$$F(k) = \int_{-\infty}^{\infty} f(x)\exp(-ikx)\mathrm{d}x \qquad (4.28)$$

Figure 4.5
Illustrating the
progression from
Fourier series to
transform.

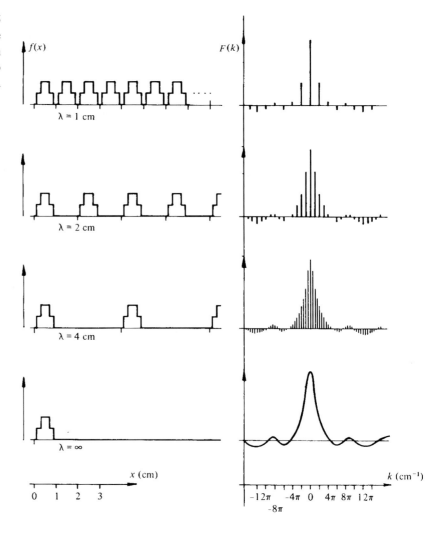

which is a continuous function of the spatial frequency k.† We shall then show that if $f(x)$ is periodic, the transform $F(k)$ is non-zero at discrete and periodic values of k only. Proof of this statement will become trivially easy once we have introduced the concept of convolution (§4.6).

An important idea illustrated by Fig. 4.5 is that of *sampling*. The set of orders of a periodic function can be regarded as equally-spaced ordinates of the Fourier transform of the unit. As the spacing is reduced by increasing the repeat distance λ, the orders sweep through the transform, sampling its value at ever closer intervals. This idea is particularly relevant to digital operations. Numerically, a function (the unit) is defined only within a certain limited region of space. The mathematics 'assumes' that this unit is repeated periodically, and uses (4.17) to calculate the Fourier series. The transform of the unit is therefore sampled digitally at closely spaced but distinct points, whose spacing is determined by the length of the repeat period.

4.4.2 Fourier transform of a square pulse

We can illustrate the calculation of a Fourier transform by using the equivalent example to that in §4.3.2, a single square pulse. This is one of the simplest, and optically most useful, functions. We define it to have height H and width h (Fig. 4.6(a)) and the integral (4.28) becomes

$$
\begin{aligned}
F(k) &= \int_{-h/2}^{h/2} H \ \exp(-ikx)\mathrm{d}x \\
&= \frac{H}{-ik}\left[\exp\left(\frac{-ikh}{2}\right) - \exp\left(\frac{ikh}{2}\right)\right]
\end{aligned}
$$

† In comparing this with (4.27), notice that the $1/2\pi$ has been dropped.

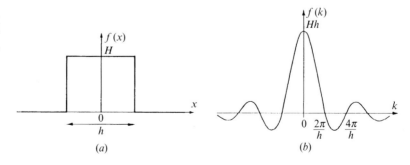

Figure 4.6 (*a*) A square pulse and (*b*) its transform.

$$= Hh \frac{\sin(kh/2)}{kh/2} . \tag{4.29}$$

The function $\sin(\theta)/\theta$ appears very frequently in Fourier transform theory, and has therefore been given the name 'sinc(θ)'. Equation (4.29) can thus be written:

$$F(k) = Hh \, \mathrm{sinc} \, (kh/2) . \tag{4.30}$$

The transform is illustrated in Fig. 4.6(b). It has a value Hh (the area under the pulse) at $k = 0$ and decreases as k increases, reaching zero when $kh = 2\pi$. It then alternates between positive and negative values, being zero at $kh = 2n\pi$ ($n \neq 0$). It should be noted that the transform is real: this follows because the function is symmetrical about the origin (see §4.4.7).

In Fig. 4.7 we can see the reciprocal property of the transform discussed in §4.3.3. As h is increased, the value of k at which the transform becomes zero decreases and the interval between successive zeros also decreases; the coarser the function, the finer is the detail of its transform. Conversely, as h decreases the transform spreads out and when h reaches zero there is no detail at all in the transform, which has become a constant, Hh.

4.4.3 The Dirac δ-function

The limiting process above has introduced a new and very useful function, the Dirac δ-function. It is the limit of a square pulse as its width h goes to zero but its enclosed area Hh remains at unity. It is therefore zero everywhere except at $x = 0$, when it has infinite value, $\lim_{h \to 0} 1/h$. The transform of the δ-function can be found by the limiting process above; we start with a square pulse of width h and height h^{-1}, which has transform

$$F(k) = \mathrm{sinc} \, (kh/2) \tag{4.31}$$

and see that as $h \to 0$ the transform becomes unity for all values of k. The transform of a δ-function at the origin in one dimension is unity.

A mathematically important property of this function is

$$\int_{-\infty}^{\infty} f(x)\delta(x-a)\mathrm{d}x = f(a) . \tag{4.32}$$

This integral *samples* $f(x)$ at $x = a$.

4.4.4 A shift of origin

To illustrate the change in phase, but not in amplitude, which occurs when the origin is shifted, we can calculate the transform of the function $f_1(x) = f(x - x_0)$. We write x' for $x - x_0$, and

$$
\begin{aligned}
F_1(k) &= \int_{-\infty}^{\infty} f(x - x_0)\exp(-ikx)dx \\
&= \int_{-\infty}^{\infty} f(x')\exp\{-ik(x' + x_0)\}dx' \\
&= \exp(-ikx_0)\int_{-\infty}^{\infty} f(x')\exp(-ikx')dx' \\
&= \exp(-ikx_0)F(k) \ .
\end{aligned} \tag{4.33}
$$

This differs from $F(k)$ only by the phase factor $\exp(-ikx_0)$. In particular, the amplitudes $|F_1(k)|$ and $|F(k)|$ are equal.

Figure 4.7
Progression from a square pulse, (*a*) and (*b*), to a δ- function (*c*). The area Hh remains constant throughout.

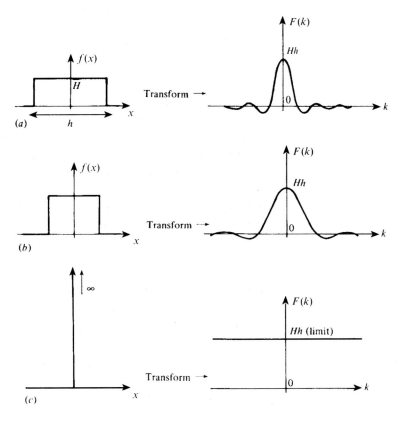

4.4.5 Multiple δ-function

An array of δ-functions at various values of x is a function we shall use repeatedly:

$$f(x) = \sum_n \delta(x - x_n) \ . \tag{4.34}$$

From (4.33) its transform is clearly

$$F(k) = \sum_n \exp(-ikx_n) \ . \tag{4.35}$$

If there are two δ-functions, for example, at $x_n = \pm b/2$ we have a transform

$$F(k) = 2\cos(kb/2) \tag{4.36}$$

which is real ($f(x)$ is even) and oscillatory (Fig. 4.8). Its importance in discussing the optical experiment of Young's fringes will be evident in §8.3.1.

The transform of a regular array of δ-functions is particularly important (Fig. 4.9):

$$f(x) = \sum_{-\infty}^{\infty} \delta(x - nb) \ . \tag{4.37}$$

It follows from (4.35) that

$$F(k) = \sum_{-\infty}^{\infty} \exp(-iknb) \ . \tag{4.38}$$

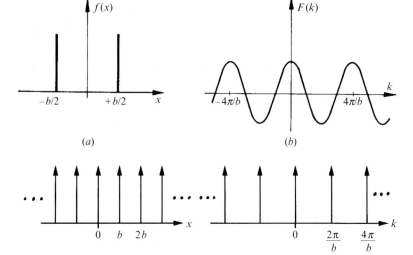

Figure 4.8 (a) Two δ-functions at ±b/2 and (b) their transform.

Figure 4.9 Periodically repeated δ-function and its transform.

Now the function $f(x)$ extends to infinity in both the positive and negative directions, and its integral is infinite. As a result it can be shown that from the purely mathematical point of view it does not have a Fourier transform. However, we know that the mathematics only represents a real physical entity, which must itself be finite in extent. We shall represent physical reality better by fading out the series during a large but finite number of terms. (It is better to fade out than to cut the series off abruptly, as will be demonstrated experimentally in §12.3.5 and §12.5.1.) We shall do this by multiplying the series by the factor $\exp(-s|n|)$, where s is a number small enough not to affect the series appreciably within any specified region of interest. But now (4.38) can be summed; we write it as the sum of two geometric series, less a single term which appears twice:

$$F(k) = \sum_{0}^{\infty} \exp(-iknb - sn) + \sum_{-\infty}^{0} \exp(-iknb + sn) - 1 , \qquad (4.39)$$

in which s can be made as small as required. Equation (4.39) gives:

$$F(k) = [1 - \exp(-ikb - s)]^{-1} + [1 - \exp(ikb - s)]^{-1} - 1 . \qquad (4.40)$$

This function has periodic peaks of height $2/(1 - e^{-s}) - 1 \approx 2/s$ whenever $kb = 2\pi m$, for any integer value of m. By expanding the exponents to second order it is easy to show that these peaks have width $2s/b$ which is proportional to s and so they approximate to δ-functions as $s \to 0$. As a result, the function (4.40) can be considered as the periodic series of δ-functions:

$$F(k) = \frac{4}{b} \sum_{m=-\infty}^{\infty} \delta(k - 2\pi m/b) . \qquad (4.41)$$

where the pre-factor $4/b$ in this expression is just the area of a peak.†

4.4.6 The Gaussian function

Another function whose Fourier transform is particularly useful in optics is the *Gaussian* (Fig. 4.10):

$$f(x) = \exp(-x^2/2\sigma^2) . \qquad (4.42)$$

---

† Since $f(x)$ does not really have a Fourier transform, different ways of calculating the pre-factor may give different answers!

From the definition of the transform, (4.28), we have

$$F(k) = \int_{-\infty}^{\infty} \exp(-x^2/2\sigma^2)\exp(-ikx)\mathrm{d}x \tag{4.43}$$

$$= \exp\left[-k^2\left(\frac{\sigma^2}{2}\right)\right]\int_{-\infty}^{\infty}\exp\left\{-\left[\frac{x}{(2\sigma^2)^{\frac{1}{2}}} + ik\left(\frac{\sigma^2}{2}\right)^{\frac{1}{2}}\right]^2\right\}\mathrm{d}x$$

by completing the square in the exponent. The integral is standard and occurs frequently in statistical theory. Its value is independent of k,

$$\int_{-\infty}^{\infty}\exp\frac{-\xi^2}{2\sigma^2}\,\mathrm{d}\xi = (2\pi\sigma^2)^{\frac{1}{2}}, \tag{4.44}$$

and therefore

$$F(k) = (2\pi\sigma^2)^{\frac{1}{2}}\exp\left[-k^2\left(\frac{\sigma^2}{2}\right)\right]. \tag{4.45}$$

The original function (4.42) was a Gaussian with variance σ; the transform is also a Gaussian, but with variance σ^{-1}. The *half-peak width* of the Gaussian (the width of the peak at half its maximum height) can be shown to be equal to 2.36σ. Because the Gaussian transforms into a Gaussian, this example illustrates particularly clearly

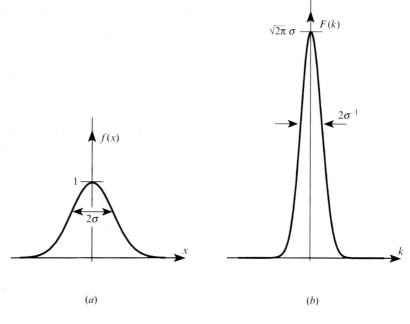

Figure 4.10 Gaussian function and its transform. The widths are shown at $e^{-\frac{1}{2}} = 0.60$ of the maximum height.

(a)

(b)

the reciprocal relationship between the scales of the function and its transform.

4.4.7 Transforms of complex functions

In §4.3.1 we discussed the relationships between F_n and F_{-n} for periodic functions having various symmetry properties. We included in the discussion the possibility that $f(x)$ was complex, and since complex functions form the backbone of wave optics we must extend our discussion of transforms to include them. If the function $f(x)$ is complex, and has transform $F(k)$ defined in the usual manner, we can write down the transform of its complex conjugate $f^*(x)$ as:

$$\int_{-\infty}^{\infty} f^*(x) \exp(-ikx)dx = \left[\int_{-\infty}^{\infty} f(x) \exp(ikx)dx\right]^* = F^*(-k) . \quad (4.46)$$

Thus the transform of $f^*(x)$ is $F^*(-k)$. It now follows that if $f(x)$ is real, then $f(x) = f^*(x)$ and so as in (4.14),

$$\text{real function:} \quad F^*(-k) = F(k). \quad (4.47)$$

By similar manipulations, in parallel to those in (4.19)–(4.22), we find:

$$\text{even function:} \quad F(k) = F(-k), \quad (4.48)$$

$$\text{odd function:} \quad F(k) = -F(-k). \quad (4.49)$$

Combining these with (4.47) for real functions, it follows that a real even function has a real transform, and a real odd function has a pure imaginary transform. For all these cases,

$$|F(-k)|^2 = |F(k)|^2 . \quad (4.50)$$

In later chapters we shall often be using complex functions to represent real physical quantities, for mathematical convenience. The *Hilbert transform* is a formal way of defining the complex function associated with a given real function, and it can be conveniently expressed in terms of their Fourier transforms. If the real function is $f^R(x)$, where

$$f^R(x) = \int_{-\infty}^{\infty} F(k) \exp(-ikx)\,dk , \quad (4.51)$$

the *associated complex function* $f(x) = f^R(x) + if^I(x)$ is

$$f(x) = 2\int_{0}^{\infty} F(k) \exp(-ikx)\,dk . \quad (4.52)$$

The reader can easily confirm from this definition that $\text{Re}[f(x)] = f^R(x)$.

4.4.8 The Fourier transform in two dimensions, and symmetry properties

All that has been said so far about Fourier transforms and series in one dimension also applies to higher dimensionalities. In particular, two-dimensional functions (screens) are very important in optics. The transform is defined in terms of two spatial frequency components, k_x and k_y, by a double integral :

$$F(k_x,\ k_y) = \int\int_{-\infty}^{\infty} f(x,\ y)\exp\left[-i(xk_x + yk_y)\right] dx\ dy . \qquad (4.53)$$

If the function $f(x,\ y)$ can be written as the product $f_1(x)f_2(y)$, the integral (4.53) can be factorized into two one-dimensional transforms:

$$
\begin{aligned}
F(k_x,\ k_y) &= \int_{-\infty}^{\infty} f(x)\exp(-ixk_x)dx \int_{-\infty}^{\infty} f_2(y)\exp(-iyk_y)dy \\
&= F_1(k_x)F_2(k_y) .
\end{aligned}
\qquad (4.54)
$$

In the same way as the components $(x,\ y)$ form a vector $\mathbf{r}$ in direct space, the components $(k_x,\ k_y)$ form a vector $\mathbf{k}$ in reciprocal space. Three-dimensional analogues of (4.53) and (4.54) can be written down with no trouble.

If $f(x, y)$ can not be expressed as a product in the above way, the integral (4.53) may be difficult to evaluate analytically. An important class of such problems in optics is that for which f has axial symmetry, and can be written in terms of polar coordinates $(r,\ \theta)$:

$$f(r,\theta) = f_1(r)\, f_2(\theta). \qquad (4.55)$$

Some examples of such problems are discussed in Appendix 1.

The symmetry properties can easily be extended to two dimensions. If $f(x, y)$ is *centrosymmetric*, i.e.

$$
\begin{aligned}
f(x,y) &= f(-x,-y), & (4.56) \\
F(k_x,k_y) &= F(-k_x,-k_y) . & (4.57)
\end{aligned}
$$

Similarly for the case

$$
\begin{aligned}
f(x,y) &= -f(-x,-y) & (4.58) \\
F(k_x,k_y) &= -F(-k_x,-k_y) . & (4.59)
\end{aligned}
$$

If $f(x, y)$ is real:

$$F(k_x,k_y) = F^*(-k_x,-k_y) , \qquad (4.60)$$

implying that $|F(k_x,k_y)|^2$ is centrosymmetric. Equations (4.56)–(4.57) imply that both the function and its transform are invariant on rotation by 180° about the origin. More generally, if the function is invariant on rotation by $360°/n$ (n-fold axial symmetry) its transform behaves likewise. Finally, consider the case of a *real* function with odd n. $|F(k_x, k_y)|^2$ has n-fold symmetry and is also centrosymmetric, implying $2n$-fold symmetry. An example is an equilateral triangle which has a transform with six-fold symmetry. Mirror-plane symmetry or anti-symmetry behave similarly; if $f(x, y) = \pm f(-x, y)$ then $F(-k_x, k_y) = \pm F(k_x, k_y)$, and in both cases $|F(k_x, k_y)|^2$ has mirror-plane symmetry.

4.5 The Fourier inversion theorem

One very useful property of Fourier transforms is that the processes of transforming and untransforming are identical. This property is not trivial, and will be proved below. Another way of stating it is to say that the Fourier transform of the Fourier transform is the original function again, which is true except for some minor details, and is known as the *Fourier inversion theorem*.

If the original function is $f(x)$, the Fourier transform $f_1(x')$ of its Fourier transform can be written down directly as a double integral:

$$f_1(x') = \int \left\{ \int_{-\infty}^{\infty} f(x) \exp(-ikx) \, dx \right\} \exp(-ikx') \, dk \qquad (4.61)$$

which can be evaluated as follows:

$$
\begin{aligned}
f_1(x') &= \int \int_{-\infty}^{\infty} f(x) \exp\{-ik(x + x')\} \, dx \, dk \\
&= \int_{-\infty}^{\infty} f(x) \left[\frac{\exp\{-ik(x + x')\}}{-i(x + x')} \right]_{k=-\infty}^{k=\infty} dx \, . \qquad (4.62)
\end{aligned}
$$

The function with the square brackets can be written as the limit:

$$\lim_{k \to \infty} \frac{2 \sin ky}{y} \, , \quad \text{where } y = (x + x') \, , \qquad (4.63)$$

which can be shown† to be equal to $2\pi\delta(y)$. The transform $f_1(x')$ is thus:

$$f_1(x') = \int_{-\infty}^{\infty} 2\pi\delta(x + x') f(x) \, dx = 2\pi f(-x') \, . \qquad (4.64)$$

† That the function

$$\lim_{k \to \infty} (2 \sin ky)/y$$

is a δ-function, can be justified by drawing the function out for a few values of k. That it has a value of 2π follows from the standard definite integral:

$$\int_{-\infty}^{\infty} \frac{\sin ky}{y} \, dy = \pi \, .$$

The magnitude of a δ-function is equal to the area underneath it.

On retransforming the transform we have therefore recovered the original function, intact except for inversion through the origin (x has become $-x$) and multiplied by a factor 2π. In the two-dimensional transform of a function $f(x, y)$ (§4.4.8), the result of retransforming the transform is to invert *both* axes, which is equivalent to a rotation of 180° about the origin.

It is *conventional* to redefine the inverse transform in a way that 'corrects' the above two deficiencies, so that the transform of the transform comes out exactly equal to the original function. One defines the *forward transform*, $f(x)$ to $F(k)$, as before (4.28):

$$F(k) = \int_{-\infty}^{\infty} f(x)\exp(-ikx)\,dx \qquad (4.65)$$

and the *inverse transform*, $F(k)$ to $f(x)$, as

$$f(x) = \frac{1}{2\pi}\int_{-\infty}^{\infty} F(k)\exp(ikx)\,dk \ . \qquad (4.66)$$

With this convention the inverse transform of the forward transform is exactly identical to the original function. Of course, physical systems are ignorant of such conventions. If we carry out the transform and its inverse experimentally, as in an imaging system (§12.2), the image is indeed inverted!

4.5.1 Examples

The Fourier inversion theorem can be illustrated by any function which can itself be transformed analytically, and whose transform can also be transformed analytically. In §4.4 we have already introduced the periodic array of δ-functions (§4.4.5) and the Gaussian (§4.4.6), which both transform into themselves (see also §9.5.3).

Another example from §4.4.5 is the pair of δ-functions. We saw that the function $\delta(x + b/2) + \delta(x - b/2)$ transforms into $2\cos(kb/2)$, (4.36). The inverse transform of the cosine can be evaluated from (4.66), using the footnote on page 89:

$$\frac{1}{2\pi} \times 2 \int_{-\infty}^{\infty} \cos(\frac{kb}{2})\exp(ikx)dk$$

$$= \frac{1}{2\pi}\int_{-\infty}^{\infty} \{\exp[ik(x + \frac{b}{2})] + \exp[ik(x - \frac{b}{2})]\}dk$$

$$= \delta(x + \frac{b}{2}) + \delta(x - \frac{b}{2}) , \qquad (4.67)$$

which is the original function. The Fourier inversion theorem is particularly useful, of course, when the transform can be carried out analytically in one direction only.

4.6 Convolution

An operation which appears very frequently in optics – and indeed in physics in general – is called *convolution*, or folding. The convolution of two real functions f and g is defined mathematically as:

$$h(x) = \int_{-\infty}^{\infty} f(x')g(x - x')\mathrm{d}x' \ . \tag{4.68}$$

The convolution operation will be represented in this book by the symbol '$\otimes$' so that (4.68) is written:

$$h(x) = f(x) \otimes g(x) \ . \tag{4.69}$$

This operation is particularly important in Fourier theory.

4.6.1 Illustration by means of a 'pinhole' camera

The convolution function is best illustrated by the simplest optical instrument, the pinhole camera. Suppose we consider the photograph of a plane object taken with a pinhole camera with a large pinhole. Because of the size of the pinhole, any one bright point on the object will produce a blurred spot in the image-plane, centred at the point x' where the image would come if focusing were sharp. In one dimension this blurred spot would be described as a function $g(x - x')$ whose origin is at $x = x'$. The intensity of the blurred spot is proportional to the intensity $f(x')$ that the sharp image would have at x'. The intensity at point x is therefore

$$f(x')g(x - x') \tag{4.70}$$

and for the complete blurred image the total intensity observed at x is the integral

$$h(x) = \int_{-\infty}^{\infty} f(x')g(x - x')\mathrm{d}x' \ . \tag{4.71}$$

The above description is illustrated in Fig. 4.11, where two dimensions have been employed and some fancy pinholes have been introduced in order to illustrate various features of convolution. In two dimensions, the convolution function is written

$$h(x, y) = \int \int_{-\infty}^{\infty} f(x', y')g(x - x', y - y')\,\mathrm{d}x'\,\mathrm{d}y' \ . \tag{4.72}$$

A quantitative analysis of this demonstration is given in Appendix 2.

Figure 4.11 Convolution of two-dimensional functions illustrated by the pinhole camera method described in §4.6.1. The objects are shown in (a), (b) and (c); the 'pinholes' used are transparencies identical to (a) and (c). (d) shows $b \otimes a$, (e) shows $c \otimes a$, (f) shows the self-convolution $a \otimes a$ and (g) shows the self-convolution $c \otimes c$. Since (a) and (c) are centrosymmetric, their self-convolutions and auto-correlations are identical.

4.6.2 Convolution with an array of δ-functions

One of the most important applications of convolution in physical optics occurs when one of the functions is an array, regular or otherwise, of δ-functions. It can be illustrated in two dimensions by an infinite sheet of postage stamps (Fig. 4.12), and it will be seen that under these conditions the idea of convolution becomes particularly simple. Basically a sheet of stamps can be described as a rectangular lattice of points (Fig. 4.12(*a*)) with separation typically $b = 2.5$ cm along the *y*-axis and $a = 2.0$ cm along the *x*-axis; at each lattice point is placed an identical unit, that of one postage stamp. We define a particular point on the postage stamp as the origin, and describe the density of ink on one stamp by a function $f(x, y)$ referred to this origin. We then describe the lattice by a series of δ-functions at the lattice points:

$$g(x, y) = \sum_{l,m} \delta(x - la)\,\delta(y - mb), \qquad (4.73)$$

where *l* and *m* are integers. The complete sheet of postage stamps can then be represented by the convolution function (Fig. 4.12(*b*)):

$$h(x, y) = \iint g(x', y')f(x - x', y - y')\,\mathrm{d}x'\,\mathrm{d}y'. \qquad (4.74)$$

4.6.3 Convolution in optics

We have devoted considerable attention to the convolution operation because it has many applications in optics. At the risk of pre-empting discussions in later chapters, we shall briefly mentions some of the situations which are considerably simplified by the use of convolutions.

(1) A diffraction grating (§9.2) can be represented by a slit or other arbitrary line-shape function convoluted with a one-dimensional array of δ-functions.

(2) The electron density in a crystal is represented by the density in a single molecular unit convoluted with the three-dimensional lattice of δ-functions representing the crystal lattice (§8.4.1)

(3) In a Fraunhofer diffraction experiment, the intensity is all that can be observed directly. The transform of the intensity function is the convolution of the diffraction mask function with its inverse (§4.7.1).

4.6.4 Fourier transform of a convolution

Not only does the convolution operation occur frequently in physics, but its Fourier transform is particularly simple. This fact makes it very convenient to use. We shall now prove the *convolution theorem*, which states that the *Fourier transform of the convolution of two functions is the product of the transforms of the original functions.*

Consider the convolution $h(x)$ of the functions $f(x)$ and $g(x)$, as defined in (4.68). Its Fourier transform is

$$
\begin{aligned}
H(k) &= \int_{-\infty}^{\infty} \left[\int_{-\infty}^{\infty} f(x')g(x-x')\mathrm{d}x' \right] \exp(-ikx)\mathrm{d}x \\
&= \iint_{-\infty}^{\infty} f(x')g(x-x')\exp(-ikx)\mathrm{d}x'\mathrm{d}x .
\end{aligned}
\tag{4.75}
$$

By writing $y = x - x'$, we can rewrite this as

$$
H(k) = \iint_{-\infty}^{\infty} f(x')g(y)\exp\{-ik(x'+y)\}\mathrm{d}x'\mathrm{d}y ,
\tag{4.76}
$$

which separates into two factors:

$$
\int_{-\infty}^{\infty} f(x')\exp(-ikx')\mathrm{d}x' \int_{-\infty}^{\infty} g(y)\exp(-iky)\mathrm{d}y = F(k)\,G(k)
\tag{4.77}
$$

Figure 4.12 Sheet of postage stamps, representing convolution of one stamp with a two-dimensional lattice of δ-functions. (By kind permission of HM Postmaster-General.)

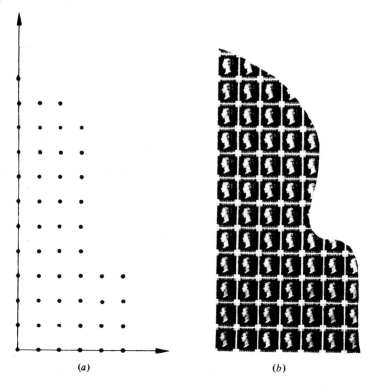

(a)   (b)

or, simply,

$$H(k) = F(k) G(k). \tag{4.78}$$

This is the required result.

We can now invoke the Fourier inversion theorem (§4.5) and deduce immediately that 2π *times the Fourier transform of the product of two functions equals the convolution of their individual transforms*, which is an alternative statement of the convolution theorem.

4.6.5 Transform of a wave-group as an example of convolution

There are many examples of functions which can most conveniently be Fourier-transformed after they have been broken down into a convolution or a product, and the reader will meet several of them in the succeeding chapters. We shall just give one simple example here, which is often employed as a convenient model for more complicated ideas (e.g. in §§2.8 and 11.2.2).

A *Gaussian wave-group* has the form $A \exp(ik_0 x)$ modified by a Gaussian envelope (§4.4.6) having variance σ (Fig. 4.13*a*). It can be written in the form:

$$f(x) = A \exp(ik_0 x) \exp(-x^2/2\sigma^2) . \tag{4.79}$$

This function will immediately be recognized as the *product* of the complex exponential $\exp(ik_0 x)$ and the Gaussian (4.42). Its transform is therefore the *convolution* of the transforms of these two functions which are, respectively $2\pi A \, \delta(k - k_0)$ and (4.45), namely, $(2\pi\sigma^2)^{\frac{1}{2}} \exp(-k^2\sigma^2/2)$. Now the first of these transforms is a δ-function at the point $k = k_0$, and on convoluting the latter transform with it, we simply shift the origin of the transform Gaussian to that point, getting:

$$F(k) = (2\pi)^{\frac{3}{2}} \sigma A \exp\left[-(k - k_0)^2 \sigma^2/2\right] , \tag{4.80}$$

as shown in Fig. 4.13(*b*). This result was used in §2.8.

4.7 Correlation functions

A form of convolution function which is of great importance in statistics and has many applications in physics is the *correlation function*, which is formally defined as †

$$h_C(x) = \int_{-\infty}^{\infty} f(x') g^*(x' + x) \, dx' . \qquad (4.81)$$

It is easy to show by changing variables this is the convolution of $f(-x)$ with $g^*(x)$. As its name implies, the function measures the degree of similarity between the functions f and g. Suppose the two tend to be similar in magnitude and phase, when referred to origins at 0 and $-x_0$ respectively. Putting $x = x_0$, $f(x')$ and $g(x' + x_0)$ will then have about the same complex values and so $f(x') g^*(x' + x_0)$ will be positive and real. Thus the integral $h_C(x_0)$ will be large and positive. We shall use this function considerably in studying coherence in Chapter 11. Its Fourier transform is

$$H_C(k) = F(-k) G^*(-k) . \qquad (4.82)$$

4.7.1 Auto-correlation function and the Wiener–Khinchin theorem

A particular case of the correlation function is the *auto-correlation function*, h_{AC} which is defined by (4.81) with $f \equiv g$, i.e. by

$$h_{AC}(x) = \int_{-\infty}^{\infty} f(x') f^*(x' + x) \, dx' . \qquad (4.83)$$

Since the functions now have the same origin, the auto-correlation function clearly has a strong peak when $x = 0$. We get for the Fourier

† Some books define the correlation function with $-x$ instead of $+x$ in the argument of g. This makes no difference to the physics. We have followed Born and Wolf (1980).

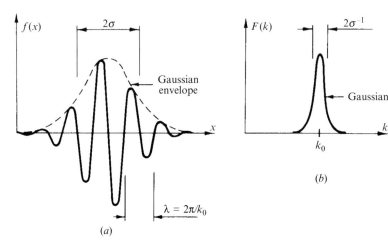

Figure 4.13
(a) A wave-group;
(b) transform of a wave-group.

transform of h_{AC}

$$H_{AC}(k) = F(-k) F^*(-k) = |F(-k)|^2, \qquad (4.84)$$

and for real f we get by using (4.50)

$$H_{AC}(k) = |F(k)|^2. \qquad (4.85)$$

In words, (4.85) states that for a real function $f(x)$ the Fourier transform of the auto-correlation function is the square modulus of the transform of the function, also known as its *power spectrum*. From the Fourier inversion theorem, the statement is also true in reverse, barring a factor 2π. It is known as the *Wiener–Khinchin theorem*, and applies similarly in more than one dimension.

Useful information in the auto-correlation function is not limited to the peak at $x = 0$. For example, suppose that a function has a strong periodicity with wavenumber K_0 and period $\Lambda = 2\pi/K_0$. The functions $f(x')$ and $f^*(x'+n\Lambda)$ will then tend to be similar and so their product will be positive; thus periodic peaks in $h_{AC}(x)$ will appear. The transform, the power spectrum H_{AC}, has a corresponding peak at $k = K_0$, and is thus useful for recognizing the existence of periodicities in the function $f(x)$.

In two and three dimensions, the correlation function finds many applications in pattern recognition, and the auto-correlation function has been widely used in the interpretation of X-ray diffraction patterns, where it is called the *Patterson function*. It is instructive to see how it is built up in a simple case in two dimensions (Fig. 4.14), where $f(x, y)$ consists of three equal real δ-functions. On each δ-point of $f(x', y')$ we put the origin of the function $f^*(-x, -y)$, which is just

Figure 4.14
Auto-correlation of a two-dimensional function consisting of three δ-functions (shown on left). The lines are inserted to guide the eye, and are not part of the function.

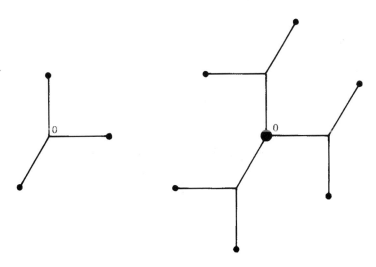

$f(x, y)$ rotated by 180°. We immediately see a strong point developing at the origin. This strong point at the origin is intrinsic to the auto-correlation of a real function. Experimental methods of determining spatial auto-correlation functions will be discussed briefly in §12.6.5.

4.7.2 Energy conservation: Parseval's theorem

The process of Fourier transformation essentially takes a certain function $f(x)$ and represents it as the superposition of a set of waves. We shall see later, in Chapter 8, that in optics Fraunhofer diffraction is described by a Fourier transform, where $f(x)$ represents the amplitude distribution leaving the diffracting obstacle and $F(k)$ represents the amplitude distribution in the diffraction pattern. No light energy need be lost in this process, and it would therefore seem necessary that the total power leaving the object be equal, or at least proportional, to that arriving at the diffraction pattern. In mathematical terms, we expect that

$$\int_{-\infty}^{\infty} |f(x)|^2 dx = C \int_{-\infty}^{\infty} |F(k)|^2 dk . \qquad (4.86)$$

This is called *Parseval's theorem*. It can be deduced easily from our discussion of the auto-correlation function in §4.7.1. Applying the Fourier inversion theorem to (4.84), the inverse transform of $|F(-k)|^2$ must be equal to $h_{AC}(x)$. Writing this out explicitly, we get from (4.66)

$$\frac{1}{2\pi} \int_{-\infty}^{\infty} |F(-k)|^2 \exp(ikx) dk = \int_{-\infty}^{\infty} f(x') f^*(x' + x) dx' . \qquad (4.87)$$

Now let $x = 0$ in this equation and substitute $-k$ for k. This gives:

$$\frac{1}{2\pi} \int_{-\infty}^{\infty} |F(k)|^2 dk = \int_{-\infty}^{\infty} |f(x')|^2 dx' \qquad (4.88)$$

which is Parseval's theorem, with $C = 1/2\pi$. This factor has arisen solely because of the convention that the same factor be introduced into the definition (4.66) of the inverse transform.

CHAPTER FIVE

Electromagnetic waves

5.1 Electromagnetism and the wave equation

This chapter will discuss the electromagnetic wave as a specific and most important example of the general treatment of wave propagation presented in Chapter 2. We shall start at the point where the elementary features of classical electricity and magnetism have been summarized in the form of Maxwell's equations, and the reader's familiarity of the steps leading to this formulation will be assumed (see, for example Grant and Phillips, 1975; Jackson, 1975).† It is well-known that Maxwell's formulation included for the first time the displacement current $\partial \mathbf{D}/\partial t$, the time-derivative of the ficticious displacement field $\mathbf{D} = \epsilon_0 \mathbf{E} + \mathbf{P}$, which is a combination of the applied electric field $\mathbf{E}$ and the electric polarization density $\mathbf{P}$. This field will turn out to be of prime importance when we come to extend the treatment in this chapter to wave propagation in anisotropic media in Chapter 6.

The development presented in this chapter emphasizes the properties of simple harmonic waves in isotropic linear media, and the way in which waves behave when they meet the boundaries between media. An *isotropic* medium is one in which all directions in space are equivalent, and there is no difference between right-handed and left-handed rotation. An example would be a monatomic liquid; in contrast, crystals are generally *anisotropic*. A *linear* medium is one in which the polarization produced by an applied electric or magnetic field is proportional to that field. Since interatomic electric fields are of the order of 10^{11} V.m^{-1}, macroscopic laboratory electric fields (usually $< 10^8$ V.m^{-1}) are small in comparison, and their effects are consequently linear, but the oscillating fields produced by intense laser beams are often orders of magnitude larger and cause *non-linear* re-

† In deference to common usage, the S.I. (M.K.S.) system will be used.

sponse. Essentially the rest of the book consists of elaboration of these ideas. The final chapter discusses the quantization of the electromagnetic field, an aspect which Maxwell could not have predicted, and will bring us close to the frontiers of modern research.

5.1.1 Maxwell's equations

Considering that Maxwell did not have the modern concepts of vector diffential operators (grad, div, curl) at his disposal, it was an almost incredible achievement that he was able to summarize the classical properties of the electric fields **E** and **D**, the magnetic fields **H** and **B**, charge density ρ and current density **j** in a set of four simple equations. It was even more remarkable that he could see that these equations led to wave propagation. Using vector operators, the derivation is much more transparent.

Gauss's law in electrostatics becomes: $\nabla \cdot \mathbf{D} = \rho,$ (5.1)

Gauss's law in magnetostatics becomes: $\nabla \cdot \mathbf{B} = 0,$ (5.2)

Ampère's law becomes: $\nabla \times \mathbf{H} = \dfrac{\partial \mathbf{D}}{\partial t} + \mathbf{j},$ (5.3)

Faraday's law becomes: $\nabla \times \mathbf{E} = -\dfrac{\partial \mathbf{B}}{\partial t}.$ (5.4)

In vacuum, **D** and **E** are identical fields; the fact that in S.I. units, $\mathbf{D} = \epsilon_0 \mathbf{E}$, where ϵ_0 has a non-unit value, only reflects the fact that **D** and **E** are measured in different *units*. The same applies to the applied magnetic field **H** and the magnetic induction, **B**, which is the measured field when magnetic polarization effects are taken into account. In a vacuum, $\mathbf{B} = \mu_0 \mathbf{H}$ where μ_0 reflects the difference in units.† In a medium, **D** really differs from **E**, and **B** from **H**. This is represented, in the case of a linear isotropic medium, by scalar dimensionless constants ϵ and μ:

$$\mathbf{D} = \epsilon\epsilon_0 \mathbf{E}, \qquad (5.5)$$

$$\mathbf{B} = \mu\mu_0 \mathbf{H}. \qquad (5.6)$$

The values of ϵ and μ may be frequency dependent, as will be discussed in Chapter 10.

---

† We emphasize the typographical distinction we shall make between μ for magnetic permeability and µ for refractive index.

5.1.2 Electromagnetic waves

The simplest case for which Maxwell's equations lead to a non-dispersive wave equation (2.6) is in an isotropic insulating medium, where charge density ρ and current density $\mathbf{j}$ are both zero. Then (5.1)–(5.4) become:

$$\nabla \cdot \mathbf{D} = \epsilon\epsilon_0 \nabla \cdot \mathbf{E} = 0, \tag{5.7}$$

$$\nabla \cdot \mathbf{B} = \mu\mu_0 \nabla \cdot \mathbf{H} = 0, \tag{5.8}$$

$$\nabla \times \mathbf{H} = \frac{\partial \mathbf{D}}{\partial t} = \epsilon\epsilon_0 \frac{\partial \mathbf{E}}{\partial t}, \tag{5.9}$$

$$\nabla \times \mathbf{E} = -\frac{\partial \mathbf{B}}{\partial t} = -\mu\mu_0 \frac{\partial \mathbf{H}}{\partial t}. \tag{5.10}$$

Taking $(\nabla\times)$ of both sides of (5.10) and substituting (5.9) we have:

$$\nabla \times (\nabla \times \mathbf{E}) = -\mu\mu_0 \frac{\partial}{\partial t}(\nabla \times \mathbf{H}) = -\mu\mu_0\epsilon\epsilon_0 \frac{\partial^2 \mathbf{E}}{\partial t^2}. \tag{5.11}$$

On expanding $\nabla \times (\nabla \times \mathbf{E}) = \nabla(\nabla \cdot \mathbf{E}) - \nabla^2 \mathbf{E}$, (5.11) becomes

$$\nabla^2 \mathbf{E} = \epsilon\mu\,\epsilon_0\mu_0 \frac{\partial^2 \mathbf{E}}{\partial t^2}. \tag{5.12}$$

In cartesians, $\nabla^2 \mathbf{E}$ is the vector

$$\nabla \cdot (\nabla \mathbf{E}) \equiv (\nabla^2 E_x, \nabla^2 E_y, \nabla^2 E_z).$$

5.1.3 Wave velocity and refractive index

Following §2.2, we immediately see that the solution to (5.12) is a vector wave with velocity

$$v = (\epsilon\mu\,\epsilon_0\mu_0)^{-\frac{1}{2}} \tag{5.13}$$

In free space, this velocity is $c = (\epsilon_0\mu_0)^{-\frac{1}{2}}$ which is an important fundamental constant, now defined as $2.997\,924\,58 \times 10^8$ m s^{-1} exactly.† (This number retains the metre and second as accurately as they have ever been defined, but makes c itself the fundamental constant – a decision made in 1986.) Following this definition, the S.I. defines μ_0 as $4\pi \times 10^{-7}$ Hm^{-1} from which ϵ_0 can be calculated as $(\mu_0 c^2)^{-1} = 8.854 \times 10^{-12}$ Fm^{-1}.

In accordance with the usual practice in optical work we shall assume in the rest of the book that the magnetic permability μ of

---

† One of the rather disappointing features of the definition of the speed of light as a fundamental constant is that its determination, by any experimental means, has thus been degraded to the measure of the length of an optical path in terms of a known constant!

media is unity at the frequencies of light waves. The ratio between the velocity of electromagnetic waves in a vacuum to that in an isotropic medium, which is the definition of *refractive index*, μ, is then

$$\mu = c/v = \epsilon^{\frac{1}{2}}, \qquad (5.14)$$

where again the value of ϵ at the right frequency must be used. To avoid confusion with the refractive index μ, μ_0 will usually be replaced by $(\epsilon_0 c^2)^{-1}$ wherever it appears.

5.2 Plane-wave solutions of the wave equation

The wave (cf §2.6)

$$\mathbf{E} = \mathbf{E}_0 \exp[i(\mathbf{k} \cdot \mathbf{r} - \omega t)] \qquad (5.15)$$

is a solution of (5.12) where $\omega/k = v$. Replacing ∇ and $\partial/\partial t$ by $-i\mathbf{k}$ and $i\omega$ respectively (§2.6.1) allows us to write (5.7)–(5.10) in the form:

$$\mathbf{k} \cdot \mathbf{D} = \epsilon \epsilon_0 \mathbf{k} \cdot \mathbf{E} = 0, \qquad (5.16)$$

$$\mathbf{k} \cdot \mathbf{B} = (\epsilon_0 c^2)^{-1} \mathbf{k} \cdot \mathbf{H} = 0, \qquad (5.17)$$

$$\mathbf{k} \times \mathbf{H} = -\omega \mathbf{D} = -\omega \epsilon \epsilon_0 \mathbf{E}, \qquad (5.18)$$

$$\mathbf{k} \times \mathbf{E} = \omega \mathbf{B} = \omega (\epsilon_0 c^2)^{-1} \mathbf{H}. \qquad (5.19)$$

These equations immediately gives us an insight into the disposition and size of the field vectors (Fig. 5.1): $\mathbf{D}$, $\mathbf{k}$ and $\mathbf{B}$ are mutually orthogonal, as are $\mathbf{E}$, $\mathbf{k}$ and $\mathbf{H}$ by virtue of the isotropy of the medium. Electromagnetic waves are therefore transverse. Morever, the magnitudes of $\mathbf{E}$ and $\mathbf{H}$ are related by

$$\frac{E}{H} = \frac{k}{\epsilon \epsilon_0 \omega} = \frac{1}{\epsilon^{\frac{1}{2}} \epsilon_0 c} \equiv Z. \qquad (5.20)$$

The constant Z is called the *impedance* of the medium to electromagnetic wave propagation and, somewhat surprisingly, has the dimensions of ohms. In free space $Z_0 = 1/\epsilon_0 c = 377$ ohms. In general

$$Z = Z_0/\mu \qquad (5.21)$$

relates the impedance of a medium to its refractive index.

The plane containing $\mathbf{D}$ and $\mathbf{k}$ is called the *plane of polarization* which will be of paramount importance in Chapter 6.

The fact that equations (5.16)–(5.19) are completely real indicates that there are no phase differences between the oscillations of the electric and magnetic fields. $\mathbf{H}$ can thus be written

$$\mathbf{H} = \mathbf{H}_0 \exp[i(\mathbf{k} \cdot \mathbf{r} - \omega t)], \qquad (5.22)$$

where $\mathbf{H}_0$ is orthogonal to $\mathbf{k}$ and $\mathbf{E}_0$ and has magnitude E_0/Z.

5.2.1 Flow of energy in an electromagnetic wave

An important feature of electromagnetic waves is that they can transport energy. The vector describing the flow of energy is the *Poynting vector*, which can be shown in general to be

$$\mathbf{\Pi} = \mathbf{E} \times \mathbf{H}. \qquad (5.23)$$

It has dimensions of energy per unit time per unit area, and its absolute value is called the *intensity* of the wave. It is easy to see that this vector lies parallel to $\mathbf{k}$ in the isotropic medium. The time-averaged value of $\mathbf{\Pi}$, when $\mathbf{E}$ and $\mathbf{H}$ have the same phase, and are mutually orthogonal as in Fig. 5.1, is

$$\langle \mathbf{\Pi} \rangle = \langle E_0 \sin \omega t \, H_0 \sin \omega t \rangle = \tfrac{1}{2} E_0 H_0 = \tfrac{1}{2} E_0^2/Z \qquad (5.24)$$

(since the average value of $\sin^2 \omega t$ is $\tfrac{1}{2}$ over a time $\gg 1/\omega$).

However, we shall later come across situations where there is a phase difference between $\mathbf{E}$ and $\mathbf{H}$, and then the mean value is different. In particular, if there is a $90°$ phase difference between the phases of $\mathbf{E}$ and $\mathbf{H}$,

$$\langle \mathbf{\Pi} \rangle = \langle E_0 \sin \omega t H_0 \cos \omega t \rangle = 0 \qquad (5.25)$$

and no energy is transported. Evanescent waves are an example of such behaviour (§2.4.2 and §5.5.4).

5.3 Radiation

Electromagnetic radiation is initiated by moving charges. Two types of source are of particular importance in optics and, although a detailed

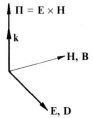

Figure 5.1 Disposition of vectors in an electromagnetic wave in an isotropic medium.

treatment is out of place in an optics textbook, we shall mention them briefly. A full treatment is given in texts on electromagnetic theory.

5.3.1 Radiation by an accelerating charge

A charged particle moving at uniform velocity in a straight line is equivalent to an electric current, and produces a constant magnetic field. This does not radiate electromagnetic waves. However, if the particle accelerates, then the field has a time derivative and radiation results. An important direct application is the X-radiation from charged particles in a synchrotron; it is emitted from the regions of the synchrotron where the particles change direction due to a magnetic field, so they are subject to centripetal acceleration (Wille, 1991). For a particle of charge q moving with velocity $\mathbf{v}(t)$ and acceleration $\dot{\mathbf{v}}$, the radiative electric field at radius vector $\mathbf{r} \equiv \hat{\mathbf{n}}r$ is

$$\mathbf{E} = \frac{q}{4\pi\epsilon_0 rc^2}[\hat{\mathbf{n}} \times (\hat{\mathbf{n}} \times \dot{\mathbf{v}})], \tag{5.26}$$

$$\text{with magnitude} \quad |E| = \frac{q|\dot{\mathbf{v}}|}{4\pi\epsilon_0 c^2 r}\sin\theta \tag{5.27}$$

where θ is the angle between $\dot{\mathbf{v}}$ and $\hat{\mathbf{n}}$. The field lies in the plane containing these vectors, transverse to $\hat{\mathbf{n}}$, as shown in Fig. 5.2(a).

The magnetic field is given by $\mathbf{H} = Z_0^{-1}(\hat{\mathbf{n}} \times \mathbf{E})$ and is thus also transverse to $\hat{\mathbf{n}}$, but polarized normal to the $(\mathbf{v}, \hat{\mathbf{n}})$ plane. The fields are *retarded*, which means that $\mathbf{v}$ is evaluated a time r/c earlier than $\mathbf{E}$ and $\mathbf{H}$ are measured. Together, they result in energy being radiated predominately in the plane normal to the direction of the acceleration. The Poynting vector is, from (5.27):

$$\boldsymbol{\Pi} = \mathbf{E} \times \mathbf{H} = \frac{\hat{\mathbf{n}}q^2\dot{\mathbf{v}}^2}{16\pi^2\epsilon_0 c^3 r^2}\sin^2\theta, \tag{5.28}$$

Figure 5.2 Radiation from an accelerating charge: (*a*) orientation of vectors, (*b*) section of the radiation polar diagram, which is a torus in three dimensions.

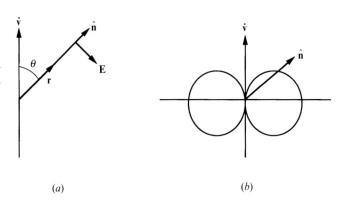

(*a*)   (*b*)

which has maximum value in the direction normal to $\dot{\mathbf{v}}$. We can represent this by a *radiation polar diagram* in which the magnitude of $\mathbf{\Pi}(\hat{\mathbf{n}})$ is represented as a polar graph with its centre at the radiating charge, Fig. 5.2(*b*). In the case of a charged particle traversing a circular orbit, as in a synchrotron, the radiation is then maximum in the plane tangential to the orbit and perpendicular to the instantaneous acceleration $\dot{\mathbf{v}}$. The total power radiated is found by integrating (5.28) over the surface of a sphere of radius r and is

$$P = \frac{q^2 |\dot{\mathbf{v}}|^2}{6\pi\epsilon_0 c^3}. \tag{5.29}$$

The above calculation has assumed that the charge motion is not relativistic, i.e. $v \ll c$. At velocities close to c, the radiation pattern has a narrower peak lobe more closely aligned to the direction of orbit and when transformed to the laboratory frame of reference it takes the form shown in Fig. 5.3.

The frequency spectrum of synchrotron radiation arises from Fourier analysis of what is essentially a short burst of radiation that is emitted every time an electron passes through the curved orbital region. In practice this contains considerable amounts of X-radiation and is indispensible for many X-ray diffraction experiments as well as X-ray imaging (see §7.5).

5.3.2 Radiation emitted by an oscillating dipole

The radiative system most frequently encountered in elementary optics is a periodically oscillating dipole. This arises, for example, in scattering theory (§13.2) when a wave is incident on a polarizable body – an atom, molecule or larger particle. The electric field of the wave polarizes the body and gives it a dipole moment which then oscillates at the wave frequency.

The radiated fields can be derived directly from §5.3.1. A point charge q has position $z(t) = a\cos\omega t$, representing a dipole with instantaneous moment $p = qz = qa\cos\omega t$. The acceleration is

$$\dot{\mathbf{v}} = -a\omega^2 \hat{\mathbf{z}} \cos\omega t. \tag{5.30}$$

Figure 5.3 Synchrotron radiation. The diagram shows the section of the radiation polar diagram in the plane of the orbit, for a charged particle at $v \approx c$, where $\gamma \equiv (1 - v^2/c^2)^{-1/2} \gg 1$, transformed relativistically to the laboratory frame of reference.

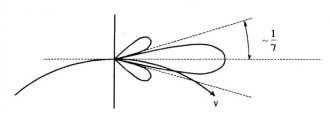

Then, at distances large† compared with a we have $\mathbf{E}$ and $\mathbf{H}$ transverse to $\hat{\mathbf{n}}$:

$$\mathbf{E} = \frac{-qa\omega^2}{4\pi\epsilon_0 c^2 r}[\hat{\mathbf{n}} \times (\hat{\mathbf{n}} \times \hat{\mathbf{z}})], \tag{5.31}$$

$$\mathbf{H} = Z_0^{-1}\hat{\mathbf{n}} \times \mathbf{E}. \tag{5.32}$$

The radiation polar diagram for the Poynting vector $\mathbf{\Pi} = \mathbf{E} \times \mathbf{H}$ from such an oscillating dipole has the same $\sin^2\theta$ dependence as in (5.28) as was shown in Fig. 5.2(*b*). It is most important to appreciate that $\mathbf{\Pi}$ has zero value along the axis $\hat{\mathbf{z}}$ of the dipole, emphasizing the fact that electromagnetic radiation is transversely polarized.

The total power radiated from the dipole is, like (5.29),

$$P = \frac{1}{6\pi\epsilon_0} \cdot \frac{p_0^2\omega^4}{c^3}\cos^2\omega t, \tag{5.33}$$

where $p_0 = qa$ is the amplitude of the dipole moment oscillations. Since the average value of $\cos^2 x$ is $\frac{1}{2}$, the mean power radiated during a period $\gg \omega^{-1}$ is

$$\langle P \rangle = \frac{p_0^2\omega^4}{12\pi c^3\epsilon_0}. \tag{5.34}$$

A noticeable feature of this expression is the strong dependence on ω; the power radiated from a dipole oscillator is proportional to the fourth power of the frequency. One practical result of this dependence is the blue colour of the sky (§13.2.1); another one is a basic limitation to the transparency of optical fibres (§10.3).

5.4 Reflexion and refraction

5.4.1 Boundary conditions at an abrupt interface

At a sharp boundary between two media, there are simple relationships which must be obeyed between the fields on the two sides. The components of the fields $\mathbf{E}$ and $\mathbf{H}$ parallel to the surface are equal on the two sides, whereas the normal components of $\mathbf{D}$ and $\mathbf{B}$ must likewise be continuous. Full proof of these conditions, which we shall use extensively in what follows, is given in the texts on electromagnetic theory.

† When the distance to the observer is comparable to a and/or the wavelength, the *near field* is more complicated. This regime has some important applications in optics, such as near-field microscopy (§12.5), but will not be considered here.

5.4.2 The Fresnel coefficients

Suppose that a plane electromagnetic wave with wave-vector **k** and electric field amplitude $\mathbf{E}_0$ is incident on a plane surface separating isotropic media with refractive indices μ_1 and μ_2. The angle of incidence between the incident wave-vector **k** and the normal to the surface $\hat{\mathbf{n}}$ is $\hat{\imath}$. Without loss of generality we can treat separately the two cases where the incident vector **E** lies in the plane defined by **k** and $\hat{\mathbf{n}}$, denoted by $\parallel$, and that where **E** is normal to this plane, denoted by $\perp$. Any other polarization, plane or otherwise (§6.2), can be considered as a linear superposition of these two cases. Other names which are commonly used are p and TM for $\parallel$, s and TE for $\perp$.

Fig. 5.4 shows the geometry of this situation. Notice that reflected and transmitted waves have been introduced. The plane containing the incident wave-vector **k**, the reflected and transmitted wave-vector and the normal $\hat{\mathbf{n}}$ is the (x, z) plane, and the vector $\hat{\mathbf{n}}$ is along the z direction. We denote the amplitudes (electric field magnitudes) of the incident, reflected and transmitted waves by I, R and T respectively. The magnitudes of the wave vectors in the two media are k_1 and k_2 and clearly $k_1/k_2 = \mu_2/\mu_1$ since both waves have the same frequency.

Consider first the $\perp$ mode, so that the incident $\mathbf{E} = (0, I, 0)$. At $t = 0$,

Figure 5.4 Incident, reflected and transmitted waves.

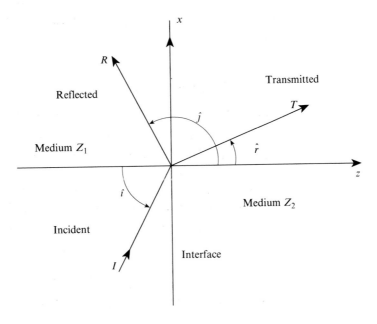

$$
\begin{array}{rl}
\text{incident wave:} & E_y = E_{yI} = I\exp[-\mathrm{i}(k_1 z\cos\hat{\imath} + k_1 x\sin\hat{\imath})]\,,\\[4pt]
\text{reflected wave:} & E_y = E_{yR} = R\exp[-\mathrm{i}(k_1 z\cos\hat{\jmath} + k_1 x\sin\hat{\jmath})]\,,\\[4pt]
\text{transmitted wave:} & E_y = E_{yT} = T\exp[-\mathrm{i}(k_2 z\cos\hat{r} + k_2 x\sin\hat{r})]\,.
\end{array}
$$

$$(5.35)$$

Any changes of phase occuring on reflexion and transmission will be indicated by negative or complex values of R and T. The magnetic fields are related by impedances $Z = E/H = Z_0/\mu$ and are perpendicular to $\mathbf{k}$ and $\mathbf{E}$. The fact that the reflected wave travels in the opposite z-direction to the others will be taken care of by the appropriate value of $\hat{\jmath}$ so that the Poynting vector, the energy flow, is in the correct direction. Given the direction of the field $\mathbf{E} = (0, E_y, 0)$, we find

$$
\begin{array}{rlcl}
\text{incident wave:} & H_z & = & E_{yI}Z_0^{-1}\mu_1\sin\hat{\imath}\\[4pt]
& H_x & = & -E_{yI}Z_0^{-1}\mu_1\cos\hat{\imath}\,;\\[4pt]
\text{reflected wave:} & H_z & = & E_{yR}Z_0^{-1}\mu_1\sin\hat{\jmath}\,,\\[4pt]
& H_x & = & -E_{yR}Z_0^{-1}\mu_1\cos\hat{\jmath}\,;\\[4pt]
\text{transmitted wave:} & H_z & = & E_{yT}Z_0^{-1}\mu_2\sin\hat{r}\,,\\[4pt]
& H_x & = & -E_{yI}Z_0^{-1}\mu_2\cos\hat{r}\,.
\end{array}
$$

$$(5.36)$$

The boundary conditions can then be applied. E_y is itself the parallel component, which is continuous, so from (5.35) at the point $x = 0$, $z = 0$ we have

$$I + R = T\,. \tag{5.37}$$

For $E_{yI} + E_{yR} = E_{yT}$ at any point in the plane $z = 0$, their oscillatory parts must be identical:

$$k_1\sin\hat{\imath} = k_1\sin\hat{\jmath} = k_2\sin\hat{r}\,, \tag{5.38}$$

from which $\hat{\jmath} = \pi - \hat{\imath}$ and Snell's law follows:

$$\sin\hat{\imath} = \frac{k_2}{k_1}\sin\hat{r} = \frac{\mu_2}{\mu_1}\sin\hat{r} = \mu_r\sin\hat{r}\,, \tag{5.39}$$

where μ_r is the relative refractive index between the two media. Continuity of the parallel component H_x at $(x, z) = (0, 0)$ gives

$$IZ_1^{-1}\cos\hat{\imath} + RZ_1^{-1}\cos\hat{\jmath} = TZ_2^{-1}\cos\hat{r}\,. \tag{5.40}$$

We define reflexion and transmission coefficients $\mathscr{R} \equiv R/I$, $\mathscr{T} \equiv T/I$. Then, for this polarization (denoted by the subscript $\perp$) we have from (5.37), (5.38) and (5.21),

$$\mathscr{R}_\perp = \frac{\mu_1\cos\hat{\imath} - \mu_2\cos\hat{r}}{\mu_1\cos\hat{\imath} + \mu_2\cos\hat{r}} = \frac{\cos\hat{\imath} - \mu_r\cos\hat{r}}{\cos\hat{\imath} + \mu_r\cos\hat{r}} \tag{5.41}$$

$$\mathcal{T}_\perp = \frac{2\mu_1 \cos\hat{\imath}}{\mu_1 \cos\hat{\imath} + \mu_2 \cos\hat{r}} = \frac{2\cos\hat{\imath}}{\cos\hat{\imath} + \mu_r \cos\hat{r}}. \tag{5.42}$$

The coefficients for the ∥ plane of polarization can be worked out similarly. When $\mathcal{R}$ and $\mathcal{T}$ refer to the component E_x, we find:†

$$\mathcal{R}_\| = \frac{\mu_1 \cos\hat{r} - \mu_2 \cos\hat{\imath}}{\mu_1 \cos\hat{r} + \mu_2 \cos\hat{\imath}} = \frac{\cos\hat{r} - \mu_r \cos\hat{\imath}}{\cos\hat{r} + \mu_r \cos\hat{\imath}}, \tag{5.43}$$

$$\mathcal{T}_\| = \frac{2\mu_1 \cos\hat{\imath}}{\mu_1 \cos\hat{r} + \mu_2 \cos\hat{\imath}} = \frac{2\cos\hat{\imath}}{\cos\hat{r} + \mu_r \cos\hat{\imath}}. \tag{5.44}$$

These functions are shown in Fig. 5.5. The two cases are sometimes combined in the convenient forms

$$\mathcal{R} = \frac{u_1 - u_2}{u_2 + u_1}, \tag{5.45}$$

$$\mathcal{T} = \frac{2u_1}{u_2 + u_1}, \tag{5.46}$$

where

$$\text{for } \perp \quad u_1 \equiv \mu_1 \cos\hat{\imath}, \quad u_2 \equiv \mu_2 \cos\hat{r}, \tag{5.47}$$

$$\text{for } \| \quad u_1 \equiv \mu_1 \sec\hat{\imath}, \quad u_2 \equiv \mu_2 \sec\hat{r}. \tag{5.48}$$

This form is particularly useful in formulating the general theory of multilayer dielectric systems (§10.4), since both polarizations and all angles of incidence can be treated with the single pair of formulae. At normal incidence, the reflexion and transmission coefficients for the two polarizations are equal and are given by

$$\mathcal{R} = \frac{1 - \mu_r}{1 + \mu_r}, \tag{5.49}$$

$$\mathcal{T} = \frac{2}{1 + \mu_r}. \tag{5.50}$$

As an example, at an air–glass interface, where $\mu_r = 1.5$, the amplitude reflexion coefficient $\mathcal{R}$ (5.49) is $-0.5/2.5 = -0.2$, and so the intensity reflexion coefficient $\mathcal{R}^2 = 4\%$.

5.4.3 Brewster angle

For the polarization plane parallel to the incidence plane, Fig. 5.5 indicates that the reflexion coefficient is zero at a particular angle $\hat{\imath}_B$.

---

† There is sometimes some confusion about the sign of $\mathcal{R}_\|$. If the coefficient were to refer to the component E_z (which might be more convenient at angles near grazing incidence), the sign of $\mathcal{R}_\|$ would be reversed. To avoid any confusion, we shall only refer the coefficient to E_x in this book.

For this condition we have

$$\mu_1 \cos \hat{r} - \mu_2 \cos \hat{\imath} = 0 ,$$
$$\frac{\cos \hat{r}}{\cos \hat{\imath}} = \frac{\mu_2}{\mu_1} = \mu_r = \frac{\sin \hat{\imath}}{\sin \hat{r}} . \qquad (5.51)$$

We leave it to the reader to confirm that this equation can be rewritten as

$$\tan \hat{\imath} = \cot \hat{r} = \mu_r . \qquad (5.52)$$

The angle $\hat{\imath} = \hat{\imath}_B$ which is the solution of this equation is called the *Brewster angle*. At this angle of incidence light of the parallel polarization is not reflected.

5.5 Incidence in the denser medium

When $\mu_r < 1$, meaning that the incidence is in the denser medium, several interesting phenomena occur. First, we should point out that the fact that $\mathcal{T}$ can be greater than unity in (5.50), and (5.42) or (5.44) for certain angles, does not contradict the conservation of energy. We must calculate Π in each case. For (5.50), putting $\Pi = E^2 Z^{-1} = E^2 \mu Z_0^{-1}$ per unit area, the proportion of the energy transmitted is

$$\left(\frac{2\mu_1}{\mu_2 + \mu_1} \right)^2 \frac{\mu_2}{\mu_1} = \frac{4\mu_1 \mu_2}{(\mu_1 + \mu_2)^2} = \frac{4\mu_r}{(1 + \mu_r)^2} \qquad (5.53)$$

Figure 5.5 Reflexion coefficient $\mathscr{R}(\hat{\imath})$ at the surface of a medium of refractive index $\mu_r = 1.5$ for the $\perp$ and $\parallel$ polarizations.

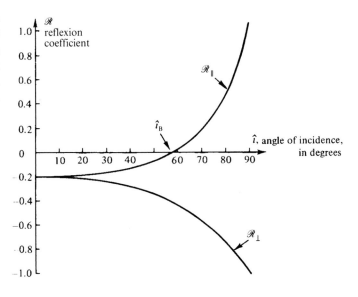

which reaches a maximum value of 1 when $\mu_r = 1$. At non-normal incidence the fact that the areas of transmitted and reflected beams are in the ratio $\cos\hat{\imath}:\cos\hat{r}$ must also be taken into account when calculating total energy flows.

5.5.1 Total internal reflexion

In the usual situation where $\mu_r > 1$, any angle of incidence results in a calculable angle of refraction $\hat{r}$. But if $\mu_r < 1$, there is no real solution for $\hat{r}$ if $\hat{\imath} > \sin^{-1}\mu_r \equiv \hat{\imath}_c$. This angle is called the *critical angle*, and above it both $|\mathcal{R}_\perp|$ and $|\mathcal{R}_\parallel|$ become unity. This phenomenon is called *total internal reflexion*. How can the wave field be described at angles $\hat{\imath} > \hat{\imath}_c$? We postulate for such angles a complex angle of refraction as a formal solution to Snell's law. It then turns out that the disturbance in the second medium is evanescent, as follows. For the equation

$$\sin\hat{r} = \frac{1}{\mu_r}\sin\hat{\imath} \equiv (1 + \beta^2)^{\frac{1}{2}} > 1 \tag{5.54}$$

we have

$$\cos\hat{r} = (1 - \sin^2\hat{r})^{\frac{1}{2}} = \pm i\beta \ (\beta \text{ real and positive}). \tag{5.55}$$

Of the two signs for $\cos\hat{r}$, the upper and lower ones will be seen to apply to waves propagating along $+z$ and $-z$ respectively. Substituting in the equations for $\mathcal{R}$ and $\mathcal{T}$ (5.41)–(5.44) we obtain:

$$\mathcal{R}_\perp = \frac{\cos\hat{\imath} \mp i\mu_r\beta}{\cos\hat{\imath} \pm i\mu_r\beta}, \tag{5.56}$$

$$\mathcal{T}_\perp = \frac{2\cos\hat{\imath}}{\cos\hat{\imath} \pm i\mu_r\beta}, \tag{5.57}$$

$$\mathcal{R}_\parallel = \frac{\pm i\beta - \mu_r\cos\hat{\imath}}{\pm i\beta + \mu_r\cos\hat{\imath}}, \tag{5.58}$$

$$\mathcal{T}_\parallel = \frac{2\cos\hat{\imath}}{\mu_r\cos\hat{\imath} \pm i\beta}. \tag{5.59}$$

As the reflexion coefficients are both of the form

$$\mathcal{R} = \frac{p - iq}{p + iq} = \exp\left[-2i\tan^{-1}\left(\frac{p}{q}\right)\right] = \exp(-i\alpha) \tag{5.60}$$

it is clear that they represent complete reflexion ($|\mathcal{R}| = 1$) but with a phase change α

$$\alpha_\perp = \pm 2\tan^{-1}\frac{\mu_r\beta}{\cos\hat{\imath}}, \tag{5.61}$$

$$\alpha_\parallel = \mp 2\tan^{-1}\frac{\mu_r\cos\hat{\imath}}{\beta}. \tag{5.62}$$

Fig. 5.6 shows the reflexion coefficient for $\mu_r = 1.5^{-1}$ over the whole range of $\hat{\imath}$ from zero to $\pi/2$.

Neither of the transmission coefficients is zero, however, and so we must investigate the transmitted wave more closely. We shall write the space-dependent part of the transmitted wave in full:

$$
\begin{aligned}
E &= E_0 \exp[-i(kz\cos\hat{r} + kx\sin\hat{r})] \\
&= E_0 \exp(\mp k\beta z)\exp[-ikx(1+\beta^2)^{\frac{1}{2}}]. \quad (5.63)
\end{aligned}
$$

When the upper signs in (5.55)–(5.63) are chosen, the wave is evanescent and decays exponentially to zero as $z \to \infty$. The characteristic decay distance is $(k\beta)^{-1}$. As an example, a material with $\mu = 1.5$ has a critical angle of 41.8°. Then at an incident angle of 42.8°,

$$
\beta = (2.25\sin^2 42.8° - 1)^{\frac{1}{2}} = 0.20
$$

Figure 5.6 Modulus and phase of the reflexion coefficient $\mathscr{R} = |\mathscr{R}|\exp(i\alpha)$ at the surface when incidence is in the denser medium, $\mu_r = 1/1.5$.

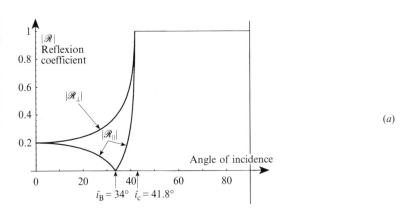

(a)

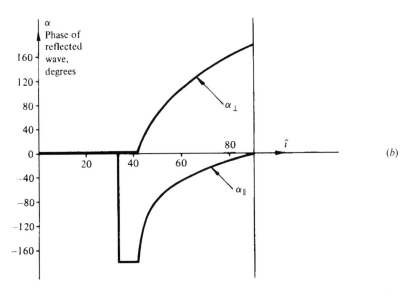

(b)

and the decay distance is thus $\lambda/2\pi\beta \approx 0.8\lambda$.

The phenomena of total internal reflexion and the consequent evanescent wave have several important uses. Various types of prism employing total internal reflexion are used in optical instruments to reflect light rays with or without inversion of an image. An important application is found in the common design of field glasses. In optical waveguides and fibres (§§10.2–10.3), repeated total internal reflexion at the wall or an interface between media is used to transfer light energy along the length of the fibre, with negligible loss. In addition, the existence of the evanescent wave outside the fibre gives rise to one of the ways in which energy can be extracted without any mechanical disturbance.

5.5.2 Phase changes on total internal reflexion

The phase changes (5.61–5.62) for the two polarizations for propagation in the z-direction have somewhat different dependence on the angle $\hat{\imath}$ in the region between $\hat{\imath}_c$ and $\pi/2$. Using the upper signs again in both equations, they can be seen to have the values 0 and π respectively at $\hat{\imath} = \hat{\imath}_c$ ($\beta = 0$), and π and 0 at $\hat{\imath} = \pi/2$. The difference $\alpha_\parallel - \alpha_\perp + \pi$ can be evaluated† for any particular value of μ_r, and is shown in Fig. 5.7 for the two values $\mu_r = 1.5^{-1}$ and $\mu_r = 2.5^{-1}$. For $\beta/\cos\hat{\imath} = 1$ the phase difference has its maximum value, 46° at $\hat{\imath} = 51.7°$ and 94° at $\hat{\imath} = 31.7°$ respectively for the two values of μ_r. A polarizing device which uses this property is the *Fresnel rhomb* (Problem 5.3).

† The need for addition of π to $\alpha_\parallel - \alpha_\perp$ arises because only the $\parallel$ polarization has a Brewster angle, at which the phase of $\mathscr{R}$ changes by π (see Fig. 5.6(b)).

Figure 5.7 Phase difference $\alpha_\parallel - \alpha_\perp + \pi$ for total internal reflexion when (a) $\mu_r = 1/2.5$ and (b) $\mu_r = 1/1.5$, plotted as a function of $\beta/\cos\hat{\imath}$.

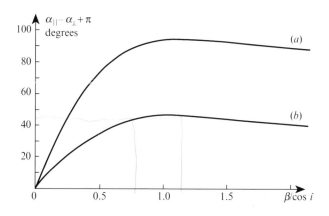

5.5.3 Optical tunnelling

If light is totally reflected from a plane surface at an angle greater
than the critical angle, but a second interface exists within the region
of the evanescent wave, the total reflexion can be frustrated, and
we have the phenomenon of *optical tunnelling* whereby a wave is
partially transmitted through a region where it would be forbidden
by geometrical optics. This is the electromagnetic equivalent to alpha-
particle or electron tunnelling in quantum mechanics. A schematic
experiment is shown in Fig. 5.8. This process has several applications,
such as beam-splitters and optical waveguide couplers.

Calculation of the transmittance through the 'forbidden layer' is
not difficult, once the effective refractive indices of the media are
expressed in terms of u_1 and u_2 (5.45, 5.46), where clearly the value
of u_2 in the air layer is imaginary. We can anticipate the technique
to be developed in §10.4 for multilayer calculations and quote the
result:

$$\mathcal{T} = [\cosh k\beta d + \tfrac{1}{2}\sinh k\beta d(\mu\cos\hat{\imath}/\beta - \beta/\mu\cos\hat{\imath})]^{-1}$$
$$\sim e^{-k\beta d} \quad \text{at large } d. \tag{5.64}$$

It is easy to demonstrate the tunnelling by the experiment shown
in Fig. 5.9(*a*). The second prism of Fig. 5.8 is replaced by a lens
with a large ($\sim 1\,\text{m}$) radius of curvature which rests lightly on the
horizontal hypotenuse of the prism, so that a variety of values of d
are sampled simultaneously. Looking at the reflected light, a dark
patch indicating frustrated reflexion around the point of contact can
be seen – Fig. 5.9(*b*). Altering the incidence in the prism to an angle
below the critical angle returns u_2 to a real value and interference
fringes (Newton's rings) replace the patch – Fig. 5.9(*c*). These can
be used to calibrate the thickness profile of the forbidden layer and
to confirm that significant tunnelling occurs up to thicknesses of
about $\tfrac{3}{4}\lambda$.

Figure 5.8
Tunnelling of a wave
through the air gap
between two media,
spaced by $d \sim (k\beta)^{-1}$.

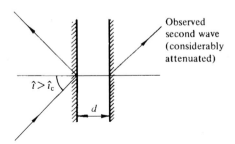

$\hat{\imath} > \hat{\imath}_c$

Observed
second wave
(considerably
attenuated)

d

5.5.4 Energy flow in the evanescent wave

The amplitude of the evanescent wave decays with increasing z, so clearly no energy can be transported in that direction, away from the interface. But on the other hand, the wave is there, and has an energy density, so we have to show that energy transport within it is restricted to the directions parallel to the interface. For the $\perp$ wave in the second medium, substitution of (5.63) in (5.36) shows the fields to be:

$$E_y = E_0 \exp(-k\beta z) \exp\{-i[kx(1+\beta^2)^{\frac{1}{2}} - \omega t]\}, \qquad (5.65)$$

$$
\begin{aligned}
H_z &= Z_0^{-1} \sin \hat{r} \, E_y = Z_0^{-1} (1+\beta^2)^{\frac{1}{2}} E_y \\
&= Z_0^{-1} (1+\beta^2)^{\frac{1}{2}} E_0 \exp\{-i[kx(1+\beta^2)^{\frac{1}{2}} - \omega t]\}, \qquad (5.66)
\end{aligned}
$$

$$
\begin{aligned}
H_x &= Z_0^{-1} \cos \hat{r} \, E_y = iZ_0^{-1} \beta E_y \\
&= Z_0^{-1} E_0 \beta \exp\{-i[kx(1+\beta^2)^{\frac{1}{2}} - \omega t + \pi/2]\}. \qquad (5.67)
\end{aligned}
$$

So the Poynting vector has components

$$\Pi_x = E_y H_z \sim E_y^2 (1+\beta)^{\frac{1}{2}}, \qquad (5.68)$$

$$\Pi_z = -E_y H_x \sim i\beta E_y^2. \qquad (5.69)$$

The imaginary value of Π_z tells us that no energy is transported normal to the surface; it is clear from (5.67) that E_y and H_x have a $\pi/2$ phase difference and so the average of their product is zero. However, there is no phase difference between E_y and H_z and so $\langle \Pi_x \rangle \neq 0$, and energy is transported in that direction.

5.5.5 Mirages

Reflexion can also occur when the interface between two media is not sharply defined. This occurs frequently; a well-known situation is the

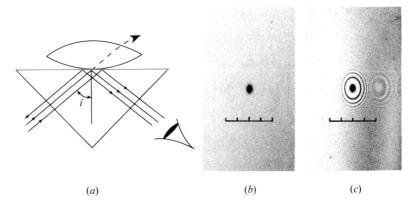

Figure 5.9 (*a*) Experiment to show optical tunnelling; (*b*) observation with $\hat{\imath} > \hat{\imath}_c$; (*c*) observation with $\hat{\imath} < \hat{\imath}_c$.

(*a*) (*b*) (*c*)

mirage, which is most commonly seen by the modern reader as he drives along an asphalt road in the summer heat, and sees the blue sky above him reflected at glancing angle by a layer of overheated air in contact with the road. It almost looks as if the road is flooded†. Another situation which we shall meet in §10.3.2 is the graded-index optical fibre. There are several ways of approaching the problem:

– by geometric optics,
– by Huygens's construction and Fermat's principle.
– by solution of the wave equation.

We assume that the the refractive index μ varies slowly as a function of height z, i.e. $d\mu/dz \ll k_0$, so that locally the medium appears uniform. Geometrical optics tells us that the angle of a ray varies with height such that

$$\mu(z) \sin \hat{\imath}(z) = K, \text{ a constant.} \qquad (5.70)$$

For a given $\hat{\imath}(0)$ at the observer's eye height, $z = 0$, total internal reflexion occurs at the level z_r where $\hat{\imath}(z) = \pi/2$:

$$\mu(z_r) = \mu(0) \sin \hat{\imath}(0). \qquad (5.71)$$

For example, at temperature T (in deg C), air has a refractive index

$$\mu(T) = 1.000\,291 - 1 \times 10^{-6} T.$$

If the observer is at 30 °C and the road at 70 °C, a ray reflected from the air immediately above the road reaches the observer at angle $\hat{\imath}(h)$ given by

$$\mu(70) = \mu(30) \sin \hat{\imath}(h)$$

whence $\hat{\imath} = \frac{\pi}{2} - 9 \times 10^{-3}$ rad, which is about 0.6° below the horizon.

Huygens's construction allows us to follow wavefronts as a slowly descending wave approaches $z = 0$ from above. The wavelength (distance between wavefronts) is always shorter at larger z (since $d\mu/dz > 0$) and so the wave bends around (Fig. 5.10). There is no well-defined height of reflexion. At first sight one might think that this type of reflexion doesn't invert the image, but deeper consideration shows that it really does.

The wave treatment allows the full complex amplitude to be calculated at each height, and shows, as might be expected, an evanescent wave below the level of geometrical reflexion z_r as defined by (5.71).

† This replaces the standard description of a mirage as seen by the thirsty Bedouin from their camels, crossing the endless desert sands scorched by the burning midday Sun...

This can be treated by a method which is much used in quantum mechanics, the WKB method due to Wentzel, Kramers and Brillouin (see for example Cohen-Tannoudji *et al.*, 1977; Gasiorowicz, 1974).†

5.6 Electromagnetic waves incident on a conductor

The media we have discussed so far have all been insulators, as a result of which we have been able to neglect the current term in equation (5.3),

$$\nabla \times \mathbf{H} = \frac{\partial \mathbf{D}}{\partial t} + \mathbf{j}. \tag{5.72}$$

If we wish to consider what happens to an electromagnetic wave incident on a conductor we must bring this term into play, as the electric field $\mathbf{E}$ will induce a non-zero current density $\mathbf{j}$ if the conductivity σ is appreciable:

$$\mathbf{j} = \sigma \mathbf{E}. \tag{5.73}$$

We can substitute (5.73) into (5.72), at the same time replacing $\mathbf{D}$ by $\epsilon \epsilon_0 \mathbf{E}$ to give:

$$\nabla \times \mathbf{H} = \epsilon \epsilon_0 \frac{\partial \mathbf{E}}{\partial t} + \sigma \mathbf{E}. \tag{5.74}$$

Now, remembering that the wave is oscillatory with frequency ω, we replace the operator $\partial/\partial t$ by $-i\omega$ and thus obtain the equation

$$\nabla \times \mathbf{H} = -i\epsilon_0 E \omega \left(\epsilon - \frac{\sigma}{i\omega\epsilon_0} \right). \tag{5.75}$$

† The full calculation appeared in the second edition of this book.

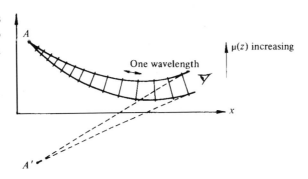

Figure 5.10 Fermat's principle applied to the mirage.

The conductivity term can be absorbed into the dielectric constant by letting it be complex:

$$\epsilon_c = \epsilon - \frac{\sigma}{i\omega\epsilon_0}. \tag{5.76}$$

This is an important result; propagation in a conductor can be treated formally as propagation in a medium with a complex dielectric constant. The reason is easy to see. In an insulator the dielectric field produces displacement current $\partial \mathbf{D}/\partial t$ in quadrature with it; in a conductor the real current density is in phase with $\mathbf{E}$, and thus the net effect is a total current at an intermediate phase angle, which is represented by a complex ϵ_c.

As the mathematics is now similar to that in §5.1 for a real dielectric, we shall take the standard result, $v/c = \epsilon^{-\frac{1}{2}}$, and substitute ϵ_c from equation (5.76) to give

$$\frac{v}{c} = \left(\epsilon + \frac{i\sigma}{\epsilon_0 \omega} \right)^{-\frac{1}{2}}. \tag{5.77}$$

Let us assume ϵ to be of the order of unity. Substitution of values for σ and ω for metallic conductors shows the imaginary term to be completely dominant even at optical frequencies. We therefore write:

$$c/v = \mu \approx (i\sigma/\epsilon_0\omega)^{\frac{1}{2}} = (\sigma/2\epsilon_0\omega)^{\frac{1}{2}}(1+i), \tag{5.78}$$

where μ is the complex refractive index (§2.4). We can then write down the effect of applying a wave of frequency ω,

$$\mathbf{E} = \mathbf{E}_0 \exp[i(kz - \omega t)] = \mathbf{E}_0 \exp[i\omega(\mu z/c - t)], \tag{5.79}$$

normally to the surface $z = 0$ of a conductor; at depth z we have from (5.78):

$$\mathbf{E}(z) = \mathbf{E}_0 \exp[-(\sigma\omega/2\epsilon_0 c^2)^{\frac{1}{2}} z] \exp\{i[(\sigma\omega/2\epsilon_0 c^2)^{\frac{1}{2}} z - \omega t]\}. \tag{5.80}$$

This is an attenuated wave, with characteristic decay length l and wavelength λ inside the conductor given by

$$l = \lambda/2\pi = (2\epsilon_0 c^2/\sigma\omega)^{\frac{1}{2}}. \tag{5.81}$$

The decay per wavelength is thus independent of the frequency and implies that a wave cannot travel more than a few wavelengths inside a conductor.

Table 5.6 shows some typical values of the *skin-depth* l at various frequencies; it is clear that at optical frequencies the penetration into copper is almost negligible, even on an atomic scale, although admittedly the theory above is not applicable at such frequencies, where the skin-depth is less than the electronic mean free path in the metal.

Table 5.1 *Skin-depths in copper at various frequencies at* $0\,°C$

Frequency (Hz)	Free-space wavelength	Skin-depth
50		10 mm
1000		2.4 mm
10^6	300 mm	0.01 mm
0.6×10^{15}	5000 Å	17 Å

5.6.1 Reflexion by a metal surface

As a result of the above one can calculate many of the optical properties of a simple conductor. Such calculations ignore band-structure effects, which give rise to the characteristic colour of copper, for example, but are accurate in the visible region for metals such as the alkali metals. One can then substitute the complex value for μ from (5.78) in (5.41) and (5.43) for the reflectivity. We shall illustrate this with an example.

Light is incident obliquely on a metal surface in the ∥ mode. The reflexion coefficient as a function of angle is, from (5.43),

$$\mathcal{R}_{\parallel} = \frac{\cos \hat{r} - s(1 + \mathrm{i}) \cos \hat{\imath}}{\cos \hat{r} + s(1 + \mathrm{i}) \cos \hat{\imath}}, \tag{5.82}$$

where

$$s = (\sigma/2\epsilon_0 \omega)^{\frac{1}{2}}. \tag{5.83}$$

Since $s = \mathrm{Re}(\mu) \gg 1$ we can assume $\cos \hat{r} = 1$, whence

$$\mathcal{R}_{\parallel} = \frac{1 - s(1 + \mathrm{i}) \cos \hat{\imath}}{1 + s(1 + \mathrm{i}) \cos \hat{\imath}}. \tag{5.84}$$

For small angles of incidence $\hat{\imath}$, $\mathcal{R}_{\parallel}$ has the value -1: perfect reflexion with a π phase change. As we approach glancing incidence, the phase of the reflected wave changes continuously reaching $\mathcal{R}_{\parallel} = +1$ at $\hat{\imath} = \pi/2$. The phase change occurs around what might be described as a 'complex Brewster angle' at which the real and imaginary parts of $\mathcal{R}$ are comparable, i.e.

$$s \cos \hat{\imath} \approx 1. \tag{5.85}$$

Here the value of $|\mathcal{R}|$ falls to a value somewhat less than unity. With an aluminium surface at room temperature, this angle is about 89°

for visible light, and the intensity reflexion coefficient $|\mathscr{R}|^2$ falls to a minimum of about 20%. At smaller angles of incidence, the behaviour of real metals is indistinguishable from that of an ideal metal with $\sigma \to \infty$.

5.6.2 Reciprocity and time-reversal: the Stokes relationships

The reader may have noticed that the reflexion coefficient (5.45) is negative when incidence is in the lower-index medium, and positive, with the same value, if the light ray is exactly reversed, so that it is incident in the denser medium. This reversal of the sign of the reflexion coefficient when the light path is reversed is a feature of non-absorbing systems which arises in a very general manner. It is the result of the *time-reversal symmetry* of Maxwell's equations in the absence of absorption mechanisms, which would be represented by a non-zero current density term **j** in (5.3).

If we change t in equations (5.9)–(5.10) to $-t$ we find no change in the resulting wave equation (5.12). So any set of related waves, such

Figure 5.11
Reflexion from
opposite sides of an
interface.

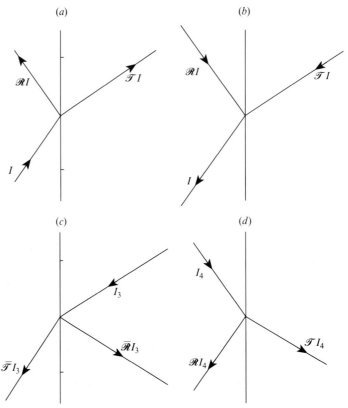

as the usual incident, reflected and transmitted trio, has an equally valid time-reversed set. But the effect of changing t to $-t$ in a wave is to reverse the direction of its propagation without affecting its amplitude, so that if we apply this procedure to the trio shown in Fig. 5.11(a), we get those in Fig. 5.11(b). The reflector is completely general here, except that it must be non-absorbing; it could be a single interface or any multilayer or other system satisfying this requirement. The incident, reflected and transmitted amplitudes I, R and T in both figures are related by coefficients $\mathscr{R} = R/I$ and $\mathscr{T} = T/I$. In Fig. 5.11(b) the amplitudes are unchanged, but it represents an unusual situation: there are *two* incident waves and *one* leaving the system. Obviously, some form of interference must be involved, but that is all within the framework of Maxwell's equations and the details don't matter to us. However, the situation in Fig. 5.11(b) can be represented by the superposition of two perfectly conventional trios, one incident from each side, which are shown in Fig. 5.11(c) and (d). The former has incidence from the reverse side, and reflexion and transmission coefficients $\bar{\mathscr{R}}$ and $\bar{\mathscr{T}}$ respectively. In the figures, the amplitudes have been labelled accordingly. Equating amplitudes of Fig. 5.11(b) to the sum of the last two gives us:

$$I = \bar{\mathscr{T}} I_3 + \mathscr{R} I_4, \tag{5.86}$$

$$\mathscr{R} I = I_4, \tag{5.87}$$

$$0 = \bar{\mathscr{R}} I_3 + \mathscr{T} I_4, \tag{5.88}$$

$$\mathscr{T} I = I_3. \tag{5.89}$$

These lead directly to the *Stokes relationships*:

$$\bar{\mathscr{R}} = -\mathscr{R}, \tag{5.90}$$

$$1 = \mathscr{T}\bar{\mathscr{T}} + \mathscr{R}^2. \tag{5.91}$$

The general result (5.90) is what we set out to prove: the reflexion coefficients from opposite sides of a non-absorbing partial reflector are *equal and opposite in sign*. Conservation of energy in the system is then expressed by (5.91). Obviously, in an absorbing system no such generalization is possible; consider, for example, the properties of metal foil painted green on one side!

The above argument assumes that $\mathscr{R}$ and $\mathscr{T}$ are real quantities, implying that neither reflexion nor transmission of the wave introduces a phase change. But there are many cases, such as total internal reflexion, §5.5.2, where this is not true. It is really quite easy to extend the argument to complex values of $\mathscr{R}$ and $\mathscr{T}$. The starting point is to realise that the time-reversed field corresponding to I

is I^*. Although we shall not prove this in general (see Altman and Suchy, 1991), one can appreciate its significance in the case of total internal reflexion, where the transmitted wave carries no energy and the reflexion coefficient is $\mathscr{R} = e^{i\alpha}$. On reflexion, the incident wave I becomes $I\mathscr{R} = Ie^{i\alpha}$. In the time-reversed system, the field $(Ie^{i\alpha})^*$ is reflected with the same reflexion coefficient into I^*, so that the relationship is consistent. Now we replace I by I^* in Fig. 5.11(b) and then follow through the same calculation as in (5.86)–(5.89). This gives us the Stokes relations in their complex form:

$$\bar{\mathscr{R}} = -\mathscr{R}^*, \tag{5.92}$$

$$1 = \bar{\mathscr{T}}\mathscr{T}^* - \bar{\mathscr{R}}\mathscr{R}^*. \tag{5.93}$$

CHAPTER SIX

Polarization and anisotropic media

6.1 Introduction

As we saw in Chapter 5, electromagnetic waves in isotropic materials are transverse, their electric and magnetic field vectors $\mathbf{E}$ and $\mathbf{H}$ being normal to the direction of propagation $\mathbf{k}$. The direction of $\mathbf{E}$ or rather, as we shall see later, the electric displacement field $\mathbf{D}$, is called the *polarization* direction, and for any given direction of propagation there are two independent such vectors, which can be in any two mutually orthogonal directions normal to $\mathbf{k}$. When the medium through which the wave travels is *anisotropic*, which means that its properties depend on orientation, the above statements meet with some restrictions. We shall see that the result of anisotropy in general is that the fields $\mathbf{D}$ and $\mathbf{B}$ remain transverse to $\mathbf{k}$ under all conditions, but $\mathbf{E}$ and $\mathbf{H}$, no longer having to be parallel to $\mathbf{D}$ and $\mathbf{B}$, are not *necessarily* transverse. Moreover, the two independent polarizations that propagate must now be chosen specifically with relation to the axes of the anisotropy. A further direct consequence of $\mathbf{E}$ and $\mathbf{H}$ no longer being necessarily transverse, is that the Poynting vector $\mathbf{\Pi} = \mathbf{E} \times \mathbf{H}$ may not be parallel to the wave-vector $\mathbf{k}$.

In this chapter, we shall first discuss the various types of polarized radiation that can propagate. We shall then go on to extend the theory of electromagnetic waves as described in Chapter 5 to take into account anisotropic media. This aim, the theory of *crystal optics*, will be achieved by means of a somewhat unconventional geometrical approach, the origin of which lies in an analogy between electromagnetic wave propagation in crystals and electron wave propagation in crystalline metals. We shall then see how the same formalism can be extended to take into account anisotropy introduced by helicity (screw-like) properties of materials, and by electric and magnetic fields.

6.2 Polarized light in isotropic media

6.2.1 Linearly-polarized light

The simplest basic periodic solution to Maxwell's equations in an isotropic medium is a wave in which **E**, **H** and **k** form a triad of mutually perpendicular vectors. This is called a *linearly-polarized* (or *plane-polarized*) wave. The vectors **E** and **k** define a plane called the *plane of polarization*. We have:

$$\mathbf{E} = \mathbf{E}_0 \exp[i(\mathbf{k} \cdot \mathbf{r} - \omega t)], \tag{6.1}$$

$$\mathbf{H} = \mathbf{H}_0 \exp[i(\mathbf{k} \cdot \mathbf{r} - \omega t)]. \tag{6.2}$$

The energy flow $\mathbf{\Pi} = \mathbf{E} \times \mathbf{H}$ is parallel to **k**. For a given direction of **k** any pair of orthogonal polarizations can be chosen to represent independent ways of fulfilling these requirements.

6.2.2 Circularly-polarized light

In a medium which responds linearly to electric and magnetic fields (i.e. we assume B is proportional to H, and D to E), any linear superposition of the above two linearly-polarized waves is also a solution of Maxwell's equations. A particularly important case is that in which the two waves are superposed with a phase difference of $\pi/2$ (either positive or negative) between them. If we take the example where **k** is in the z-direction, and the two linearly-polarized waves have equal amplitudes

$$\mathbf{E}_{01} = E_0\hat{\mathbf{x}}, \qquad \mathbf{E}_{02} = E_0\hat{\mathbf{y}}, \tag{6.3}$$

in which $\hat{\mathbf{x}}$ and $\hat{\mathbf{y}}$ are unit vectors along the x- and y-axes, we have the superposition with $\pi/2$ phase difference

$$\mathbf{E} = E_0\hat{\mathbf{x}} \exp[i(kz - \omega t)] + E_0\hat{\mathbf{y}} \exp[i(kz - \omega t + \pi/2)]. \tag{6.4}$$

Remembering that the real electric field is the real part of this complex vector, we have

$$\mathbf{E}^R = E_0\hat{\mathbf{x}} \cos(kz - \omega t) + E_0\hat{\mathbf{y}} \sin(kz - \omega t). \tag{6.5}$$

At given z, this represents a vector of constant length $E^R = E_0$ which rotates around the z-axis at angular velocity ω. The sense of rotation is clockwise if viewed as an observer would see the wave coming towards him; this is called *right-handed circularly-polarized light*. Rather awkwardly, the vector then draws out a left-handed screw-thread![†]

[†] This convention is not universal; see for example Yariv and Yeh (1984).

Alternatively, if we were to freeze the wave at time $t = 0$, the vector $\mathbf{E}^R$ has the form

$$\mathbf{E}^R = E_0(\hat{\mathbf{x}} \cos kz + \hat{\mathbf{y}} \sin k\hat{z}). \tag{6.6}$$

This vector, when drawn out as a function of z, traces out a right-handed screw (Fig. 6.1). The magnetic field traces a similar screw, $\pi/2$ out of phase with $\mathbf{E}$. If the phase difference is $-\pi/2$, a second independent polarization, in which the sense of rotation of $\mathbf{E}$ is anti-clockwise, is created. This is called *left-handed circularly-polarized light*. At given t, the vector traces out a left-handed screw.

6.2.3 Elliptically-polarized light

The superposition described in §6.2.2 need not involve two linearly-polarized waves of equal amplitude. If the two waves have amplitudes E_{0x} and E_{0y}, then it is easy to see that the vector in Fig. 6.1 traces out a screw of elliptical cross section. Similarily, at constant z, the vector $\mathbf{E}^R$ traces out an ellipse. This type of light is called *elliptically-polarized light* and also has left- and right-handed senses.

6.2.4 Fundamental significance of polarized types

When we introduce quantum optics in Chapter 14, we shall see that the quantum statistics of the electromagnetic field are equivalent to those of an ensemble of identical particles with Bose statistics. These

Figure 6.1 Electric field vector at time $t = 0$ for a circularly-polarized wave.

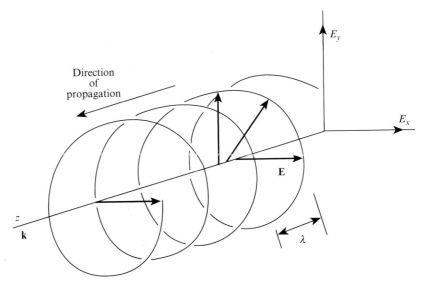

are called *photons*. In order to have such statistics, they must have integral spin. Moreover, we shall see that in order to conserve angular momentum when light interacts with atoms, this must be ± 1 units of $\hbar$. It therefore emerges that the closest equivalent to a single photon is a wave with circular polarization, right-handed for spin=+1 and left-handed for spin=−1. Linearly-polarized light should then be rightly considered as the superposition of two circularly-polarized waves with opposite handedness; the direction of polarization achieved then depends on the phase difference between the two circularly-polarized waves.

6.2.5 Partially-polarized and unpolarized light

Light as generated by a discharge or filament lamp is generally not polarized. What this really means is that such light can be described as a superposition of many linearly-polarized waves, each having its own random plane of polarization and individual phase. Moreover, because the light is not strictly monochromatic, the phase relation between the waves changes with time. Such a chaotic collection of waves has no discernable polarization properties and is called *unpolarized*. Sometimes, such light has some statistical preference for a particular plane of polarization because of some anisotropy in the medium and is *partially-polarized*. An example is light from blue sky, where scattering provides the anisotropy (§13.2.2).

It is sometimes necessary to describe the degree of polarization of a light wave. This can be done in several ways, which are described in detail in specialized texts (e. g. Clarke and Grainger, 1971; Azzam and Bashra, 1989). Basically, incoherent partially-polarized light (the most general practical case) can be described by an unpolarized intensity plus a polarized intensity, the latter of which has an axis, degree of elliptical polarization and sense of rotation. It is possible to express these properties in terms of four parameters, which form a four-element vector called the *Stokes vector*. A polarizing element or mirror, which changes the polarization state of a wave, can then be described by a 4×4 matrix called a *Müller matrix*, which multiplies the Stokes vector of the incident wave to give that of the outgoing one. If the light is coherent, fewer parameters are needed to describe it, since an unpolarized coherent component does not exist. We shall not use these descriptions in the rest of this book, and therefore will not discuss them further.

6.2.6 Orthogonal polarization states

Two modes of polarization are called *orthogonal* if their electric field vectors are orthogonal in the conventional manner:

$$\mathbf{E}_1 \cdot \mathbf{E}_2^* = 0. \qquad (6.7)$$

The electric field vector intended here is the complex amplitude which multiplies $\exp[i(\mathbf{k} \cdot \mathbf{r} - \omega t)]$. For example, for linearly-polarized waves, these amplitudes are two real vectors which are orthogonal in the usual geometric sense; $\mathbf{E}_1$ normal to $\mathbf{E}_2$ implies that

$$E_{1x}E_{2x} + E_{1y}E_{2y} = 0. \qquad (6.8)$$

Two circularly-polarized waves with opposite senses are likewise orthogonal. From (6.4) we have

$$\mathbf{E}_1 = E_0(\hat{\mathbf{x}} + i\hat{\mathbf{y}}), \qquad \mathbf{E}_2 = E_0(\hat{\mathbf{x}} - i\hat{\mathbf{y}})$$
$$\mathbf{E}_1 \cdot \mathbf{E}_2^* = E_0^2(\hat{\mathbf{x}} \cdot \hat{\mathbf{x}} + i^2\hat{\mathbf{y}} \cdot \hat{\mathbf{y}}) = 0. \qquad (6.9)$$

Any elliptically-polarized mode has an orthogonal companion; the two can be shown to have the same ellipticity but with major and minor axes interchanged and opposite senses.

6.3 Production of polarized light

Any dependence of the propagation properties of light on its polarization can be used, in principle, to produce polarized light. Two well-known phenomena having this property are reflexion at the surface of a dielectric (§5.4.2) and scattering by small particles (§13.2.2). Other methods, which will be discussed in more detail later, involve crystal propagation (§6.8.4) and selective absorption (dichroism – §6.3.2).

The action of 'polarizing light', essentially means taking unpolarized light and extracting from it a beam of polarized (linear, circular or elliptical) light. The rest of the light, which is sometimes just as well-polarized in an orthogonal orientation, is wasted, or may be used for some other purpose. There is no way of reorganizing the light so as to get a single beam of polarized light from an unpolarized source without making the beam either broader or more divergent. Such a process, were it to exist, could be used to defy the second law of thermodynamics (Problem 6.9)!

6.3.1 Polarization by reflexion

One of the easiest ways to polarize light is to reflect it from a plane dielectric surface at Brewster's angle (§5.4.3):

$$\hat{\imath}_B = \tan^{-1} \mu_r. \tag{6.10}$$

At this angle, the reflexion coefficient for the ∥ component is zero; thus the reflected light is completely polarized in the ⊥ direction (Fig. 6.2). However, even for this component the reflexion coefficient is small (typically 5–6%) so that this method of polarization is quite inefficient. The polarization is also complete only for a specified angle of incidence. By stacking several plates in series, each one of which reflects some of the ⊥ component, the transmitted ∥ component can be polarized reasonably well, with less angular sensitivity (Problem 5.6). Polarization by Brewster's reflexion has one important property; it is automatically calibrated in the sense that the geometry alone defines exactly the plane of polarization. It is also extremely sensitive to surface quality and cleanliness, a property which is exploited in the technique of *ellipsometry* for investigating interfaces (Azzam and Bashra, 1989).

Crystal polarizers use total internal reflexion to separate polarized components from unpolarized light, and must be used if the highest quality of polarization is required. We shall discuss these in more detail in §6.8.4.

6.3.2 Polarization by absorption

Several materials, both natural and synthetic, absorb different polarizations by different amounts. This behaviour is called *dichroism*. It is

Figure 6.2
Production of
linearly-polarized
light by reflexion at
the Brewster angle.

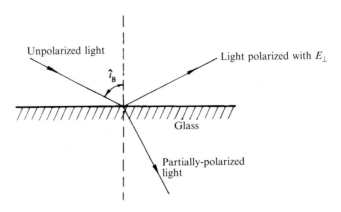

Unpolarized light

$\hat{\imath}_B$

Light polarized with $E_\perp$

Glass

Partially-polarized light

widely used to polarize light linearly, and may also be used to polarize it circularly.

A simple mechanism which polarizes light linearly by this effect is a parallel grid of conducting wires separated by somewhat less than one wavelength. This system transmits most of the light when its electric polarization vector is normal to the wires. No diffraction orders are created (§8.3.4) if the separation is less than λ. But if the electric field vector is parallel to the wires, currents are induced in them and the wave energy is absorbed. So an incident unpolarized beam emerges polarized fairly well normal to the wires.† In practice, at infra-red wavelengths (1 μm and above) polarizers of this sort are constructed by microfabrication or ion-implantation of gold or silver strips on a transparent dielectric substrate.

The most common polarizing material, 'Polaroid', also uses this mechanism. This consists of a streched film of polyvinyl alcohol dyed with iodine. The oriented conducting polymeric chains behave similarly to the wire grid. This material is cheap to produce and can be made in thin sheets of almost unlimited size.

6.3.3 Extinction ratio

A measure of the effectiveness of a polarizer is gained by passing unpolarized light through two identical devices in series. If the two transmit along parallel axes, a single component is transmitted by both, and the output intensity is I_1. If one is now rotated, so that the transmitting axes are perpendicular, ideally no light would be transmitted, but in practice a small intensity I_2 passes. Careful orientation of the polarizers minimizes this value. The ratio I_1/I_2 is called the *extinction ratio*. In good crystal polarizers this may be as high as 10^7, and a similar value can be obtained for clean reflectors exactly at Brewster's angle. Polaroid typically gives a value of 10^3.

6.4 Wave propagation in anisotropic media

In the next two sections we shall discuss the way in which the electromagnetic wave propagation theory in Chapter 5 must be extended to take into account the anisotropic (orientation-dependent) properties of the material. This will be done at a purely phenomenological level; no account will be given of the atomic or molecular origin of the anisotropy, which is a subject well outside the scope of this book.

---

† This is popular and easy to demonstrate using centimetre microwaves.

6.4.1 Huygens's construction

We shall first discuss the general relationship between propagation of waves and rays and anisotropic properties. This was first done in about 1650 by Huygens, who did not understand the origins of the anisotropy. Consider 'wavelets' originating from points on a given wavefront *AB* of *limited extent*, as in Fig. 6.3 (see also Fig. 2.6(*b*)). If the velocity is a function of the direction of propagation, the wavelets are not spherical; we shall see that they are in fact ellipsoidal. The new wavefront A_1B_1 is then the common tangent to the wavelets, as shown. But the *extent* of the wavefront moves sideways, showing that the light ray which it represents is at an angle to the wave-vector. This is a common feature of crystal propagation: the Poynting vector **Π**, which represents the direction of the light ray, is not in general parallel to **k**, as we mentioned in §6.1.

6.4.2 The refractive-index surface

The exact relationship between the velocity anisotropy and **Π** can be seen by a geometrical method.† We consider monochromatic light

† This approach is parallel to the treatment of electron waves in metal crystals. The refractive-index surface is analogous to the Fermi surface of the metal.

Figure 6.3 Huygens's principle applied to propagation of a limited beam in an anisotropic medium.

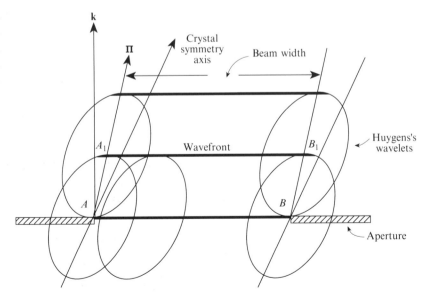

with given frequency ω_0. The refractive index, since it is a function of the direction of propagation, can be written as a vector:

$$\mu = \frac{c\mathbf{k}}{\omega_0}, \tag{6.11}$$

where $\mathbf{k}$ has the value of the wave-vector measured in the medium for propagation in that direction, and the vector μ has the same direction.

Now the phase velocity is $v = \omega/k$. The group velocity $\mathbf{v}_g$, which is the velocity of propagation of *energy* and therefore corresponds in direction to $\boldsymbol{\Pi}$, has components, by extension of §2.5,

$$v_{gx} = \frac{\partial \omega}{\partial k_x}, \quad v_{gy} = \frac{\partial \omega}{\partial k_y}, \quad v_{gz} = \frac{\partial \omega}{\partial k_z}. \tag{6.12}$$

This is written in vector terminology:

$$\mathbf{v}_g = \nabla_k \omega \quad \equiv \quad \left(\frac{\partial \omega}{\partial k_x}, \frac{\partial \omega}{\partial k_y}, \frac{\partial \omega}{\partial k_z} \right) \tag{6.13}$$

$$= \frac{c}{\omega_0} \left(\frac{\partial \omega}{\partial \mu_x}, \frac{\partial \omega}{\partial \mu_y}, \frac{\partial \omega}{\partial \mu_z} \right) = \frac{c}{\omega_0} \nabla_\mu \omega. \tag{6.14}$$

We shall now represent the propagation properties in the medium by the vector μ. At a particular value $\omega = \omega_0$ this has a given value for each direction of propagation and therefore can be represented by a closed surface.† The radius vector, from the origin to the surface in each direction, is equal to the value of μ for propagation in that direction. We shall call this the *refractive-index surface*, or *μ-surface*. Since (6.14) is analogous to the well-known electrostatic relationship $\mathbf{E} = -\nabla V$ which shows that lines of electric field E are normal to the equipotential surfaces $V =$ constant, it follows similarly that the vector $\mathbf{v}_g$ (6.14) is normal to the surface of constant ω, i.e. *normal to the μ-surface*. In general, by construction of the μ-surface for a given material at frequency ω_0, we can therefore deduce geometrically the direction of the Poynting vector $\boldsymbol{\Pi}$ (the group velocity direction) for any wave by drawing the normal to the μ-surface at the point representing its $\mathbf{k}$ (Fig. 6.4). Clearly, in an isotropic material the μ-surface is a sphere, and $\boldsymbol{\Pi} \parallel \mathbf{k}$.

† If there is no propagation possible in a certain direction, the surface does not exist in that orientation, and is therefore not closed. This happens if, for example, the propagation is evanescent in that direction. It occurs frequently when waves propagate in a magneto-plasma; examples were given in the second edition of this book; see also Budden (1966).

6.5 Electromagnetic waves in an anisotropic medium

We shall now solve Maxwell's equations when the dielectric properties of the medium are anisotropic. As in Chapter 5, we assume no magnetic polarization ($\mu = 1$) since this is usually the case in transparent media at optical frequencies. For harmonic waves of the form

$$E = E_0 \exp[i(\mathbf{k} \cdot \mathbf{r} - \omega t)] \tag{6.15}$$

we once again use the operator substitutions (§2.3) as we did in §5.1:

$$\frac{\partial}{\partial t} = -i\omega, \quad \nabla = i\mathbf{k}. \tag{6.16}$$

Maxwell's equations in an uncharged insulator then emerge as:

$$\nabla \cdot \mathbf{B} = 0 \quad \Rightarrow \quad i\mathbf{k} \cdot \mathbf{B} = 0, \tag{6.17}$$

$$\nabla \cdot \mathbf{D} = 0 \quad \Rightarrow \quad i\mathbf{k} \cdot \mathbf{D} = 0, \tag{6.18}$$

$$\nabla \times \mathbf{H} = \frac{\partial \mathbf{D}}{\partial t} \quad \Rightarrow \quad i\mathbf{k} \times \mathbf{H} = -i\omega \mathbf{D}, \tag{6.19}$$

$$\nabla \times \mathbf{E} = -\frac{\partial \mathbf{B}}{\partial t} \quad \Rightarrow \quad i\mathbf{k} \times \mathbf{E} = i\omega \mathbf{B}. \tag{6.20}$$

These equations should be compared with (5.16)–(5.19). Notice that **D** and **B** are transverse (normal to **k**). Substituting $\mu_0 \mathbf{H}$ for **B** (non-magnetic material), we take ($\mathbf{k} \times$) equation (6.20) and get

$$\mathbf{k} \times (\mathbf{k} \times \mathbf{E}) = \mu_0 \omega \mathbf{k} \times \mathbf{H} = -\mu_0 \omega^2 \mathbf{D}. \tag{6.21}$$

This equation relates the vectors **k**, **E** and **D** and can easily be seen to revert to (5.13) in the isotropic case, $\mathbf{D} = \epsilon_0 \epsilon \mathbf{E}$.

First we look at the disposition of the vectors in (6.21). The vector $\mathbf{k} \times (\mathbf{k} \times \mathbf{E})$ lies in the plane of **k** and **E**, normal to **k**. For the equation to have a solution at all, **D** *must therefore lie in the plane of* **k** *and* **E**. We also know, from (6.18) that **D** is normal to **k**. This is illustrated in Fig. 6.5(*a*). This condition defines what we shall call a *characteristic*

Figure 6.4
Construction of the
μ-surface.

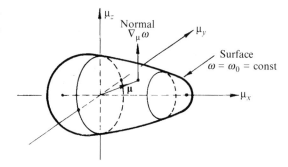

wave, which is a propagation mode for the material. For such a wave, there is an angle θ between **E** and **D** and

$$|\mathbf{k} \times (\mathbf{k} \times \mathbf{E})| = k^2 E \cos\theta = \mu_0 \omega^2 D. \qquad (6.22)$$

The wave velocity is thus given by

$$v^2 = \frac{\omega^2}{k^2} = \frac{E \cos\theta}{\mu_0 D}. \qquad (6.23)$$

Finally, from (6.17)–(6.19), the magnetic fields **B** and **H** are normal to **k** and to **D**, so that a full picture of the disposition of the vectors **D**, **E**, **k**, **H** and the Poynting vector $\mathbf{\Pi} = \mathbf{E} \times \mathbf{H}$ can be drawn, Fig. 6.5(*b*).

The problem we have to solve for a particular medium is, having chosen the direction of the wave-vector **k**, to identify the characteristic waves, which means finding those directions of **D** which result in coplanar **D**, **E** and **k**. Then the wave velocity and refractive index $\mu = c/v$ can be found from (6.23). There will be in general two distinct solutions for each direction of **k** (under some circumstances they may be degenerate). The polarizations of the two characteristic waves will be found to be orthogonal (§6.2.6). Thus when we construct the μ-surface (§6.4.2) we shall find it to be doubly-valued (i.e. two values of μ in each direction), which has many interesting and important properties.

6.6 Crystal optics

6.6.1 The dielectric tensor

Crystals are anisotropic because of their microscopic structure (see e.g., Hecht, 1987). Here we shall only consider the anisotropy as a

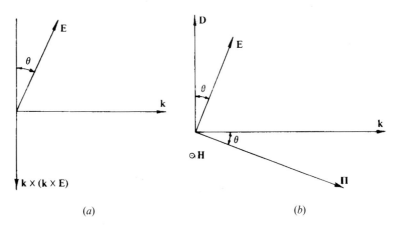

Figure 6.5 (*a*) The vectors **k**, **E** and $\mathbf{k} \times (\mathbf{k} \times \mathbf{E})$; (*b*) vectors **D**, **E**, **Π**, **k** and **H** for a wave.

(*a*)　　　　　　　　　　(*b*)

continuum phenomena, because interatomic distances are orders of magnitude smaller than the light wavelength. In an anisotropic linear dielectric medium, the vectors $\mathbf{D}$ and $\mathbf{E}$ are proportional in magnitude, but not necessarily parallel, so that we write a tensor relationship

$$\mathbf{D} = \epsilon_0 \epsilon \mathbf{E} \qquad (6.24)$$

where ϵ, the dielectric tensor, represents the matrix

$$\epsilon = \begin{pmatrix} \epsilon_{11} & \epsilon_{12} & \epsilon_{13} \\ \epsilon_{21} & \epsilon_{22} & \epsilon_{23} \\ \epsilon_{31} & \epsilon_{32} & \epsilon_{33} \end{pmatrix}. \qquad (6.25)$$

Its meaning is straightforward. If an electric field $\mathbf{E} = (E_1, E_2, E_3)$ is applied, the resulting displacement field $\mathbf{D}$ has components (D_1, D_2, D_3) where

$$D_i = \epsilon_0(\epsilon_{i1} E_1 + \epsilon_{i2} E_2 + \epsilon_{i3} E_3). \qquad (6.26)$$

The theory of linear algebra shows that there always exist three *principal axes* ($i = 1, 2, 3$) for which $\mathbf{D}_i$ and $\mathbf{E}_i$ are parallel. We then have three *principal dielectric constants* ϵ_i defined by

$$\frac{D_{1i}}{E_{1i}} = \frac{D_{2i}}{E_{2i}} = \frac{D_{3i}}{E_{3i}} = \epsilon_0 \epsilon_i. \qquad (6.27)$$

The three principal axes are mutually orthogonal and, for a non-absorbing crystal, the ϵ_is are real. By using these three axes as x, y and z, the tensor (6.25) can be written in a simpler form, which we shall use as far as possible:

$$\epsilon = \epsilon_0 \begin{pmatrix} \epsilon_1 & 0 & 0 \\ 0 & \epsilon_2 & 0 \\ 0 & 0 & \epsilon_3 \end{pmatrix}. \qquad (6.28)$$

In a non-absorbing medium, ϵ_1, ϵ_2 and ϵ_3 are real; this can be shown to be equivalent to $\epsilon_{ji} = \epsilon_{ij}^*$ in (6.25), and is the definition of a Hermitian tensor. The process of rotating the tensor so that (x, y, z) become principal axes is called *diagonalizing the tensor* and the technique for doing it is discussed in every book on linear algebra. The most general crystal, called a *biaxial crystal* for reasons which will be apparent later, has three distinct values for $\epsilon_1, \epsilon_2, \epsilon_3$; crystals with higher symmetry, *uniaxial crystals*, have two of the values (say, ϵ_1 and ϵ_2) equal. If all three are equal, the material is isotropic and the discussion in this section is irrelevant.

6.6.2 The index ellipsoid, or optical indicatrix

To carry out our plan of presenting crystal optics geometrically we need to represent the tensor as an ellipsoid. In general, the ellipsoid with semi-axes a, b, c is the surface

$$\frac{x^2}{a^2} + \frac{y^2}{b^2} + \frac{z^2}{c^2} = 1. \tag{6.29}$$

In formal terms this can be written

$$(x, y, z) \begin{pmatrix} a^{-2} & 0 & 0 \\ 0 & b^{-2} & 0 \\ 0 & 0 & c^{-2} \end{pmatrix} \begin{pmatrix} x \\ y \\ z \end{pmatrix} = 1, \tag{6.30}$$

or, in shorthand $\qquad \mathbf{r} \cdot \mathsf{M} \cdot \mathbf{r} = 1. \tag{6.31}$

The inverse to (6.24) is:

$$\epsilon_0 \mathbf{E} \;=\; \epsilon^{-1} \cdot \mathbf{D} \tag{6.32}$$

where, for the diagonal form (6.28),

$$\epsilon^{-1} \equiv \begin{pmatrix} \epsilon_1^{-1} & 0 & 0 \\ 0 & \epsilon_2^{-1} & 0 \\ 0 & 0 & \epsilon_3^{-1} \end{pmatrix}. \tag{6.33}$$

We can now study the geometrical meaning of the formal equation $\mathbf{D} \cdot \mathbf{E} = 1$ which becomes, using (6.32),

$$\mathbf{D} \cdot \epsilon^{-1} \cdot \mathbf{D} = \epsilon_0 ; \tag{6.34}$$

this is represented by the ellipsoid (6.30) if

$$(x, y, z) = \mathbf{D} \epsilon_0^{-\frac{1}{2}} \tag{6.35}$$

$$\text{and} \quad \epsilon_1 = a^2, \; \epsilon_2 = b^2, \; \epsilon_3 = c^2. \tag{6.36}$$

Thus the ellipsoid, Fig. 6.6(a) has semi-axes $\epsilon_1^{\frac{1}{2}}, \epsilon_2^{\frac{1}{2}}, \epsilon_3^{\frac{1}{2}}$ which we shall see to be three principal values of the refractive index μ (remember that in an isotropic medium $\mu = \epsilon^{\frac{1}{2}}$: §5.1.3). To understand the meaning of the ellipsoid, we imagine $\mathbf{D}$ varying in direction, its length being calculated at each point so that $\mathbf{D} \cdot \mathbf{E} = 1$ (in units of energy density). From (6.35), the tip of the vector $\mathbf{D} \epsilon_0^{-\frac{1}{2}}$ then traces out the ellipsoid. The vector $\mathbf{E}$ can be shown to have the direction of the normal to the ellipsoid† at the tip of $\mathbf{D}$ (Fig. 6.6).

† Proof: the tangent plane to the ellipsoid (6.29) at (x_1, y_1, z_1) is

$$\frac{x x_1}{a^2} + \frac{y y_1}{b^2} + \frac{z z_1}{c^2} = 1 \tag{6.37}$$

A vector normal to this plane is $(x_1/a^2, y_1/b^2, z_1/c^2)$. Replacing (x_1, y_1, z_1) by $\mathbf{D}$ and a^2 by ϵ_1, etc., shows this normal to be in the direction of $\mathbf{E}$.

The refractive index μ for the wave with polarization vector **D** then follows simply. From (6.23) we have

$$v^2 = \frac{c^2}{\mu^2} = \frac{E\cos\theta}{\mu_0 D} = \frac{ED\cos\theta}{\mu_0 D^2} = \frac{1}{\mu_0 D^2} \tag{6.38}$$

since $ED\cos\theta = \mathbf{E}\cdot\mathbf{D} = 1$ at all points on the ellipsoid. Thus $\mu = c\mu_0^{\frac{1}{2}}D = D/\epsilon_0^{\frac{1}{2}}$. In other words, *the radius vector of the ellipsoid in each direction equals the refractive index of the medium for a wave with polarization vector* **D** *in that direction.* This ellipsoid is called the *index ellipsoid* or *optical indicatrix.* Notice, by the way, that the values have come come out correctly in a principal direction, x for example; the ellipsoid has semi-axis $\epsilon_1^{\frac{1}{2}}$, which is just the refractive index for a wave polarized in that direction. It is most important to realize that it is the *polarization direction*, not the propagation direction, which determines the velocity of the wave. Waves propagating in different directions, but with the same polarization vector, travel at the same velocity.

6.6.3 Characteristic waves

We now have to determine the polarizations and velocities of the characteristic waves for a given propagation direction **k**. We saw, in §6.5, that the requirement for characteristic waves is that **D**, **E** and **k** have to be coplanar. We now have the means to find them. We proceed as follows, given the propagation vector **k**.

Figure 6.6 The optical indicatrix and the relationship between **D** and **E**. **E** is normal to the surface of the indicatrix.

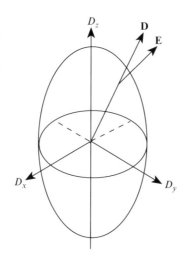

(1) We find all possible polarizations **D**. These lie in a plane normal to **k** since **D** is always transverse (6.18).

(2) We construct **E** for each **D** using the indicatrix. Recall that **E** is normal to the surface of the indicatrix at the tip of **D** (§6.6.2).

(3) We look for coplanar **D**, **E** and **k**.

Fig. 6.7 illustrates these stages. In stage (1), we construct a plane normal to **k** through the origin. It intersects the indicatrix in an ellipse. In stage (2), we construct **E** normal to the indicatrix at each point on the intersection ellipse. By symmetry, stage (3) selects the points which are on the major and minor axes of the intersection ellipse as those for which **E** lies in the (**k**, **D**) plane.† Thus there are always *two characteristic waves* for propagation with a particular **k**, and OP and OQ in the figure represent their polarization vectors, which must be orthogonal. Their refractive indices are given by the lengths of OP and OQ. The existence of two waves with different refractive indices and polarizations is shown by the photograph in Fig. 6.8.

6.6.4 The μ-surface for a crystal

We can get a good idea of the shape of the μ-surface in three dimensions by working out its sections in the (x, y), (y, z) and (z, x) planes. Without loss of generality, assume the indicatrix to have minor, intermediate and major axes of lengths μ_1, μ_2 and μ_3 along the x, y and z axes respectively.

Start with **k** along x and consider what happens as it rotates in the (x, y) plane, as in Fig. 6.9. When **k** is along the x axis, the ellipse $PQRS$ of Fig. 6.7 has its major and minor axes $OZ = \mu_3$ and $OY = \mu_2$. When **k** rotates around z, OZ is always the major axis, but the minor

† This is easily seen by considering the projection of **E** on the plane of the ellipse.

Figure 6.7 (*a*) The elliptical section of the indicatrix *PQRS* which is normal to **k**. (*b*) The tangent planes are indicated at two points on *PQRS* to show that on the axes (at *P*, for example) **D**, **E** and **k** are coplanar, but not at other points.

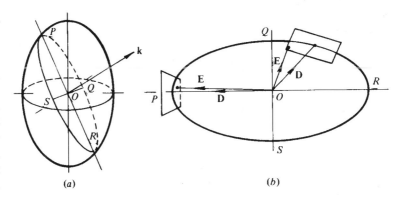

(*a*)   (*b*)

axis changes gradually from OY to $OX = \mu_1$, which value it reaches when **k** is along y. Plotting the two values of μ on a polar plot as a function of the direction of k gives us the section of the μ-surface in the (x, y) plane. Fig. 6.10 shows the result. There is a circle, radius μ_3, corresponding to polarization in the z-direction and an ellipse (μ_2, μ_1)‡ corresponding to the polarization orthogonal to z and **k**.

In the same way we construct the sections in the (y, z) plane (Fig. 6.11) and (z, x) plane (Fig. 6.12). The latter figure shows the circle of radius μ_2 to intersect the ellipse (μ_3, μ_1) at four points A. These correspond to two circular sections of the indicatrix (Fig. 6.13),

‡ Meaning the ellipse with major axis μ_2 and minor axis μ_1.

Figure 6.8 Photograph of refraction of an unpolarized laser beam by a crystal of NaNO$_3$, whose natural growth facets form a prism. A reflected beam and two refracted beams are visible, showing the existence of two refractive indices. At P the refracted beams meet a polarizer, which transmits only one beam.

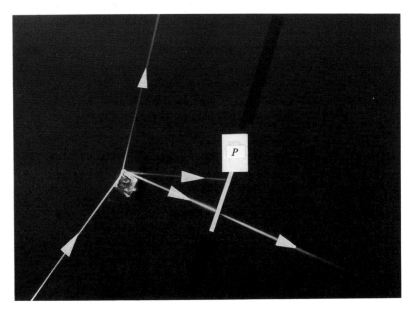

Figure 6.9 Section of the indicatrix when **k** lies in the (x, y) plane.

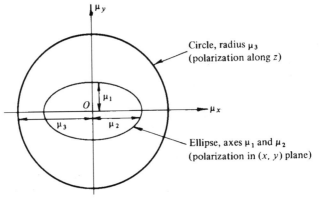

Circle, radius μ_3
(polarization along z)

Ellipse, axes μ_1 and μ_2
(polarization in (x, y) plane)

and the directions **k** corresponding to the *OA*s are called *optic axes*. For propagation in these directions, the two characteristic waves are degenerate (and so in fact any two orthogonal polarizations could be chosen for characteristic waves). Since there are two such orientations *OA*, the general crystal ($\mu_3 \neq \mu_2 \neq \mu_1$) is called *biaxial*. It is quite easy to see that there are no other circular sections of the general ellipsoid, so there are no other optic axes.

Construction of the complete μ-surface can now be done, qualitatively, by interpolation, and this will serve us sufficiently for understanding the physics of crystal optics. One octant of the surface is shown in Fig. 6.14, together with a photograph of a model. The surface clearly has two branches, which we shall call 'outer' and 'inner'. They touch along the optic axis, which is the only direction for which the refractive indices of the two characteristic waves are equal. The other octants are constructed by reflexion.

Figure 6.10 Section of the μ-surface in the (x, y) plane.

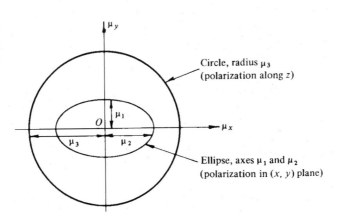

Circle, radius μ_3 (polarization along z)

Ellipse, axes μ_1 and μ_2 (polarization in (x, y) plane)

Figure 6.11 Section of the μ-surface in the (y, z) plane.

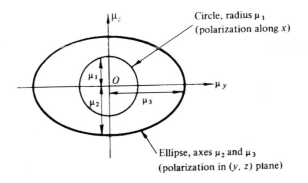

Circle, radius μ_1 (polarization along x)

Ellipse, axes μ_2 and μ_3 (polarization in (y, z) plane)

6.6.5 Ordinary and extraordinary rays

Once we have constructed the μ-surface, it is in principle a simple matter to deduce the polarizations and Poynting vectors **Π** of the two characteristic waves in any given direction (Fig. 6.15). We associate with each direction of **k** and characteristic polarization a *ray* which travels in the direction of **Π**. The ray is what is actually seen when a wave travels through a crystal (Fig. 6.16), and the existence of two rays for any given **k** direction gives rise to the well-known phenomenon of the double image (see Fig. 1.3). Two types of ray can be defined:

– an *ordinary ray*, for which **Π** and **k** are parallel;
– an *extraordinary ray*, for which **Π** and **k** are not parallel.

Since Snell's law applies to the directions of **k**, it applies to **Π** as well only for ordinary waves. In general we find one general and two special cases which are illustrated by Fig. 6.16.

(1) In an arbitrary direction $\mathbf{k}_1$ both surfaces give rise to extraordinary rays. Once **Π** and **k** are known, the magnetic field vector **H** is determined by being their common normal, and the polarization vector **D** by being the common normal to **k** and **H**.

Figure 6.12 Section of the μ-surface in the (z, x) plane.

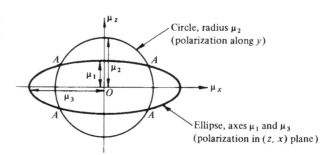

Figure 6.13 Circular sections of the indicatrix.

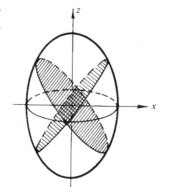

(2) If $\mathbf{k}_2$ lies in a symmetry plane (x, y), (y, z) or (z, x) there is one ordinary and one extraordinary ray.

(3) If $\mathbf{k}_3$ lies along one of the axes x, y or z, both rays are ordinary, despite their having different values of μ.

Examples of important biaxial crystals are – mica, with $\mu_1 = 1.552$, $\mu_2 = 1.582$, $\mu_3 = 1.587$; – LiB_3O_5, with $\mu_1 = 1.578$, $\mu_2 = 1.606$, $\mu_3 = 1.621$.

Figure 6.14 The μ-surface for a biaxial crystal.

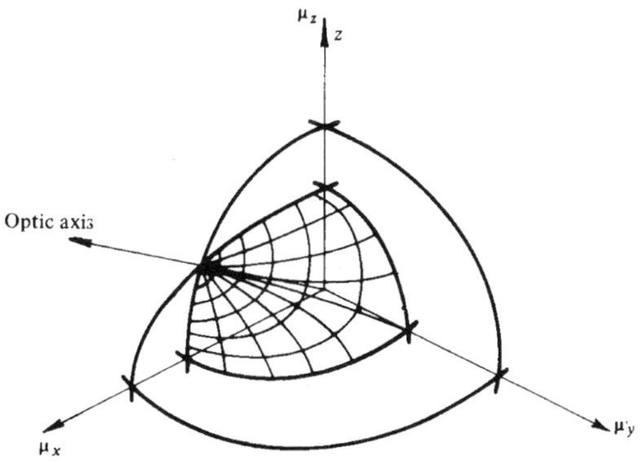

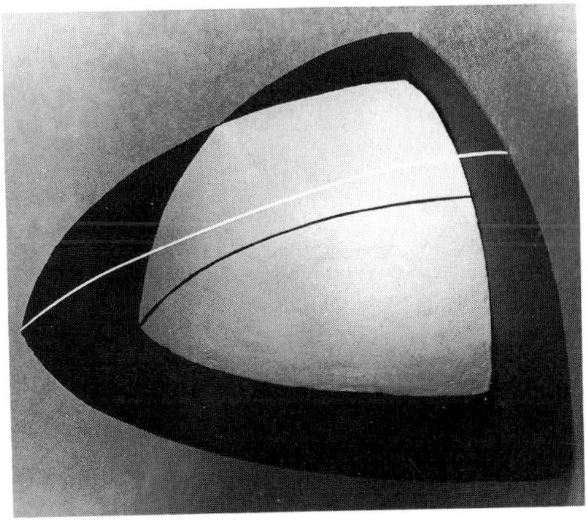

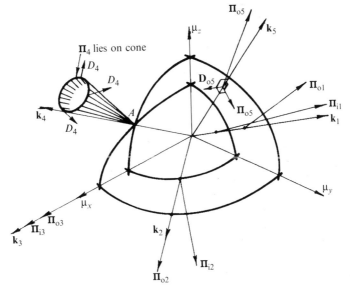

Figure 6.15 Relationships between the propagation vectors at various points on the surface. Suffices 'o' and 'i' refer to outer and inner branches.

6.6.6 Conical propagation

A peculiar form of propagation occurs when $\mathbf{k}$ is along the optic axis ($\mathbf{k}_4$ in Fig. 6.15). Because of the degeneracy of μ, any polarization can be chosen (§6.6.4), but each one gives rise to a different $\boldsymbol{\Pi}$. The various $\boldsymbol{\Pi}$s possible lie on a cone, one edge of which is along the optic axis. If we have a plate of a biaxial crystal, and an unpolarized light beam is incident on it so that it is refracted into the optic axis, the light spreads out into a cone inside the crystal. This phenomenon is called *external conical refraction.*

6.7 Uniaxial crystals

Many crystals have a dielectric tensor which has only two distinct principal values. Then (6.28) becomes

$$\epsilon = \begin{pmatrix} \epsilon_1 & 0 & 0 \\ 0 & \epsilon_1 & 0 \\ 0 & 0 & \epsilon_3 \end{pmatrix}. \tag{6.39}$$

It follows that the indicatrix is a spheroid (ellipsoid of revolution) with one semi-axis of length μ_3 and circular section of radius μ_1. Visualizing the μ-surface via its sections gives once again a two-branched surface. One branch is a sphere of radius μ_1; the other is a spheroid with semi-axis μ_1 and section of radius μ_3. It is immediately apparent that the two branches touch along the μ_z-axis, which is the only optic axis (Fig. 6.17(*a*)). Hence the name for such crystals, *uniaxial crystals.* It is usual to refer to μ_1 as the *ordinary* index (μ_o) and μ_3 as the *extraordinary* index (μ_e); if $\mu_e > \mu_o$, the crystal is said to be *positive uniaxial*, and if $\mu_e < \mu_o$, *negative uniaxial.* Many important optical

Figure 6.16 An unpolarized light ray splits into two as it traverses a crystal plate. The angles between extraordinary rays and the interface clearly do not satisfy Snell's law. The optic axis is in direction *OA*.

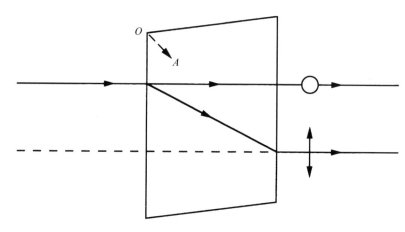

crystals are uniaxial, such as calcite, or Iceland spar (CaCO₃), with $\mu_o=1.66$ and $\mu_e=1.49$; KDP, (KH$_2$PO$_4$) with $\mu_o=1.51$ and $\mu_e=1.47$; quartz (SiO$_2$) with $\mu_o=1.54$ and $\mu_e=1.55$; calomel (HgCl) with $\mu_o=1.97$ and $\mu_e=2.66$.

6.7.1 Propagation in a uniaxial crystal

It follows from the form of the μ-surface, as the reader will easily verify, that:

(1) for a general **k** there is one ordinary and one extraordinary ray;
(2) for **k** along the optic axis, there are two degenerate ordinary rays;
(3) for **k** normal to the optic axis, two ordinary rays propagate with indices μ_o and μ_e. The former is polarized along the optic axis (z) and the latter in the orthogonal direction in the (x, y) plane;
(4) conical propagation does not occur.

6.7.2 Optical activity

When a plane-polarized wave enters a quartz crystal along its optic axis, it is found that the plane of polarization rotates at about 22° per mm of propagation. Quartz is a uniaxial crystal, so this is not consistent with the behaviour that we have described so far. The continuous rotation of the plane of polarization is known as *optical activity* and can occur in any material, crystalline or non-crystalline, having a helical structure – such as quartz which occurs naturally in both right- and left-handed versions (Fig. 6.18). Sugar solutions are

Figure 6.17 Axial sections of the μ-surface of uniaxial crystals: (*a*) such as calcite, (*b*) such as quartz, which is also optically-active.

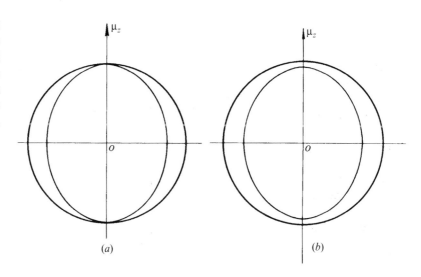

(*a*)　　　　　(*b*)

also well-known examples of non-crystalline optically-active media; dextrose rotates the polarization in a right-handed sense, whereas laevulose is the opposite.

Phenomenologically the dielectric properties of an optically-active uniaxial crystal can be described by a Hermitian dielectric tensor with imaginary off-diagonal components.† We write

$$\epsilon = \begin{pmatrix} \epsilon_1 & ia & 0 \\ -ia & \epsilon_1 & 0 \\ 0 & 0 & \epsilon_3 \end{pmatrix}. \tag{6.40}$$

This tensor satisfies $\epsilon_{ij} = \epsilon_{ji}^*$, i.e. is Hermitian, and can be diagonalized to give principal values $\epsilon_1 + a, \epsilon_1 - a$ and ϵ_3. The principal polarizations are, respectively, $\mathbf{D}_1 = (1, i, 0), \mathbf{D}_2 = (1, -i, 0), \mathbf{D}_3 = (0, 0, 1)$. The first two represent circularly-polarized waves propagating along the z-axis, since they show $\pi/2$ phase differences between the oscillations of their x and y components.

When a wave propagates parallel to z in such a medium we can now see why its plane of polarization rotates. A linearly-polarized wave can be constructed from the superposition of two circularly-polarized waves of opposite senses:

$$\mathbf{D}_r = D_0(1, i, 0) \exp[i(\mu_r k_0 z - \omega t)] \tag{6.41}$$

$$\mathbf{D}_l = D_0(1, -i, 0) \exp[i(\mu_l k_0 z - \omega t)] \tag{6.42}$$

in which the refractive indices μ_r and μ_l are, respectively, $(\epsilon_1 \pm a)^{\frac{1}{2}}$. Their mean is $\bar{\mu}$ and difference $\delta\mu$. Combined:

$$\mathbf{D} = \mathbf{D}_r + \mathbf{D}_l = 2D_0 (\hat{\mathbf{x}} \cos \tfrac{1}{2}\delta\mu k_0 z + \hat{\mathbf{y}} \sin \tfrac{1}{2}\delta\mu k_0 z) \exp[i(\bar{\mu} k_0 z - \omega t)] \tag{6.43}$$

The angle of the plane of polarization, $\tan^{-1}(D_x/D_y) = \tfrac{1}{2}\delta\mu k_0 z$, increases continuously with z. The rate of 22° per mm gives, for green

† As an example, we shall derive such a tensor for a magneto-optical medium in §13.3.5.

Figure 6.18 Positions of silicon atoms in right- and left-handed quartz, projected on a plane normal to the optic axis. The broken line outlines the unit cell, within which there are atoms at levels of $0, \frac{1}{3}$ and $\frac{2}{3}$ of the cell height, indicated by open, shaded and filled circles respectively. These form helices of opposite sense in the two diagrams.

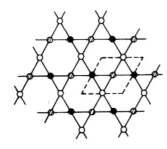
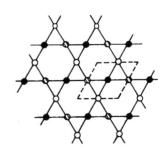

light, $\delta\mu \approx 7.10^{-5}$. Because this value is so small, quartz behaves as a normal uniaxial crystal for propagation in all directions except very close to the optic axis. Its μ-surface is shown schematically in (Fig. 6.17(b)).

An isotropic optically active material, such as sugar solution, can be described by a μ-surface consisting of two concentric spheres, with radii μ_r and μ_l.

6.8 Applications of propagation in anisotropic media

To follow the propagation of a wave of given $\mathbf{k}$ and polarization state in an anisotropic medium we first have to express its $\mathbf{D}$ vector as a superposition of those of the two characteristic waves for the same $\mathbf{k}$. We then follow each one according to its refractive index, and recombine the two at a later stage. If there is an interface to another medium, it is necessary to ensure continuity of the fields, which involves applying Snell's law separately to each characteristic wave (§6.8.3).

6.8.1 Quarter- and half-wave plates

A linearly-polarized plane wave is incident normally (z direction) on a parallel-sided crystal slab of thickness l, such that its plane of polarization bisects those of the two characteristic waves propagating in the same direction; their $\mathbf{D}$ vectors will define the x and y axes. The two characteristic waves have refractive indices μ_1 and μ_2, whose mean is $\bar{\mu}$ and difference $\delta\mu$. We have

$$\mathbf{D} = D_0\hat{\mathbf{x}}\exp[i(\mu_1 k_0 z - \omega t)] + D_0\hat{\mathbf{y}}\exp[i(\mu_2 k_0 z - \omega t)] \qquad (6.44)$$

which, at $z = 0$, combine to give the incident wave $\mathbf{D} = D_0(1,1,0)$ $\exp(-i\omega t)$. At non-zero z, (6.44) can be written

$$\mathbf{D} = D_0[\hat{\mathbf{x}}\exp(-\tfrac{1}{2}i\delta\mu k_0 z) + \hat{\mathbf{y}}\exp(\tfrac{1}{2}i\delta\mu k_0 z)]\exp[i(\bar{\mu}k_0 z - \omega t)]. \qquad (6.45)$$

Some particular cases are of great importance.

(1) When $\tfrac{1}{2}\delta\mu k_0 z = \pi/4$ the phase difference between the $\hat{\mathbf{x}}$ and $\hat{\mathbf{y}}$ components has the value $\pi/2$. The incident linearly-polarized wave has become circularly-polarized. A plate with this thickness, $l = \pi/2k_0\delta\mu = \lambda/4\delta\mu$, is called a *quarter-wave plate*. If the plane of polarization of the incident wave is in the second bisector of $\hat{\mathbf{x}}$ and $\hat{\mathbf{y}}$ the opposite sense of rotation is obtained, and if it does not

exactly bisect $\hat{\mathbf{x}}$ and $\hat{\mathbf{y}}$, the outgoing wave is elliptically-polarized. In the reverse situation, a quarter-wave plate converts circularly-polarized light into linearly-polarized. According to the values given in §6.6.5, a quarter-wave plate for $\lambda=5900\,\text{Å}$ made from mica has thickness about 0.025 mm.

(2) A plate of the same material with twice the above thickness can easily be seen to reflect the plane of polarization in the (x, z) and (y, z) planes, and therefore also reverses the sense of rotation of an incident circularly- or elliptically-polarized wave. It is called a *half-wave* plate.

6.8.2 Compensators

The phase difference between the principal waves can be changed continuously by a *compensator*, which is made from a wedge of bire-fringent material having an angle of about $1°$ between its faces. Its thickness is then a function of position, and its effect on a beam can be varied by sliding it transversely. However, it is difficult to get phase differences close to zero by this means. The *Babinet compensator* is constructed from two opposed wedges cut from a uniaxial crystal and cemented so that the optic axes of the two halves are normal to one another (Fig. 6.19), thus providing a device with parallel outside sur-faces which provides a phase difference varying continuously through zero. A circular compensator, giving a variable rotation of the plane of polarization, is constructed in a similar manner from a pair of wedges of left- and right-handed quartz.

6.8.3 The Pöverlein construction

The μ-surface construction lends itself easily to the graphical solution of refraction problems at interfaces between crystals. We should remember that Snell's law, arising as it does from the continuity of E and H at the interface, is *always* true for the $\mathbf{k}$ vector directions. Now suppose that we consider the refraction of light from, say, a homogeneous material of index μ_1 into the crystal. Fig. 6.20 shows the section of the μ-surface in the plane of incidence, which contains the $\mathbf{k}$ vectors of the incident and refracted rays as well as the normal

Figure 6.19 Babinet compensator. The parallel lines indicate the directions of the optic axes of the two wedges.

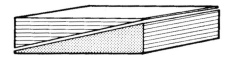

to the surface. The construction described by the figure equates the projections $k \sin \theta = k_0 \mu(\theta) \sin \theta$ (OA for the incident wave and OB for the two refracted waves) on the interface. It is known as *Pöverlein's construction* and can also be used for refraction at an interface between different birefringent materials.

6.8.4 Crystal polarizers

Crystal polarizers separate the two orthogonal polarizations by using the fact that the critical angle is a function of the refractive index and therefore depends on the polarization state. A typical example is the *Glan-air* prism which is generally constructed from calcite, although calomel, HgCl, has even more favourable properties. The construction is illustrated by Fig. 6.21, the optic axes of both halves of the device being normal to the plane of the diagram. Clearly, only the ordinary polarization is reflected at the air layer when the angle of incidence at it is between the critical angles $\sin^{-1} \mu_o$ and $\sin^{-1} \mu_e$. The crystal is cut so that this interface lies half way between the two critical angles when light is incident normally on the input surface.

If the two halves are cemented with a glue having refractive index μ_B between μ_o and μ_e, variants on this idea – the *Glan–Thompson* and *Nicol prism* – are obtained. In this case there is a critical angle only for the ordinary wave. The Nicol prism is constructed around natural cleavage angles of the calcite.

Figure 6.20
Illustrating the Pöverlein construction. OB is the projection of both $\mathbf{k}_o$ and $\mathbf{k}_e$ on the interface, and is equal to the projection OA of the incident $\mathbf{k}$ vector.

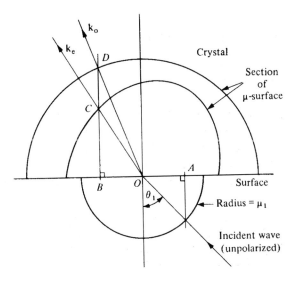

6.9 Induced anisotropic behaviour

The discussion so far has assumed that the anisotropy of a crystal is the result of its structure. There are, however, many instances in which the properties of an isotropic material (such as a liquid, polymer or a cubic crystal) become anisotropic because of some external field; in addition, the optical properties of many anisotropic materials can be changed by applied fields. We shall describe below a few examples of such behaviour, but the description should not be considered in any way as exhaustive. Some other aspects of induced dielectric effects will be discussed in Chapter 13. A much more detailed description may be found in the book by Yariv (1989).

6.9.1 The electro-optic effect

Application of an external electric field can cause induced anisotropy. Two types of effect are common. First, many isotropic materials such as glass, and liquids such as nitrobenzene, become uniaxial with their optic axis along the direction of the electric field. Because there is no way that an isotropic material could be sensitive to the sign of the field, the effect has to be proportional to the square (or a polynomial

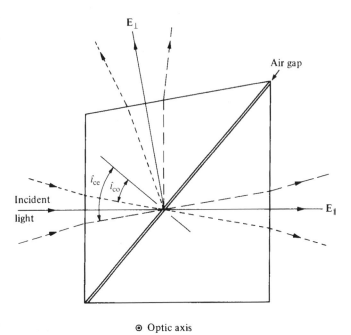

Figure 6.21 Glan-air polarizing prism.

$E_\perp$

Air gap

$\hat{\imath}_{ce}$ $\hat{\imath}_{co}$

Incident light

$E_\parallel$

⊙ Optic axis

including only even powers) of the applied field E_0:

$$\mu_e - \mu_o \propto E_0^2. \tag{6.46}$$

This is called the *Kerr effect*.

On the other hand, crystals without a centre of symmetry in the atomic arrangement of their unit cell are able to distinguish between positive and negative fields and so the electro-optic effect can depend on any power of the field; in particular, a linear effect is possible. Its magnitude can also be a function of the orientation of the field and so a complete description of the effect, even at a phenomenological level, becomes quite involved. It is usual to describe the electro-optic effect in terms of parameters that relate directly to the distortion of the indicatrix, (6.34), which is written explicitly:

$$\frac{D_x^2}{\mu_1^2} + \frac{D_y^2}{\mu_2^2} + \frac{D_z^2}{\mu_3^2} = \epsilon_0. \tag{6.47}$$

We shall consider here only one example, the *Pockels effect* in a uniaxial crystal such as KH_2PO_4. In this effect the application of the field E_0 parallel to the optic axis makes equal and opposite linear changes in μ_1, the refractive indices for the two polarizations perpendicular to the applied field. It is usual to write the distorted indicatrix in the form

$$D_x^2 \left(\frac{1}{\mu_1^2} + rE_0 \right) + D_y^2 \left(\frac{1}{\mu_1^2} - rE_0 \right) + \frac{D_z^2}{\mu_3^2} = \epsilon_0, \tag{6.48}$$

in which we have assumed that the changes in μ are very small. Then the actual changes in the x and y axes of the ellipsoid are $\pm\delta\mu_1 \approx \mp rE_0\mu_1^3/2$. Clearly, the crystal is now biaxial. It follows that a wave propagating along z, for example, can become elliptically-polarized in the same way as we discussed in §6.8.1 and a slab crystal of thickness l in the z direction will act as a quarter-wave plate when

$$lE_0 = \lambda/4r\mu_1^3. \tag{6.49}$$

The product lE_0 is a voltage which is independent of the thickness of the slab and is called the *quarter-wave voltage*, typically $500\,V$. Both the Kerr effect and the linear electro-optic effects such as the Pockels effect can be used to make an electrically operated optical shutter by placing the crystal between crossed polarizers.

6.9.2 The photo-elastic effect

A strain field can affect the indicatrix of an isotropic medium such as glass, Perspex (lucite or poly-methyl-methacrylate) or various epoxy resins. One can imagine the indicatrix being distorted as the medium stretches and this is, qualitatively, the basis of the effect. The material becomes uniaxial with its axis along that of the strain. The effect is of considerable importance as a method of visualizing strain fields in complicated two-dimensional bodies (Fig. 6.22), models of which can be constructed from the above materials.

6.9.3 The magneto-optic effect

Many isotropic diamagnetic materials, including glass and water, become optically active when a magnetic field is applied to them, the induced optic axis being parallel to the applied field $\mathbf{B}_0$. Then, if a wave propagates with $\mathbf{k} \parallel \mathbf{B}_0$, its plane of polarization rotates, in one sense if $\mathbf{k}$ is in the same direction as $\mathbf{B}_0$, and in the other sense if they are opposite. We describe this in a manner identical to (6.40), in which the parameter a is proportional to B_0, $a = r_B B_0$. A microscopic model which illustrates this effect for an electron plasma will be described in §13.3.5. It then follows from (6.42) that the two refractive indices, for left- and right-handed polarizations, satisfy $\delta\mu = \mu_l - \mu_r = \mu^3 r_B B_0$ when $r_B B_0 \ll \mu^{-2}$. The angle of rotation of the plane of polarization per unit propagation distance is then $\frac{1}{2}k_0\delta\mu = \frac{1}{2}k_0\mu^3 r_B B_0$. The constant $\frac{1}{2}k_0\mu^3 r_B$ is called *Verdet's constant*, and is approximately proportional to λ^{-2}. A typical value, that for water, is $20° \, \text{T}^{-1}\text{m}^{-1}$.

There is an important difference between optical activity in a crystal and the magneto-optic effect which emphasizes the properties of the magnetic field as a pseudo-vector.† If a wave propagates through an *optically-active* crystal, is reflected normally by a mirror and then returns through the crystal, the net rotation of the plane of polarization is zero, because the mirror interchanges left- and right-handed circularly polarized components, and so the rotation resulting from the first passage is cancelled in the second. However, if the same experiment

Figure 6.22 An example of the photo-elastic effect. A piece of strained Perspex is observed in monochromatic light between crossed polarizers, oriented (a) at $\pm45°$ and (b) $0°$ and $90°$ to the edges of the strip.

† A screw remains left- or right-handed from whichever end you look at it. But the helix traced out by an electron in a magnetic field as it comes towards you reverses its helicity if you reverse the field direction.

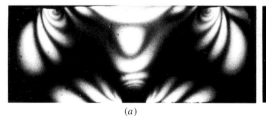

(a)　　　　　　　　　　　　　　　(b)

is carried out with a *magneto-optic* material, the propagation being parallel to $\mathbf{B}_0$, not only is the handedness of the wave reversed on reflexion, but also the sign of the magneto-optic effect, because after the reflexion the direction of $\mathbf{k}$ is reversed with respect to $\mathbf{B}_0$. So the net result of the two passages is twice the effect of the single one. A similar analysis of the electro-optic effect *vis-à-vis* birefringence (which will be left as an exercise to the reader), shows no such distinction. This property of the magneto-optic effect allows us to construct a one-way light valve, or *isolator*. If we have a magneto-optic plate in a field such that the rotation obtained is $\pi/4$, and it is sandwiched between polarizer and analyzer with their axes separated by this angle, a wave in one direction will be transmitted by both polarizers. A wave in the other direction, however, find itself with polarization exactly orthogonal to the polarizer when it leaves the device, and so is absorbed. This type of device is widely used in microwave equipment.

CHAPTER SEVEN

Diffraction

7.1 Occurrence of diffraction

As we saw in Chapter 1, the wave theory of light was not at first generally accepted because light did not appear to have any obviously wave-like properties; for example, it did not bend round obstacles as water waves are clearly seen to do. The reason why this difficulty no longer prevents our acceptance of the wave theory is that we are now aware of the relative scales of the two sorts of waves: water waves are coarse and we can see that they only bend round obstacles that have dimensions of the same order of magnitude as the wavelength; larger objects merely stop the waves in the sense that the waves bending round the edge produce negligible effects. But the wavelength of light is about 5×10^{-7}m ($0.5\,\mu$m) and an object of about a hundred waves in size – sufficient to stop a light wave – is still very small by ordinary standards. Nevertheless some bending of the light waves round the edges of obstacles does occur and can be observed over a range of conditions. For particles of the order of a few wavelengths in size, no special apparatus is needed; for example, the water droplets that condense on a car window are surprisingly uniform in size and show beautiful halos round the street lights as the car passes by. For objects that are much larger, special apparatus is needed. The effects are called *diffraction* phenomena.

7.1.1 Interference and diffraction

There is another class of phenomena that is closely related to diffraction. These are produced by the superposition of several spatially-distinguishable waves and are called *interference* phenomena. Diffraction and interference are sometimes not clearly distinguishable, and different writers attach different meanings to the two words. We shall try to maintain the convention that interference involves the deliberate production of two or more separate beams and that diffraction occurs naturally when a single wave is limited in some way. We shall not always succeed in maintaining this convention because some names – the diffraction grating, for example – do not fit in with it and are too well established to change. But we shall try to preserve the distinction wherever we can. It is, in fact, similar to the distinction between the Fourier series (corresponding to interference) and the Fourier transform (diffraction); it will be remembered from §4.4.1 that the series can be deduced as a special case of the transform, and in the same way all interference can be explained on the same basis as diffraction effects.

7.1.2 Approaches to the theory of diffraction

The formulation of a diffraction problem essentially considers an incident free-space wave whose propagation is interrupted by an obstacle or mask which changes the amplitude and/or phase of the wave locally by a well determined factor. The observer at a given point, or set of points on a screen (the eye's retina, for example), measures a wave field corresponding to the superposition of the incident field and other fields generated in order to satisfy Maxwell's equations at points on the obstacle, according to appropriate boundary conditions. An example of a problem that has been solved this way is the diffraction of a plane wave by a perfectly-conducting sphere; this is called *Mie scattering* and a detailed account of it is given by Born and Wolf (1980), and by van de Hulst (1984). Unfortunately the class of soluble problems of this type is too small for general use and a considerably simpler approach has been developed, based on Huygens's principle (§2.7.1) which describes most diffraction phenomena in a satisfactory, if not completely quantitative, manner. It makes the basic approximation that the amplitude and phase of the electromagnetic wave can be adequately described by a scalar variable, and that effects arising from the polarization of waves can be neglected. It is called the *scalar-wave approximation*. We shall develop this approach at two levels; we shall start with an intuitive approach and afterwards discuss its mathematical justification (§7.2.2).

7.2 The scalar-wave approximation

In principle, a scalar-wave calculation should be carried out for each component of the vector wave, but in practice this is rarely necessary.

On the other hand, we can see the type of conditions under which the direction of polarization might be important by considering how we would begin the problem of diffraction by a slit in a perfectly-conducting sheet of metal. Considering each point on the plane of the sheet as a potential radiator, we see that

(*a*) points on the metal sheet will not radiate at all, because the field **E** must be zero in a perfect conductor;

(*b*) points well into the slit will radiate equally well in all polarizations, because the field can equally well be in any direction in free space;

(*c*) points close to the edge of the slit will radiate better when **E** is perpendicular to the edge of the slit than when **E** is parallel. This occurs because $\mathbf{E}_\parallel$ changes continuously from zero in the metal to a non-zero value in the slit, whereas $\mathbf{E}_\perp$ is not continuous across a surface (§5.4.1). The slit thus produces a diffraction pattern appropriate to a rather smaller width when the illumination is polarized parallel to its length. Because such differences are limited to a region within only about one wavelength of the edge of the obstacle, they become most noticeable for objects with much fine detail on the scale of a few wavelengths. For example the efficiency of blazed diffraction gratings (§9.2.5) is almost always polarization dependent, and closely-spaced wire grids with spacings of the order of $2\,\mu\text{m}$ are efficient polarizers at infra-red wavelengths (§6.3).

The reader is therefore invited, for the time being, to forget that light consists of two oscillating vector fields, and imagine the vibration to be that of a single complex scalar variable ψ with angular frequency ω and wave-vector $\mathbf{k}_0$ which will be assumed to have its free-space magnitude ω/c in the direction of travel of the wave. Because ψ represents a complex scalar field it has both amplitude and phase. The time-dependent wave factor $\exp(-i\omega t)$ is of no importance in this chapter, since it is carried through all the calculations unchanged. It will therefore be omitted.

7.2.1 Diffraction by Huygens's principle

Let us try intuitively to build a theory of diffraction based on Huygens's principle of the re-emission of scalar waves by points on a surface spanning the aperture. A more rigorous, but still scalar-wave, derivation of the same theory has been given by Kirchhoff and is discussed in §7.2.2. But most of the parts of the integral formulation can be written down intuitively, and we shall first derive it in such a manner. We shall consider the amplitude observed at a point P

arising from light emitted from a point source Q and scattered by a plane mask $\mathscr{R}$ (Fig.7.1). We shall suppose that if an element of area dS at S on $\mathscr{R}$ is disturbed by a wave ψ_1 this same point acts as a coherent secondary emitter of strength $f_S \psi_1 dS$, where f_S is called the *transmission function* of $\mathscr{R}$ at point S. In the simplest examples f_S is zero where the mask is opaque and unity where it is transparent, but it is easy to imagine intermediate cases, including complex values of f_S which change the phase of the incident light by a given amount. The coherence of the re-emission is important; the phase of the emitted wave must be exactly related to that of the initiating disturbance ψ_1, otherwise the diffraction effects will change with time.

The scalar wave emitted from a point source Q of strength a_Q can be written as a spherical wave of wavenumber $k_0 = 2\pi/\lambda$ (§2.6.3),

$$\psi_1 = \frac{a_Q}{d_1} \exp(ik_0 d_1) , \qquad (7.1)$$

and consequently S acts as a secondary emitter of strength a_S,

$$a_S = f_S \psi_1 \, dS , \qquad (7.2)$$

so that the contribution to ψ received at P is

$$\begin{aligned} d\psi &= f_S \psi_1 d^{-1} \exp(ik_0 d) dS \\ &= f_S a_Q (dd_1)^{-1} \exp[ik_0(d + d_1)] dS . \end{aligned} \qquad (7.3)$$

The total amplitude received at P is therefore the integral of this expression over the plane $\mathscr{R}$:

$$\psi_P = a_Q \iint_{\mathscr{R}} \frac{f_S}{dd_1} \exp[ik_0(d + d_1)] dS . \qquad (7.4)$$

The quantities f_S, d and d_1 are all functions of the position S. It will be shown in §7.2.4 that expression (7.2) should really contain an inclination factor too; i.e. the strength of a secondary emitter

Figure 7.1 Definition of quantities for the diffraction integral.

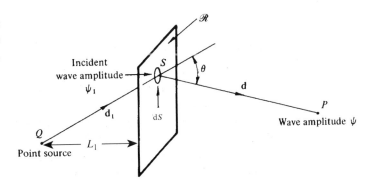

depends on the angle between the incident and scattered radiation, θ in Fig. 7.1. Huygens's principle considers this factor to be unity in the forward direction and zero in the reverse, as shown in §2.7.1 and Fig. 7.4. In addition we shall see that (7.2) should be multiplied by the constant factor ik_0 in order to lead to mathematically self-consistent results.

7.2.2 The Huygens–Kirchhoff diffraction integral

Kirchhoff reformulated the diffraction problem for a scalar wave in a more rigorous manner as a boundary-value problem, which essentially justifies the use of Huygens's principle above. In the next two sections, (7.4) will be rederived, with an explicit form for the inclination factor.

As an electromagnetic field everywhere in a bounded region of space can be uniquely determined by the boundary conditions around this region, it is of interest to see to what extent such an approach is consistent with the idea of re-radiation by points on a wavefront through the aperture. As a result of substituting a time-dependence $\exp(-i\omega t)$ in the wave equation (2.49) we have:

$$\nabla^2 \psi = -\frac{\omega^2}{c^2}\psi = -k_0^2 \psi \ , \tag{7.5}$$

which refers to any component of the electric or magnetic wave-field. We shall see that the field $\psi(0)$ at a point inside the bounded region can be written in terms of ψ and its derivatives on the boundary of the region. In simple cases where these are determined by external waves originating from a point source the result is very similar to the one which we have already found intuitively.

7.2.3 The exact mathematics for the diffraction integral

In problems involving boundaries it is often convenient to study the properties of the differences between two solutions of an equation rather than of one solution alone, since the boundary conditions become simpler to handle. The diffraction integral provides one such example, and we shall compare the required solution of (7.5) with a trial solution

$$\psi_t = \frac{a_t}{r} \exp(ik_0 r) \tag{7.6}$$

which is a spherical wave (2.48) radiating from the origin. This wave satisfies (7.5) except at $\mathbf{r} = 0$. This origin we shall define as the point of observation P, at which ψ has the value $\psi(0)$. The two wave-fields

ψ (to be calculated) and ψ_t (the convergent reference wave) satisfy the equation

$$\psi\nabla^2\psi_t - \psi_t\nabla^2\psi = -\psi k_0^2\psi_t + \psi_t k_0^2\psi = 0 \qquad (7.7)$$

at all points except $\mathbf{r} = 0$, because both ψ and ψ_t are solutions of (7.5). We shall now integrate expression (7.7) throughout a volume $\mathscr{V}$ bounded by a surface $\mathscr{S}$. The volume integral can be changed by Green's theorem to a surface integral:

$$\iiint_{\mathscr{V}} (\psi\nabla^2\psi_t - \psi_t\nabla^2\psi)\, dV = \iint_{\mathscr{S}} (\psi\nabla\psi_t - \psi_t\nabla\psi) \cdot \mathbf{n}\, dS , \qquad (7.8)$$

$\mathbf{n}$ being the outward normal to the surface $\mathscr{S}$ at each point. Because the integrand (7.7) is zero, the integrals (7.8) are also zero, provided that the region $\mathscr{V}$ does not include the origin $\mathbf{r} = 0$. The surface $\mathscr{S}$ is therefore chosen to have two parts, as illustrated in Fig. 7.2: an arbitrary outer surface $\mathscr{S}_1$ and a small spherical surface $\mathscr{S}_0$ of radius δr (much less than one wavelength) surrounding the origin. Volume $\mathscr{V}$ lies between the two surfaces, and $\mathbf{n}$, being the outward normal from $\mathscr{V}$, is therefore inward on $\mathscr{S}_0$ and outward on $\mathscr{S}_1$.

Over this two-sheet surface we thus have, for (7.8)

$$\left[\iint_{\mathscr{S}_0} + \iint_{\mathscr{S}_1} \right] (\psi\nabla\psi_t - \psi_t\nabla\psi) \cdot \mathbf{n}\, dS = 0 . \qquad (7.9)$$

We can evaluate the gradient of ψ_t from (7.6):

$$\nabla\psi_t = \frac{a_t\mathbf{r}}{r^2} ik_0 \exp(ik_0 r) - \frac{a_t\mathbf{r}}{r^3}\exp(ik_0 r) = \frac{a_t\mathbf{r}}{r^3}(ik_0 r - 1)\exp(ik_0 r) \quad (7.10)$$

and substitute in (7.9) to obtain

$$\iint_{\mathscr{S}_0 + \mathscr{S}_1} \frac{a_t}{r^3}\exp(ik_0 r)[\psi(ik_0 r - 1)\mathbf{r} + r^2\nabla\psi] \cdot \mathbf{n}\, dS = 0 . \qquad (7.11)$$

The $\mathscr{S}_0$ contribution can be evaluated directly, since over the small sphere of radius δr we can consider ψ to be constant, equal to $\psi(0)$.

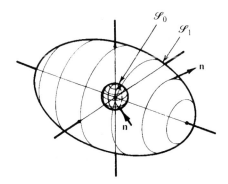

Figure 7.2 The surface for integration. $\mathscr{V}$ lies between $\mathscr{S}_0$ and $\mathscr{S}_1$.

Also since **n** is then the unit vector parallel to $-\mathbf{r}$, we have $\mathbf{r} \cdot \mathbf{n} = -r$ and can substitute $r^2 \mathrm{d}\Omega$ for $\mathrm{d}S$. Thus

$$\iint_{\mathscr{S}_0} \frac{a_t}{r^3} \exp(\mathrm{i}k_0 r)[\psi(0)(\mathrm{i}k_0 r - 1)\mathbf{r} + r^2 \nabla\psi(0)] \cdot \mathbf{n} r^2 \, \mathrm{d}\Omega$$

$$= -\iint_{\mathscr{S}_0} a_t \exp(\mathrm{i}k_0 r)[\psi(0)(\mathrm{i}k_0 r - 1)\mathbf{r} - r\nabla\psi(0) \cdot \mathbf{n}] \, \mathrm{d}\Omega, \qquad (7.12)$$

evaluated at $r = \delta r$, $\mathrm{d}\Omega$ being the element of solid angle. In the limit as $\mathrm{d}r \to 0$ there is only one term which does not approach zero and that is

$$-\iint_{\mathscr{S}_0} a_t \exp(\mathrm{i}k_0 \delta r)\psi(0)\mathrm{d}\Omega \to -4a_t\pi\psi(0), \qquad (7.13)$$

since $k_0 \, \delta r \ll 1$. Equation (7.11) therefore gives, cancelling a_t ,

$$\iint_{\mathscr{S}_1} \frac{1}{r^3} \exp(\mathrm{i}k_0 r)[\psi(\mathrm{i}k_0 r - 1)\mathbf{r} + r^2 \nabla\psi] \cdot \mathbf{n} \mathrm{d}S = 4\pi\psi(0). \qquad (7.14)$$

This expression is the analytical result of the wave equation (7.5) and applies to any solution ψ and to any surface $\mathscr{S}_1$ which surrounds the origin. (For a surface not enclosing the origin the introduction of $\mathscr{S}_0$ was unnecessary and the right-hand side of (7.14) would be zero.) We shall now consider a particular system of practical importance, for which ψ and $\nabla\psi$ are calculable at all points on $\mathscr{S}_1$.

7.2.4 Illumination by a point source

Suppose that the disturbance on $\mathscr{S}_1$ originates from a point source Q. We consider a point S on $\mathscr{S}_1$ at $\mathbf{r} = \mathbf{d}$ which also lies a distance $\mathbf{d}_1$ from Q (Fig. 7.3). The incident wave at S has amplitude $(a_Q/d_1)\exp(\mathrm{i}k_0 d_1)$ and if the transmission function at this point is f_S, the re-radiated values of $\psi(\mathbf{d})$ and its gradient are then

$$\psi(\mathbf{d}) = \frac{f_S a_Q}{d_1} \exp(\mathrm{i}k_0 d_1), \qquad (7.15)$$

$$\nabla\psi = \frac{f_S a_Q \mathbf{d}_1}{d_1^3}(\mathrm{i}k_0 d_1 - 1)\exp(\mathrm{i}k_0 d_1), \qquad (7.16)$$

as in (7.10). Substituting these values into (7.14) gives

$$a_Q \iint_{\mathscr{S}_1} f_S \exp[\mathrm{i}k_0(d+d_1)][\frac{\mathbf{d} \cdot \mathbf{n}}{d_1 d^3}(\mathrm{i}k_0 d - 1) - \frac{\mathbf{d}_1 \cdot \mathbf{n}}{d d_1^3}(\mathrm{i}k_0 d_1 - 1)]\mathrm{d}S = 4\pi\psi(0). \qquad (7.17)$$

The scalar products can be seen from the diagram to be $\mathbf{d} \cdot \mathbf{n} = d\cos\theta$ and $\mathbf{d}_1 \cdot \mathbf{n} = -d_1 \cos\theta_1$. When d and d_1 are both very much greater than the wavelength, we can neglect 1 with respect to $k_0 d$ and then, with the angles θ and θ_1 defined as in Fig. 7.3,

$$\psi(0) = \frac{ik_0 a_Q}{2\pi} \int\!\!\int_{\mathscr{S}_1} \frac{fs}{dd_1} \exp[ik_0(d + d_1)] \left(\frac{\cos\theta + \cos\theta_1}{2}\right) dS. \quad (7.18)$$

This is the justification for the expression (7.4) which we have already used in our diffraction calculations. But it contains three extra pieces of information. The first is a definite form $\frac{1}{2}(\cos\theta + \cos\theta_1)$ for the inclination factor, which is shown in Fig. 7.4 where it is compared with the Huygens guess (§2.7.1). For paraxial conditions, $\cos\theta = \cos\theta_1 \simeq 1$, the factor is unity as assumed in (7.4). It is discussed further in §7.2.5 below. The second point of interest is a phase factor i and the third is the factor $k_0/2\pi$, both of which will turn out to be physically important in the examples to follow.

7.2.5 The inclination factor

The form of the inclination factor

$$\frac{\cos\theta + \cos\theta_1}{2} \quad (7.19)$$

is of considerable interest. It is consistent with Huygens's construction in neglecting backward radiation, (§2.7.1), but curiously seems to depend on θ and θ_1 separately, so it appears to depend on the local orientation of the particular surface $\mathscr{S}$ that has been chosen. It would seem more reasonable for it to depend on $\theta - \theta_1$, so that the direction of the surface would not matter, but on closer inspection we see that this is not so. The whole picture of re-radiation is a fiction of course; it is not as if we have a *real* screen of material at $\mathscr{S}$ that is scattering light uniformly in all directions. The important property of the integral is not that any one contribution to the integrand should be independent of $\mathscr{S}$, but that the whole integral should be so, because $\mathscr{S}$ is arbitrary. But this is exactly what is assured by the particular

Figure 7.3 Part of the surface $\mathscr{S}_1$ showing normal and vectors.

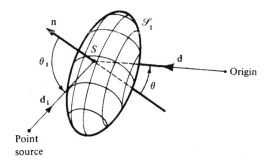

form of inclination factor. For suppose that we were to change the surface $\mathscr{S}_1$ to another surface $\mathscr{S}_2$, also satisfying the condition that it surround the origin. Then the value of $\psi(0)$ is given by

$$4\pi\,\psi(0) = \iint_{\mathscr{S}_2} = \iint_{\mathscr{S}_2} + \left(\iint_{\mathscr{S}_1} - \iint_{\mathscr{S}_2} \right), \qquad (7.20)$$

where the integrals are each of the form (7.14). The last term (bracketed) in (7.20) is the surface integral over a surface not enclosing the origin; its surface normals are outward over the part which is $\mathscr{S}_2$ and inward over $\mathscr{S}_1$ and thus the volume $\mathscr{V}$ lies between the two (Fig. 7.5).

The integral over such a surface is zero, as explained in §7.2.3. Therefore the value of $\psi(0)$ calculated from any surface $\mathscr{S}$ is independent of the surface, and the particular form of inclination factor is necessary to maintain the total integral invariant, despite the fact that the integrand itself changes.

In practice, the surface is usually chosen to span a given aperture and, when this lies in a single wavefront, θ_1 is conveniently zero. Only the aperture itself is included in the integral, as f_S is zero everywhere

Figure 7.4 The inclination factor for $\theta_1 = 0$ plotted in polar coordinates.

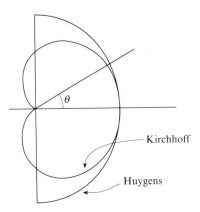

Figure 7.5 Surface consisting of two neighbouring surfaces $\mathscr{S}_1$ and $\mathscr{S}_2$.

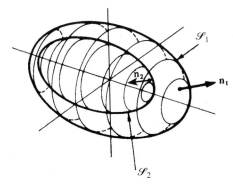

else on the surface, and evaluation of (7.18) is not difficult. We shall now follow with some examples of this sort, which are of practical interest.

7.2.6 Fraunhofer and Fresnel diffraction

Diffraction calculations involve integrating the expression (7.4), or more properly (7.18), under various conditions representing real experiments. We shall consider a classification which will help to make the principles clearer. First let us restrict our attention to a system illuminated by a plane-wave. We do this by taking the source Q to to a very distant point and making it very bright; we therefore make d_1 and a_Q very large while maintaining their ratio constant:

$$a_Q/d_1 = A. \tag{7.21}$$

Now we shall consider the situation where $\mathscr{S}_1$ coincides with the plane $\mathscr{R}$ over the region of interest, which is the diffracting mask, outside of which $f_S = 0$. Moreover, to make things simpler initially, we let $\mathscr{R}$ coincide with a wavefront of the incident wave. The axis of the system is defined as the normal to $\mathscr{R}$ through its origin O. If we denote the position of S by vector $\mathbf{r}$ in the plane of $\mathscr{R}$, f_S is replaced by $f(\mathbf{r})$ and (7.18) becomes (ignoring the inclination factor)

$$\psi = \frac{\mathrm{i}k_0 A}{2\pi} \exp(\mathrm{i}k_0 L_1) \iint_{\mathscr{R}} \frac{f(\mathbf{r})}{d} \exp(\mathrm{i}k_0 d) \mathrm{d}^2\mathbf{r}. \tag{7.22}$$

where L_1 is the normal distance from Q to $\mathscr{R}$. The factor $\exp(\mathrm{i}k_0 L_1)$, being constant over the plane $\mathscr{R}$, will henceforth be absorbed into A. The intensity observed at P is:

$$I = |\psi|^2 \equiv \psi\psi^*. \tag{7.23}$$

The classification of diffraction effects into Fresnel and Fraunhofer types depends on the way in which the phase $k_0 d$ changes as we cross the mask $\mathscr{R}$. This depends on three factors: the distance d between the point S and the point of observation, the extent of $\mathscr{R}$ for which $f(\mathbf{r})$ is not zero (i.e. the size of the mask's transmitting region), and the wavelength $\lambda = 2\pi/k_0$. If $k_0 d$ varies linearly with $\mathbf{r}$, the diffraction is called *Fraunhofer diffraction*; if the variation has non-linear terms of size comparable with $\pi/2$, the diffraction is called *Fresnel diffraction*. We can translate this statement into quantitative terms if we define a circle of radius ρ which just includes all the transmitting regions of $\mathscr{R}$ (Fig. 7.6). We now observe the diffraction in the plane $\mathscr{P}$ normal to the axis at distance L from the mask. Then at a point P in this plane, at vector distance $\mathbf{p}$ from the axis, the phase $k_0 d$ of the wave from $\mathbf{r}$ is

$$k_0 d = k_0 (L^2 + |\mathbf{r} - \mathbf{p}|^2)^{1/2} \simeq k_0 L + \tfrac{1}{2} k_0 L^{-1} (r^2 - 2\mathbf{r} \cdot \mathbf{p} + p^2) + \cdots, \quad (7.24)$$

where we have assumed that r and p are small compared with L. This expression contains the constant term $k_0(L + \tfrac{1}{2}p^2/L)$, the term $k_0 \mathbf{r} \cdot \mathbf{p}/L$, which is linear in r, and the quadratic term $\tfrac{1}{2}k_0 r^2/L$. Now since the largest value of r which contributes to the problem is ρ, we have a quadratic term of maximum size $\tfrac{1}{2}k_0 \rho^2/L$. This means that Fresnel or Fraunhofer conditions are obtained depending on whether $\tfrac{1}{2}k_0 \rho^2/L$ is considerably greater or less than about $\pi/2$. In terms of wavelength λ this gives us:

$$\text{Fresnel diffraction:} \ \rho^2 \geq \lambda L; \qquad (7.25)$$

$$\text{Fraunhofer diffraction:} \ \rho^2 \ll \lambda L. \qquad (7.26)$$

For example, if a hole of diameter 2 mm is illuminated by light of wavelength 5×10^{-4} mm, Fresnel diffraction patterns will be observed at distances L less than 2 m, and Fraunhofer diffraction at much greater distances. Calculation of the patterns will show that the transition from one type to the other is gradual (see Problem 7.8).

We can remark at this stage that when the mask is illuminated by a point source at a finite distance L_1, as in Fig. 7.1, (7.24) can easily be modified.† The phase of the wave at P is then

$$k_0(d_1 + d) = k_0[(L_1^2 + r^2)^{1/2} + (L^2 + |\mathbf{r} - \mathbf{p}|^2)^{1/2}]$$

$$\simeq k_0(L + L_1) + \frac{k_0 r^2}{2}(L^{-1} + L_1^{-1}) + \frac{k_0 p^2}{2L} - \frac{k_0}{L}\mathbf{r} \cdot \mathbf{p} + \cdots \qquad (7.27)$$

† When the object is one-dimensional, for example a slit or series of slits, it is possible to replace the point source Q by a line or slit source. Each point of the line source produces a diffraction pattern from the obstacle, and provided these are identical and not displaced laterally they will lie on top of one another and produce an intensified version of the pattern from a point source. This requires the line source and slit obstacle to be accurately parallel, but no new physical ideas are involved.

Figure 7.6 Elements of a diffraction calculation.

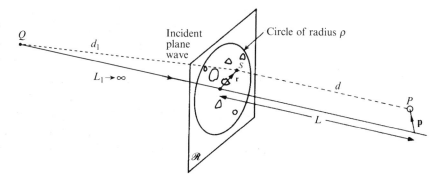

An equivalent differentiation between Fresnel and Fraunhofer diffraction classes then arises with L replaced by $1/(L^{-1}+L_1^{-1})$ in (7.25–7.26). This substitution applies to all the results in §§7.3–7.5.

7.2.7 Experimental observation of diffraction patterns

Using a point source of monochromatic light or a coherent wavefront from a laser, it is easy to observe diffraction patterns of both types. When using a point source, it is important to make sure that it is really small enough for the radiated wave to be a true spherical wave. In other words, the spherical waves emitted by various points in the source, assuming it to have finite extent D, must coincide to an accuracy of about $\frac{1}{4}\lambda$ over the transmitting part of $\mathscr{R}$, which is the circle of radius ρ. The requirement for this is easily seen to be

$$Dp/L_1 < \frac{1}{4}\lambda . \tag{7.28}$$

As we shall see in Chapter 11, this amounts to saying that the radiated wave is coherent across the transmitting part of the mask. For our 2 mm circular hole, at a distance $L_1 = 1$ m the source must have dimensions $D < 0.1$ mm; at a distance of 1 km a street lamp of 10 cm diameter will suffice.

To observe Fresnel patterns, it is only necessary to put a screen at the required distance L. To observe Fraunhofer patterns, we must make the quadratic term in r small enough to be satisfy the condition (7.26) for $r = \rho$. One way of doing this is to make both L and L_1 very large; more conveniently we can make $L = -L_1$ by using a lens to put the observing screen in a plane conjugate to the source. For example, we can look directly at the source, so that the retina of the eye is conjugate to the plane of the source, and by inserting the obstacle anywhere along the line of sight (usually close to the pupil) the Fraunhofer pattern can be observed. Defocusing the eye converts the pattern into a Fresnel pattern. For quantitative work one uses a point source with a lens giving a parallel beam, or else an expanded collimated laser beam, either of which is equivalent to infinite L_1. The Fraunhofer pattern is then observed on a screen at infinite L or else in the focal plane of a converging lens, which is conjugate to the infinite L. Many of the photographs in this book were taken with an optical diffractometer, Fig. 8.1(b), which is constructed on the above principle.

7.3 Fresnel diffraction

The following sections are to be devoted to examples of Fresnel diffraction in a few simple systems. Fraunhofer diffraction and its applications will be discussed in Chapters 8 and 9 since they are far more important as analytical tools. But Fresnel diffraction does have some applications (for example, §7.5), and historically was of crucial importance in clinching the validity of the wave theory of light (§§7.3.3 and 1.2).

The basic integral to be evaluated is equation (7.22):

$$\psi = \frac{ik_0 A}{2\pi} \int\!\!\int_{\mathscr{R}} \frac{f(\mathbf{r})}{d} \exp(ik_0 d) d^2 r \ . \tag{7.29}$$

For any given situation this integral can obviously be evaluated numerically, but we gain little physical intuition from numerical solutions, so we shall leave them as a last resort, or when accurate values are required. The problems in the first class that we shall deal with have axial symmetry, and the integral can be performed analytically for points on the axis; the second class has translational symmetry and can be investigated by graphical methods.

For small apertures ($\rho \ll L$) we can expand d in the exponent of (7.29) by the binomial theorem, as in §7.2.6. In the reciprocal, variations of d have little effect under these conditions and will be ignored and it will be replaced by L:

$$\begin{aligned}
\psi &= \frac{ik_0 A}{2\pi} \int\!\!\int_{\mathscr{R}} \frac{f(\mathbf{r})}{L} \exp[ik_0(L + \frac{r^2}{2L})]d^2 r \\
&= \frac{ik_0 A}{2\pi L} \exp(ik_0 L) \int\!\!\int_{\mathscr{R}} f(\mathbf{r}) \exp(ik_0 \frac{r^2}{2L})d^2 r \ . \tag{7.30}
\end{aligned}$$

The phase factor $\exp(ik_0 L)$ will now be absorbed into A (which, you will recall from §7.2.6, has already absorbed $\exp(ik_0 L_1)$.)

7.3.1 Circular systems, where the integral can be evaluated analytically

In a system with axial symmetry, the value of ψ *on the axis* $p = 0$ can often be evaluated by direct integration of (7.30). For such a mask, $f(\mathbf{r})$ can be written as $g(r^2) = g(s)$, where $s \equiv r^2$. The element of area is then written in terms of s:

$$d^2 r = 2\pi r dr = \pi ds \ , \tag{7.31}$$

and therefore the integral (7.30) becomes:

$$\psi = \frac{ik_0 A}{2L} \int_0^\infty g(s) \exp(\frac{ik_0 s}{2L}) ds . \tag{7.32}$$

The integral is clearly of Fourier transform type, although the limits of integration are from zero (not $-\infty$) to ∞, which can be shown to make negligible difference in physical situations. Unfortunately there is no simple transformation of this type for off-axis points!

We shall consider three important examples:

(a) the circular hole of radius R, for which $g(s) = 1$ when $s < R^2$, otherwise 0;
(b) the circular disc of radius R, for which $g(s) = 1$ when $s > R^2$, otherwise 0;
(c) the zone plate, for which $g(s)$ is periodic.

7.3.2 The circular hole

The integral becomes, under the conditions (a) above,

$$\psi = \frac{ik_0 A}{2L} \int_0^{R^2} \exp(\frac{ik_0 s}{2L}) ds \tag{7.33}$$

$$= A \left[\exp(ik_0 R^2 / 2L) - 1 \right] . \tag{7.34}$$

The observed intensity is

$$|\psi|^2 = 2A^2 [1 - \cos(k_0 R^2 / 2L)] . \tag{7.35}$$

As the point of observation moves along the axis, the intensity at the centre of the pattern alternates periodically with L^{-1} between zero and four times the incident intensity A^2 (Fig. 7.7).

7.3.3 The circular disc

In a similar way to the above, we must evaluate the integral for case (b):

$$\psi = \frac{ik_0 A}{2L} \int_{R^2}^\infty \exp(\frac{ik_0 s}{2L}) ds \tag{7.36}$$

$$= A \left[\exp(i\infty) - \exp(ik_0 R^2 / 2L) \right] . \tag{7.37}$$

The exponential $\exp(i\infty)$ can safely be taken as zero for the following reason. We have, it will be remembered, approximated the d^{-1} term in (7.29) by L^{-1}; this approximation will be invalid in the limit $s \to \infty$, where $d^{-1} \to 0$ makes it permissible to neglect this term. Thus the intensity

$$|\psi|^2 = A^2 \tag{7.38}$$

for all values of L. This surprising result, that there is *always* a bright spot at the centre of the diffraction pattern of a disc (Fig. 1.2), was the argument used finally to convince the opponents to the wave theory of light (§1.2.4). Fresnel diffraction has thus been of vital importance to the development of optics.

7.3.4 The zone plate

For a long time, the zone plate was little more than an amusing physical toy to illustrate Fresnel diffraction. Its significance has been enhanced as providing a simple model for understanding holograms (§12.6) and it has recently found applications in X-ray microscopy (§7.5). It is usually made by photo-reducing a large drawing of alternate black and white rings (Fig. 7.8) of diameters such as to make $g(s)$ a periodic function with period $2R_0^2$ (Fig. 7.9):

$$g(s) = 1 \quad [2nR_0^2 < s < (2n+1)R_0^2]$$
$$g(s) = 0 \quad [(2n+1)R_0^2 < s < (2n+2)R_0^2] \tag{7.39}$$

for integer n. This describes a square wave, for which the Fourier integral (7.32) and (4.3.2) has δ-functions of value

$$(-1)^{\frac{m-1}{2}} \frac{ik_0 A}{2m\pi L} \tag{7.40}$$

Figure 7.7 Fresnel diffraction patterns of a circular hole; (*a*) when $kR^2/2L = 2n\pi$ and (*b*) when $kR^2/2L = (2n+1)\pi$.

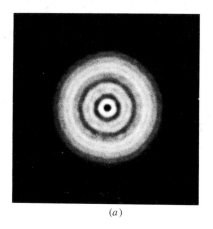

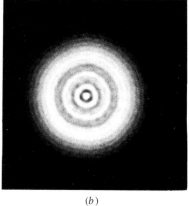

(*a*)   (*b*)

for odd m, at discrete values of L satisfying

$$\frac{k_0}{L} = \frac{2m\pi}{R_0^2}.$$ (7.41)

We therefore find at these values of L a series of 'foci', points at which the light is concentrated. On substituting L from (7.41) into the amplitudes (7.40) we find them all to be equally strong. This is a *particular* result which arises from the values of the harmonics of the square wave; it is not true for other functions. In particular, a zone plate with sinusoidal profile $g(s) = \frac{1}{2}[1 - \sin(\pi s/R_0^2)]$, which is the first component of the square wave, has a single Fourier component at positive frequency and therefore produces a single focus at $L = k_0 R_0^2/2\pi$ which will be seen in §12.6 to be the real holographic reconstruction of a point. It also has a single Fourier component at negative frequency which gives its virtual reconstruction.

The zone plate can be used in a similar way to a lens. If we concentrate on one particular order of diffraction, $n = 1$ say, we can see from §7.2.6 that if illumination is provided by a point source at

Figure 7.8
Zone plate.

distance L_1 (Fig. 7.1) the position of the image moves out to satisfy

$$\frac{1}{L} + \frac{1}{L_1} = \frac{2\pi}{k_0 R_0^2} = \frac{\lambda}{R_0^2} \tag{7.42}$$

which is equivalent to a lens of focal length R_0^2/λ. It clearly suffers from serious chromatic aberration (see Problem 7.4). Such 'lenses' are now used for X-ray microscopy (§7.5).

7.4 Fresnel diffraction by linear systems

There is no simple analytical method to evaluate the Fresnel integral (7.30) for systems without circular symmetry, and either numerical or graphical methods must be used. The latter give us some physical insight into the rather beautiful forms of Fresnel diffraction patters, and we shall discuss their application to linear systems briefly. In Cartesian coordinates, (7.30) becomes:

$$\psi = \frac{ik_0 A}{2\pi L} \int\int_{\mathscr{R}} f(x, y)\exp[\frac{ik_0}{2L}(x^2 + y^2)]\mathrm{d}x\,\mathrm{d}y \ . \tag{7.43}$$

In this section we shall consider only systems in which $f(x, y)$ can be expressed as the product of two functions, $f(x, y) = g(x)h(y)$, so that (7.43) becomes:

$$\psi = \frac{ik_0 A}{2\pi L} \int_{-\infty}^{\infty} g(x)\exp(\frac{ik_0}{2L}x^2)\mathrm{d}x \int_{-\infty}^{\infty} h(y)\exp(\frac{ik_0}{2L}y^2)\mathrm{d}y \ . \tag{7.44}$$

The two integrals can then be evaluated separately.

Figure 7.9 Functions $f(r)$ and $g(s)$ (where $s = k_0 r^2/2L$) for a zone plate.

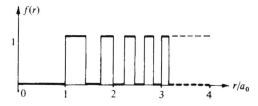

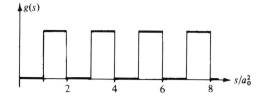

7.4.1 Graphical integration by amplitude–phase diagrams

Integrals of the type

$$\psi = \int_{x_1}^{x_2} f(x) \exp[i\phi(x)] \, dx \tag{7.45}$$

can be evaluated by drawing an amplitude–phase diagram in the complex plane. We represent each infinitesimal increment of ψ

$$d\psi = f(x) \exp[i\phi(x)] \, dx \tag{7.46}$$

by a vector in the complex plane of length $f(x) \, dx$ at angle $\phi(x)$ to the real axis. The value of ψ is then the vector sum of the increments, which is the vector joining the x_1 and x_2 ends of the curve formed by all the increments head-to-tail. This is called an *amplitude–phase diagram*, and the problem is to calculate the form of the curve that is drawn schematically in Fig. 7.10. If we move a length $\delta\sigma$ along the curve, and the angle ϕ changes by $\delta\phi$ as a result, its curvature is clearly, in the infinitesimal limit,

$$\kappa = \frac{\delta\phi}{\delta\sigma} \rightarrow \frac{d\phi}{d\sigma} \tag{7.47}$$

and if we know κ as a function of σ the complete curve can be drawn out. Now $d\sigma = f(x) \, dx$ and therefore the curve can be described in terms of σ and

$$\kappa = \frac{1}{f(x)} \frac{d\phi}{dx} . \tag{7.48}$$

7.4.2 Diffraction by a slit

Let us consider the problem of diffraction by a single long slit defined in the plane $\mathscr{R}$ from x_1 to x_2, so that $g(x) = 1$ between these limits and is zero elsewhere; $h(y) = 1$ everywhere. The integral for ψ then gives

Figure 7.10 Complex plane diagram of the integral $\psi = \int_{x_1}^{x_2} f(x) \exp[i\phi(x)] \, dx$.

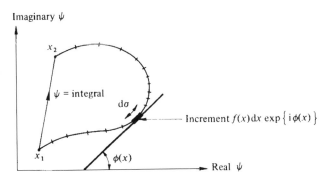

the amplitude and phase of the disturbance at P, which is opposite $(x, y) = (0, 0)$ in the plane $\mathcal{R}$ (Fig. 7.11). To build up the whole of the pattern we must repeat the calculation for P opposite various points on the slit by varying x_1 and x_2 so that $(x_1 - x_2)$ remains constant. The integral is therefore

$$\psi = \frac{ik_0 A}{2\pi L} \int_{x_1}^{x_2} \exp(\frac{ik_0}{2L} x^2) dx \int_{-\infty}^{\infty} \exp(\frac{ik_0}{2L} y^2) dy \ . \tag{7.49}$$

We shall evaluate the first integral in (7.49). Its amplitude and phase are given by the vector between the points representing x_1 and x_2 on the curve defined by

$$\sigma = \int_0^x dx = x \ , \tag{7.50}$$

$$\kappa = \frac{d\phi}{dx} = \frac{k_0 x}{L} \equiv \beta^2 x \ . \tag{7.51}$$

Thus x is the distance from the origin measured along the curve, and the curvature at that point is $\beta^2 x$. This curve can be written in terms of the dimensionless quantities:

$$X = \beta x \ , \quad K = \beta^{-1} \kappa \ . \tag{7.52}$$

The curve satisfying (7.50) and (7.51), called the *Cornu spiral*, is then defined by the simple equation

$$K = X \tag{7.53}$$

and is illustrated in Fig. 7.12.† It has curvature which increases linearly with distance from the origin, measured along the curve. To calculate the diffraction pattern from the slit we take a series of values of X_1 and X_2 such that $(X_1 - X_2)/\beta$ is the width of the slit, and measure the vector length between the points on the spiral at X_1 and X_2. This gives the amplitude and phase of ψ at P, which is opposite $x = 0$. It will be seen, then, that diffraction patterns become quite intricate when both X_1 and X_2 are in the 'horns' of the spiral – i.e. when $(X_1 - X_2)$ is typically of the order of, or greater than, 10. As an example, we have calculated the diffraction pattern for a value of $(X_1 - X_2) = 8.5$. Using light of wavelength 0.6 µm from (7.51) this corresponds to $(x_1 - x_2)^2/L \simeq 7$ µm – for example, a slit 2.7 mm wide at $L = 1$ m. Fig. 7.13 shows (*a*) the calculated intensity as a function of position, and (*b*) the diffraction pattern observed. The numerical

---

† In the Cornu spiral, the x and y axes are, respectively, $\int_0^X \cos(t^2/2) \, dt$ and $\int_0^X \sin(t^2/2) \, dt$. These integrals are tabulated by Hecht (1987).

calculations were made with the help of an accurate scale drawing of the Cornu spiral in the classic, *Physical Optics* (Wood, 1934).

We must not, however, neglect the y integral in (7.49). For this integral, the Cornu spiral gives the vector $C_+C_- = (2\pi iL/k_0)^{\frac{1}{2}}$, which is at 45° to the real axis of the diagram.† The phase is, of course, unobservable, but it is satisfying to confirm that the diffraction pattern of an infinite aperture, for which the integrals in both x and y are infinite in extent, is

$$\psi = \frac{ik_0A}{2\pi L} \int_{-\infty}^{\infty} \exp(\frac{ik_0}{2L}x^2)dx \int_{-\infty}^{\infty} \exp(\frac{ik_0}{2L}y^2)dy = A. \qquad (7.54)$$

which is what one would expect for an unobstructed wave, recalling that A already contains the phase factor $\exp[ik_0(L + L_1)]$.

† The integral is that of a Gaussian function with imaginary variance, $\sigma^2 = L/ik_0$, so that its value is $(2\pi iL/k_0)^{1/2}$.

Figure 7.11 Parameters for Fresnel diffraction by a slit.

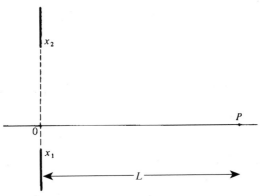

Figure 7.12 The Cornu spiral.

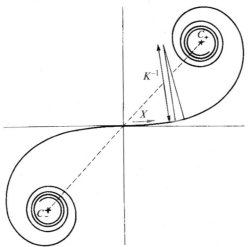

7.4.3 Diffraction by a single edge: the edge wave

Some diffraction patterns have certain characteristics that can easily be recognized as geometrical properties of the Cornu spiral. One obvious property is that for large values of $|X|$ the spiral becomes almost circular, with radius $|X|^{-1}$, which converges very slowly towards zero at the limits C_+ and C_-. The diffraction pattern of a straight edge, which is an aperture extending from a finite value of X to infinity, is expressed (Fig. 7.14). Then the vector representing $\psi(X)$ joins the point X to C_+. When X is positive, so that $X = 0$ is in the geometrical shadow, the vector simply rotates about C_+, becoming monotonically shorter with continuously increasing phase as $X \to \infty$ (Fig. 7.15a):

$$\frac{\mathrm{d}\phi}{\mathrm{d}X} = K = X. \tag{7.55}$$

This is identical with the wave coming from a line source, since the phase variation for such a wave is given by:

$$\psi = \exp[ik_0(x^2 + L^2)^{1/2}] = \exp[i\phi(x)] \tag{7.56}$$

for which, on expanding the exponent for $x \ll L$,

$$\frac{\mathrm{d}\phi}{\mathrm{d}x} \simeq \frac{k_0 x}{L}; \quad \text{or} \quad \frac{\mathrm{d}\phi}{\mathrm{d}X} \simeq X. \tag{7.57}$$

The amplitude of ψ is almost independent of X. As the phase and amplitude variations in the geometrical shadow are thus practically identical with those that would arise from a line source coincident with it, the edge appears to be a bright line; this is known as the *edge wave* construction. Well into the bright part of the pattern a similar analysis can be applied, by representing the vector between X

Figure 7.13
(a) Amplitude of the Fresnel diffraction pattern calculated for a slit of width 0.9 mm observed with $L = 20$ cm, $L_1 = 28$ cm and $\lambda = 0.6$ μm. The geometrical shadow is indicated by the broken lines. (b) Photograph of the diffraction pattern observed under the same conditions.

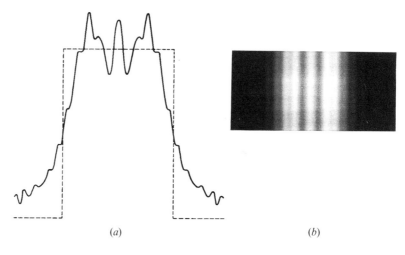

(a)   (b)

and C_+ as the sum of the two vectors C_+C_- and C_-X (Fig. 7.15(b)). The appearance is thus of uniform illumination (C_+C_-) and a bright edge again. Interference between these two waves, plane and diverging cylindrical edge wave, gives rise to a set of light and dark bands with spacing and contrast which decreases as $X \to \infty$. The edge wave is not observable in the region near $X = 0$.

7.5 Advanced topic: X-ray microscopy

Until recently, the zone plate was little more than a 'physical toy' which demonstrated the principle of Fresnel diffraction but had no serious applications. Today, resulting from improvements in microfabrication, zone plates are being used for X-ray microscopy (Howells *et al.*, 1991; Michette, 1988). We should recall (see §13.3.3) that the refractive index of materials for X-rays is very slightly less than unity (order $1 - 10^{-7}$) so that lenses cannot be constructed, although glancing-angle mirrors

Figure 7.14
(*a*) Intensity of the
Fresnel diffraction
pattern of a single
straight edge. The
geometrical shadow
is indicated by
broken lines.
(*b*) Photograph of
the observed pattern.

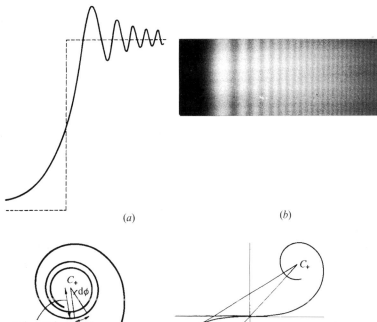

(a) (b)

Figure 7.15
Cornu-spiral
constructions for the
edge wave observed
(*a*) in the shadow
region, (*b*) in the
bright region.

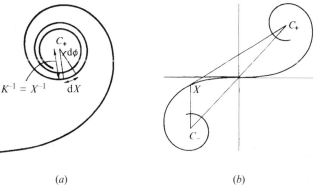

(a) (b)

using total external reflexion are a practical possibility. Because of the difficulties involved in accurate forming of the ellipsoidal surfaces of such mirrors, as required for imaging, they have not so far been successful in achieving high-resolution microscopy. However, in the zone plate we have an alternative focusing element that consists of a series of transparent and opaque rings, which can be constructed from a planar metal film and has recently been successfully used with X-ray radiation.

Supposing that we use a wavelength of 50 Å, and require a focal length of 1 mm. The scale of the rings is then given by (7.42):

$$R_0^2 = f\lambda = 5\,\mu m^2. \tag{7.58}$$

The nth ring has a radius $\sqrt{n}R_0$, and its thickness is approximately $R_0/\sqrt{n}$, so that several hundred rings require fabrication on a scale of 0.1 μm. Such structures have been made by electron-beam writing on a photoresist film protecting a layer of gold on a 1200 Å thick X-ray-transparent silicon nitride substrate. Subsequent dissolution of the unexposed photoresist and etching of the underlying gold resulted in a zone plate with structure down to about 200 Å scale.

We can calculate the resolution limit of such a 'lens' using the methods which will be developed in §12.3. First of all, the transverse resolution limit is equivalent to that of a lens with outer diameter D equal to that of the zone plate. We shall assume it to have N rings so that, from (7.39), $s_{max} = D^2/4 = 2NR_0^2$. Then the resolution limit δx_{min}, using (7.42) for the focal length f is

$$\delta x_{min} = f\theta_{min} = \frac{1.22 f\lambda}{D} = \frac{1.22 f\lambda}{2\sqrt{2N}R_0} = \frac{1.22 R_0}{2\sqrt{2N}}. \tag{7.59}$$

This can usefully be compared with the thickness of the outermost ring, which is $\sqrt{2N}R_0 - \sqrt{2N-1}R_0 \simeq R_0/\sqrt{2N}$ for large N. In other words, the transverse resolution limit is approximately equal to the scale of the finest ring, which is determined by the fabrication technique (about 200 Å). The longitudinal resolution is wavelength dependent. The images, at positions given by (7.41) of which the primary image $n = 1$ is the only one of interest (the others are blocked by appropriately-placed baffles), are ideally sharp only when $N \rightarrow \infty$. Otherwise, from the Fourier theory for a finite number N of oscillations (§4.4.5), we have the relative width of the image

$$\frac{\delta L}{L} = \frac{1}{N} \tag{7.60}$$

$$\delta L = \frac{R_0^2 k_0}{2N\pi} = \frac{1}{N}\left(\frac{R_0^2}{\lambda}\right) = \frac{D^2}{8N^2\lambda}. \tag{7.61}$$

As is usual in imaging systems, the longitudinal and transverse resolutions are related by

$$\frac{\delta L}{\lambda} \simeq \left(\frac{\delta x_{\min}}{\lambda}\right)^2, \tag{7.62}$$

i.e. the longitudinal resolution limit approximately equals the square of the transverse resolution limit, when both are measured in units of the wavelength. The transverse resolution limit is about 4λ, so the longitudinal limit is about 16λ. The poor depth discrimination indicated by the last figure, coupled with the high X-ray dosages necessary for imaging, are major problems which still await solution.

CHAPTER EIGHT

Fraunhofer diffraction and interference

8.1 Introduction

The difference between Fresnel and Fraunhofer diffraction has been discussed in Chapter 7, where we showed that Fraunhofer diffraction is characterized by a linear change of phase over the diffracting obstacle. In practice this linear change is achieved only in special circumstances, such as those occurring when the object is illuminated by a beam of parallel light. It is therefore necessary to use lenses, both for the production of the parallel beam and for the observation of the resultant diffraction pattern. An equivalent statement is that Fraunhofer diffraction is the limit of Fresnel diffraction when the source and the observer are infinitely distant from the obstacle.

8.1.1 Achieving a linear phase change across the mask

In (7.27) we showed the optical path from an axial point source Q to the point P in the observation plane, via a general point S in the mask plane $\mathcal{R}$ (as in Fig. 7.6), to be

$$\overline{QSP} \simeq L + L_1 + \tfrac{1}{2}(L^{-1} + L_1^{-1})r^2 + \tfrac{1}{2}L^{-1}(p^2 - 2\mathbf{r} \cdot \mathbf{p}) + \dots \quad (8.1)$$

where S is at $\mathbf{r} \equiv (x, y)$ and P at $\mathbf{p} \equiv (p_x, p_y)$ in their respective planes, with origins on the axis of illumination. Let P now be defined by the direction cosines (ℓ, m, n) of the line OP joining the origin of the mask to P. Then, when $p \ll L$ we can write $\mathbf{p} = (L\ell, Lm)$, and

$$
\begin{aligned}
\overline{QSP} &\simeq L + L_1 + \tfrac{1}{2}(L^{-1} + L_1^{-1})r^2 + \tfrac{1}{2}L(\ell^2 + m^2) - x\ell - ym \\
&\simeq L + L_1 - x\ell - ym + \dots
\end{aligned}
\quad (8.2)
$$

where all second and higher-order terms have been neglected. It is this linear dependence on x and y which is the origin of the great

176

importance of Fraunhofer diffraction. As was pointed out in §7.2.7, experimental conditions can easily be devised so that the second-order terms are zero, even for quite large $\mathbf{r}$. The everyday situation is when $L_1 = -L$ (observing screen conjugate to the point source Q); this arises when we look at a distant point source, with the diffracting mask close to the eye. Quantitative laboratory experiments are carried out with the aid of lenses or lens combinations, as in Fig. 8.1(a). The point source (nowadays a laser beam focused on a pinhole) is situated at the focal point B of the first lens, C, so that a plane wave is incident on the mask, $\mathscr{R}$, and thus $L_1 \to \infty$. The light leaving the mask passes through a second lens D, and the observation plane is the focal plane $\mathscr{F}$ of that lens, so that $L \to \infty$. Each point in this plane corresponds to a vector $(L\ell, Lm)$. Clearly the observation plane is conjugate to the point source, irrespective of the distance between the two lenses.

Figure 8.1 (a) Setup for observing Fraunhofer diffraction; (b) optical diffractometer. A is the light source; B is the pin-hole; C and D are the lenses; and E is an optically flat mirror. The diffraction pattern of a mask at $\mathscr{R}$ is seen in the plane $\mathscr{F}$.

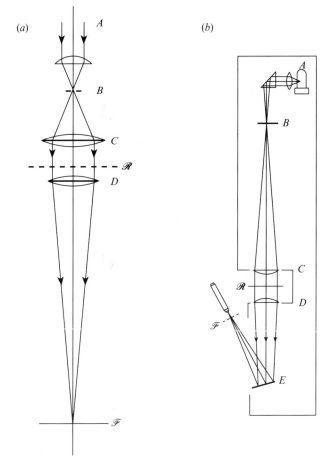

Many of the photographs in this book were taken with an *optical diffractometer* shown in Fig. 8.1(*b*) which was designed for accurate experiments of this sort. It was developed in particular as an 'analogue computer' (Taylor and Lipson, 1964) for solving X-ray crystal diffraction problems, which are of the Fraunhofer type, and will be discussed in §8.4.

8.2 Fraunhofer diffraction and Fourier transforms

We shall now examine the argument in §8.1.1 in more detail, for the case illustrated by Fig. 8.2 in which the incident light is a plane wave parallel to the optical axis and observation is in the focal plane of the second lens.

Consider a plane-wave travelling along the z-axis (Fig. 8.2) and incident at $z = 0$ on a mask with amplitude transmission function $f(x, y)$. The diffracted light is collected by a lens of focal length F situated in the plane $z = U$.

All light waves leaving the screen in a particular direction are focused by the lens to a point in the focal plane. In the figure XB, OA, YC are all parallel and are focused at P. The amplitude of the light at P is therefore the sum of the amplitudes at X, O, Y,

Figure 8.2
(*a*) Illustrating
Fraunhofer
diffraction by a
two-dimensional
object; (*b*) detail of
the region OZX.

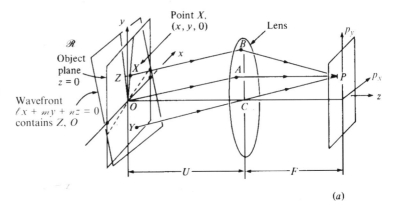

(*a*)

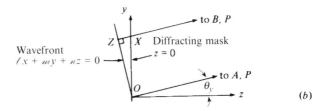

(*b*)

etc., each with the appropriate phase factor $\exp(ik_0\overline{XBP})$, etc., where $\overline{XBP}$ indicates the optical path from X to P via B, including the path through the lens.

Now the amplitude at X, the general point (x, y) in the plane $z = 0$, is simply the amplitude of the incident wave, assumed unity, multiplied by the transmission function $f(x, y)$. To calculate the optical path $\overline{XBP}$ we remember that according to Fermat's principle (§2.7.2) the optical paths from the various points *on a wavefront* to its focus are all equal. The direction of XB, OA,... is represented by direction cosines (ℓ, m, n). Then the wavefront normal to them, through O, which focuses at P is the plane

$$\ell x + my + nz = 0 \tag{8.3}$$

and the optical paths from the wavefront to P, i.e. $\overline{OAP}$ and $\overline{ZBP}$ are equal. Now ZX is just the projection of OX onto the ray XB, and this can be expressed as the component of the vector $(x, y, 0)$ in the direction (ℓ, m, n), namely:

$$ZX = \ell x + my. \tag{8.4}$$

Thus:

$$\overline{XBP} = \overline{OAP} - \ell x - my. \tag{8.5}$$

The amplitude at P is obtained by integrating $f(x, y)\exp(ik_0\overline{XBP})$ over the screen:

$$\psi_P = \exp(ik_0\overline{OAP}) \iint f(x, y)\exp[-ik_0(\ell x + my)]\,dx\,dy. \tag{8.6}$$

We can use $u \equiv \ell k_0$, $v \equiv mk_0$ to represent the position of P, and write

$$\psi(u, v) = \exp(ik_0\overline{OAP}) \iint f(x, y)\exp[-i(ux + vy)]\,dx\,dy. \tag{8.7}$$

The Fraunhofer diffraction pattern amplitude is therefore given by the two-dimensional Fourier transform of the mask transmission function $f(x, y)$.

The coordinates (u, v) can also be related to the angles of diffraction θ_x and θ_y between the vector (ℓ, m, n) and the horizontal and vertical planes, respectively, containing the axis. Then $\ell = \sin\theta_x$, $m = \sin\theta_y$, and

$$u = k_0 \sin\theta_x, \qquad v = \sin\theta_y. \tag{8.8}$$

The coordinates (p_x, p_y) of P can be related exactly to u and v only if the details of the lens are known. For Gaussian optics, $n \approx 1$ and

$$p_x = F\ell/n \approx uF/k_0, \qquad p_y = Fm/n \approx vF/k_0. \tag{8.9}$$

It would be useful if the linear approximation could be preserved out to larger angles, and lenses with this property have been designed. In general, however, one has to work at small angles for the $(p_x, p_y) : (u, v)$ relationship to be linear.

When we observe a diffraction pattern, or photograph it, we measure the intensity $|\psi(u, v)|^2$, and the exact value of $\overline{OAP}$ is irrelevant. Although there are some experiments in which the phase of the diffraction pattern is important, and which we shall consider in the next section, it is usual to ignore the phase factor and write for (8.7)

$$\psi(u, v) = \int\int f(x, y) \exp[-i(ux + vy)] \, dx \, dy. \qquad (8.10)$$

8.2.1 The phase of the Fraunhofer diffraction pattern

The intensity of the diffraction pattern is independent of the exact position of the mask relative to the lens, in that the distance OC only affects the phase factor $\exp(ik_0\overline{OAP})$. For some purposes, it is necessary to know the phase of the diffraction pattern also, for example if the diffracted wave is to be allowed to interfere with another coherent light wave as in some forms of pattern recognition or holography (§12.6).

Now the factor $\exp(ik_0\overline{OAP})$ is quite independent of $f(x, y)$, since it is determined by the geometry of the optical system. It is very easy to calculate it for the particular case where $f(x, y)$ is $\delta(x)\,\delta(y)$. This represents a pinhole in the mask at O. Then the diffraction pattern is, from (8.7),

$$
\begin{aligned}
\psi(u, v) &= \exp(ik_0\overline{OAP}) \int \delta(x) \exp(-iux) \, dx \int \delta(y) \exp(-ivy)] \, dy \\
&= \exp(ik_0\overline{OAP}). \qquad (8.11)
\end{aligned}
$$

However, we know that the action of the lens in general is to focus the light from the pinhole. Of particular interest is the case where the mask is in the front focal plane ($OC = F$). Then the wave leaving the lens is a plane-wave:

$$\psi(u, v) = \exp(ik_0\overline{OAP}) = \text{constant}. \qquad (8.12)$$

Therefore, *when the object is situated in the front focal plane*, the Fraunhofer diffraction pattern represents the *true complex Fourier transform* of $f(x, y)$. For all other object positions the intensity of the diffraction pattern is that of the Fourier transform, but the phase is not.

8.2.2 Fraunhofer diffraction in obliquely incident light

If the plane-wave illuminating the mask in Fig. 8.2 does not travel along the z-axis, the foregoing treatment can be adjusted in a rather simple manner. Specifically, when the incident wave-vector has direction cosines (ℓ_0, m_0, n_0) the phase of the wave reaching the point (x, y) on the mask is advanced by $k_0(\ell_0 x + m_0 y)$ with respect to that at the origin. Thus the retardation of the component from (x, y) with respect to that from $(0, 0)$ is $k_0[(\ell - \ell_0)x + (m - m_0)y]$. The integral (8.6) is now:

$$
\psi_P = \exp(ik_0\overline{OAP}) \iint f(x, y)
$$
$$
\times \exp\{-ik_0[(\ell - \ell_0)x + (m - m_0)y]\}\, dx\, dy. \quad (8.13)
$$

This can still be written in the form (8.7)

$$
\psi(u, v) = \exp(ik_0\overline{OAP}) \iint f(x, y) \exp[-i(ux + vy)]\, dx\, dy, \quad (8.14)
$$

provided that u and v are redefined:

$$
u = k_0(\ell - \ell_0), \quad v = k_0(m - m_0). \quad (8.15)
$$

Now remembering that (ℓ, m) are defined as sines of the angles (θ_x, θ_y), we see that u, for example, can be written

$$
u = k_0(\sin\theta_x - \sin\theta_{x0}). \quad (8.16)
$$

Suppose that a given feature of the diffraction pattern, corresponding to a certain value of u, appears at an angle of deviation $(\theta_x - \theta_{x0})$ with respect to the incident wave-vector. Then, putting u in (8.16) equal to a constant, we can write (8.16) in the form

$$
u = \text{const} = 2k_0 \sin\left(\frac{\theta_x - \theta_{x0}}{2}\right) \cos\left(\frac{\theta_x + \theta_{x0}}{2}\right), \quad (8.17)
$$

from which it is clear that $\theta_x - \theta_{x0}$ has its minimum value when $\cos[(\theta_x + \theta_{x0})/2] = 1$. This shows that the angle of deviation is minimized when $\theta_x = -\theta_{x0}$, i.e. when the diffracting screen bisects the obtuse angle (Fig. 8.3) between the incident and diffracted rays. This condition of *minimum deviation* can be met for only one component u at a time, and is often used for diffraction gratings (§8.3.4) and holograms (§12.6), where the diffraction order and the reference-wave direction respectively define the value of u very closely.

8.2.3 Diffraction by a slit

We represent a slit of width a by the function

$$f(x,y) = \text{rect}(x/a) \quad = \quad 1 \quad (|x| \le a/2)$$
$$= \quad 0 \quad (|x| > a/2). \qquad (8.18)$$

Notice that the slit is considered to be infinitely long in the y-direction. The function $f(x,y)$ separates trivially into a product of functions of x and y only (the latter being the constant 1) and so from §4.4.2,

$$\psi(u,v) \quad = \quad \int_{-a/2}^{a/2} \exp(-iux)\,dx \int_{-\infty}^{\infty} \exp(-ivy)\,dy$$
$$= \quad a\,\text{sinc}(au/2)\,\delta(v). \qquad (8.19)$$

The intensity of the Fraunhofer diffraction pattern along the axis $v = 0$ is:

$$|\psi(u,0)|^2 = a^2 \,\text{sinc}^2(au/2). \qquad (8.20)$$

This function, shown in Fig. 8.4(a), is important; it has a maximum of a at $u = 0$ and is zero at regular intervals where $au/2 = m\pi$, where m is a non-zero integer. The heights of the resulting maxima are approximately proportional to $(2m+1)^{-2}$; this result arises if we assume that these maxima lie half-way between the zeros; their *exact* positions are not easy to find.

Fig. 8.4(b) shows the observed intensity, $|\psi(u,0)|^2$. The zeros in this function occur at angles given by

$$\tfrac{1}{2}ka\sin\theta = m\pi; \qquad a\sin\theta = m\lambda. \qquad (8.21)$$

Figure 8.3 The condition of minimum deviation.

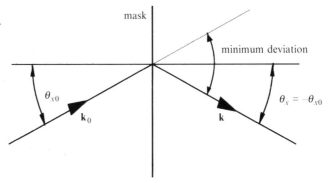

8.2.4 Diffraction by a blurred slit, represented by a triangular function

We now assume that the transmission function at the edges of the slit varies continuously and linearly with x so as to define a slit of the same width a as before, but with blurred edges. This will show that the effect of blurring the sharp edges is to reduce the prominence of the side-bands in the diffraction pattern. Consider

$$
\begin{aligned}
f(x, y) &= 1 - |x|/a \quad (|x| \leq a) \\
&= 0 \quad (|x| > a).
\end{aligned}
$$

This slit has *effective width* (defined as $\int f(x)\,dx \,/f_{max}$) equal to a, the same as that of the previous slit (§8.2.3). Then, integrating by parts, one finds

$$
\begin{aligned}
\psi(u, v) &= \left[\frac{1}{a} \int_0^a (a - x) \exp(-iux)\,dx \right. \\
&\qquad \left. + \frac{1}{a} \int_{-a}^0 (a + x) \exp(-iux)\,dx \right] \int_{-\infty}^{\infty} \exp(-ivy)\,dy \\
&= a \operatorname{sinc}^2(au/2)\,\delta(v).
\end{aligned}
\tag{8.22}
$$

The form of $\psi(u, 0)$ is the same as shown in Fig. 8.4(b). It is everywhere positive, reaching zero at values of u given by

$$
au/2 = m\pi; \quad a \sin \theta = m\lambda.
\tag{8.23}
$$

The positions of these zeros are thus exactly the same as for the uniform slit; since their effective widths are the same this result is not surprising. But the maxima of the side-bands produced are much less; their intensities are proportional to $(2m + 1)^{-4}$. Further smoothing of the function $f(x, y)$ at the edges of the slit results in even weaker side-bands. A very smooth function is the Gaussian discussed in §4.4.6 whose transform has no side-bands at all.

Figure 8.4 (*a*) Form of function $a \operatorname{sinc}(\frac{1}{2}au)$; (*b*) form of function $a^2 \operatorname{sinc}^2(\frac{1}{2}au)$.

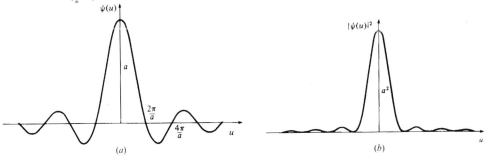

8.2.5 Diffraction by an object with phase variation only

There are many objects, particularly natural ones, that do not absorb light appreciably but change its phase on transmission. Any ordinary piece of window glass will do this; it is transparent, but its thickness is not uniform and light passing through different parts of it suffers a varying amount of phase retardation. If the refractive index of the glass is μ, the optical difference between two paths including different thicknesses t_1 and t_2 is

$$(\mu - 1)(t_1 - t_2) \tag{8.24}$$

and consequently the wavefront emerging from the glass sheet is no longer a plane as it was when incident (Fig. 8.5). Since waves of different phases but the same amplitudes are represented by complex amplitudes with the same modulus, such a state of affairs as this *phase object* can be represented by a complex transmission function $f(x, y)$ with constant modulus. We shall take as an example a thin prism of angle α and refractive index μ. The thickness of the prism at position x is αx (Fig. 8.6) and its transmission function is thus

$$\begin{aligned} f(x, y) &= \exp[ik_0(\mu - 1)t] \\ &= \exp[ik_0(\mu - 1)\alpha x]. \end{aligned} \tag{8.25}$$

The prism is assumed to be infinite in extent along both x and y directions. The diffraction pattern corresponding to $f(x)$ is then

$$\begin{aligned} \psi(u, v) &= \int_{-\infty}^{\infty} \exp[ik_0(\mu - 1)\alpha x] \exp(-iux)\,dx \int_{-\infty}^{\infty} \exp(-ivy)\,dy \\ &= \delta[u - (\mu - 1)k_0\alpha]\,\delta(y). \end{aligned} \tag{8.26}$$

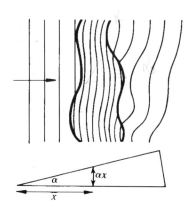

Figure 8.5 Distortion of plane-wavefront by non-uniform glass plate.

Figure 8.6 A thin prism of angle α.

The diffracted wave thus travels in the direction represented by

$$u = k_0(\mu - 1)\alpha, \quad v = 0 \tag{8.27}$$

Substituting for u this gives $\theta \approx (\mu - 1)\alpha$ for small θ. The light thus remains concentrated in a single direction, but is deviated from the incident by the same angle as deduced from geometrical optics.

8.2.6 Diffraction pattern of a rectangular hole

If we consider a rectangular hole of sides a and b, parallel to the x- and y-axes respectively, the two integrals have independent limits and (8.10) can be written as a product:

$$\psi(u,v) = \int_{-a/2}^{a/2} \exp(-iux)\,dx \int_{-b/2}^{b/2} \exp(-ivy)\,dy, \tag{8.28}$$

the function $f(x,y)$ being taken as unity over the area of the aperture. Since the origin is at the centre of the aperture the function is even and so has a real transform. Thus

$$\psi(u,v) = ab\,\mathrm{sinc}(\tfrac{1}{2}ua)\,\mathrm{sinc}(\tfrac{1}{2}vb), \tag{8.29}$$

each factor being similar to that derived for a uniform slit (8.19). The diffraction pattern has zeros at values of ua and vb equal to non-zero multiples of 2π. Thus the zeros lie on lines parallel to the edges of the slit, given by the equations

$$u = m_1\frac{2\pi}{a} \quad \text{and} \quad v = m_2\frac{2\pi}{b}. \tag{8.30}$$

Figure 8.7 (*a*) Lines of zero intensity in the diffraction pattern of the rectangular aperture shown on the left. (*b*) The observed diffraction pattern.

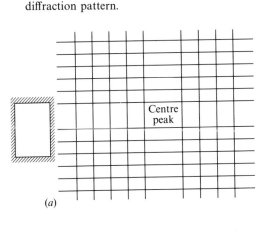

(*a*)

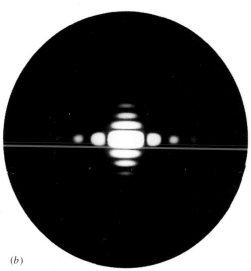

(*b*)

The centre peak, for example, is bounded by lines given by $m_1 = \pm 1$ and $m_2 = \pm 1$, which form a rectangle whose dimensions are inversely proportional to those of the diffracting aperture (Fig. 8.7). The peaks off the u- and v-axes are very weak, and are difficult to bring out in a photograph.

8.2.7 Diffraction pattern of a circular hole

For a circular hole of radius R the integral (8.10) is more difficult to evaluate since the limits are not now independent. It is best to use polar coordinates for points in the aperture and in the diffraction pattern. If (ρ, θ) are the polar coordinates in the aperture

$$x = \rho \cos \theta \quad \text{and} \quad y = \rho \sin \theta, \tag{8.31}$$

and if (ζ, ϕ) are the polar coordinates in the diffraction pattern,

$$u \equiv \zeta \cos \phi \quad \text{and} \quad v \equiv \zeta \sin \phi. \tag{8.32}$$

Thus equation (8.10) becomes

$$
\begin{aligned}
\psi(u, v) &= \int_0^R \int_0^{2\pi} \exp[-\mathrm{i}(\rho\zeta \cos \phi \cos \theta + \rho\zeta \sin \phi \sin \theta)]\rho \, \mathrm{d}\rho \, \mathrm{d}\theta \\
&= \int_0^R \int_0^{2\pi} \exp[-\mathrm{i}\rho\zeta \cos(\theta - \phi)]\rho \, \mathrm{d}\rho \, \mathrm{d}\theta.
\end{aligned}
\tag{8.33}
$$

This integral can be performed in terms of Bessel functions (Ap-

Figure 8.8 (*a*) Form of function $J_1(x)/x$, the radial amplitude distribution in the diffraction pattern of a circular aperture; (*b*) Fraunhofer diffraction of circular hole.

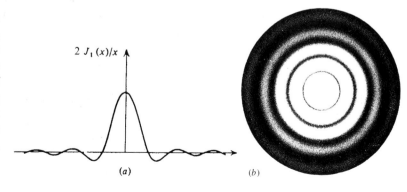

$2 J_1(x)/x$

(*a*)

(*b*)

pendix 1) and shown to be

$$\psi(\zeta, \phi) = \frac{2\pi R J_1(\zeta R)}{\zeta} = \pi R^2 \left[\frac{2J_1(\zeta R)}{\zeta R} \right]. \tag{8.34}$$

The pattern has circular symmetry; $\psi(\zeta, \phi)$, as one would expect, has no dependence on ϕ.

The form of the function $2J_1(x)/x$ is interesting. $J_1(x)$ is zero at $x = 0$ but, like sinc x, the function has a finite value of unity there. It then decreases to zero, becomes negative and continues to oscillate with a gradually decreasing period that tends to a constant as in Fig. 8.8(a).

The diffraction pattern is shown in Fig. 8.8(b). The central peak is known as the *Airy disc*, and it extends to the first zero, which occurs at $x = 3.83$, or at angle $\zeta/k_0 = 0.61\lambda/R$. As one would expect from the properties of Fourier transforms, the radius of the Airy disc is inversely proportional to the radius of the hole.

It should also be noted from equations (8.29) and (8.34) that the amplitude, not the intensity, at the centre of the diffraction pattern is proportional to the area of the hole. This result makes sense when we realize that the linear dimensions of the diffraction pattern are inversely proportional to those of the hole, and thus that the total energy flow in the diffraction pattern is proportional to the area of the hole. As the hole varies in size, the intensity at the centre varies proportionately to the square of the area.

8.2.8 A crude derivation of the size of the Airy disc

The derivation of the form of the diffraction pattern in terms of a Bessel function does not really throw any light upon the physics of the problem. If it had not been that Bessel functions appear in other physical problems, the function $J_1(x)$ would not have been worth tabulating and we should be no nearer an acceptable solution when the equation (8.34) had been derived. It is, however, possible to see a rough solution in terms of the concepts discussed in §§8.2.3–8.2.4.

Suppose we have a function $f(x, y)$ whose diffraction pattern is $\psi(u, v)$. Along the axis $v = 0$, we have in general

$$\begin{aligned} \psi(u, 0) &= \iint f(x, y) \exp(-iux) \, dx \, dy \\ &= \int_{-\infty}^{\infty} \left[\int_{-\infty}^{\infty} f(x, y) \, dy \right] \exp(-iux) \, dx. \end{aligned} \tag{8.35}$$

This means that the axial value $\psi(u, 0)$ is the Fourier transform of the integral $\int f(x, y)\,\mathrm{d}y$. For the circular aperture, this integral is a semicircular function (Fig. 8.9). We saw in §8.2.3 that the zeros of the triangular function corresponded to those of the slit if their effective widths were equal. Using the same idea, we construct the slit function, of width b, which has the same area and height as the semicircle. This requires $b = \pi R/2$. Its diffraction pattern along the u-axis is

$$\psi(u, 0) = b\,\mathrm{sinc}(\pi u R/4) \qquad (8.36)$$

which has its first zero when $uR = 4$. This agrees quite well with the exact value, $uR = 3.83$, from the Bessel function. Although this approach is not exact, it gives some impression of the way a crude approximation can sometimes give a reasonably good value for a physical quantity.

8.2.9 Addition of diffraction patterns

The additive property of transforms sometimes enables the diffraction patterns of relatively complicated objects to be derived if their shapes can be expressed as an algebraic sum of simpler ones. The separate components of the object must be expressed with respect to the same origin, and the complete complex transform is then obtained by summing the real parts and the imaginary parts of the component transforms separately. The process is particularly simple if the separate components are centro-symmetric about the same point; then their transforms are all real. For example, it is possible to derive the diffraction pattern of three slits by adding the transform of the two outer ones to that of the inner one; the diffraction pattern of four slits can be obtained by regarding them as two pairs, one with three times the spacing of the other. Some examples of this sort are included in the Problems.

An opaque obstacle can be regarded as giving a negative transform. For example, a thick rectangular frame may be regarded as the difference between the outer rectangle and the inner one. (Note, in working

Figure 8.9 Rectangle and circle of equal area.

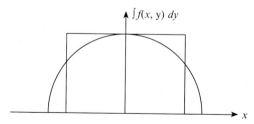

out such an example, that the height of the central peak of the transform of a rectangle is proportional to its area, as shown in §8.2.6.) The diffraction pattern of an annular ring is given by the difference between the diffraction patterns of the outer and inner circles; the result has practical implications that will be discussed later in §12.5.

8.2.10 Complementary screens; Babinet's theorem

An important theorem in optics is concerned with the interference patterns of two complementary screens. Two screens are said to be *complementary* if they each consist of openings in opaque material, the openings in one corresponding exactly to the opaque parts of the other. Babinet's theorem says that the interference patterns of two such screens are exactly the same except for a small region near the centre. For example, the pattern of a set of opaque discs should be the same as that of a set of equally-sized holes similarly arranged. The theorem is illustrated by the masks and diffraction patterns shown in Fig. 8.10.

The theorem can be proved on general grounds using the scalar theory of diffraction. Suppose that the amplitudes of the diffraction patterns of two complementary screens when illuminated by a certain beam are ψ_1 and ψ_2. Now, the diffraction function for a combination of apertures can be obtained by adding the separate (complex) functions. If we add ψ_1 and ψ_2 we should obtain the diffraction function for the unobstructed beam. If this is large and circular, the sum of ψ_1 and ψ_2 is thus the Airy-disc pattern, which is confined to a small region round the centre; the rest is blank. Therefore the sum of ψ_1 and ψ_2

Figure 8.10 Diffraction patterns of two complementary masks (shown as insets) when illuminated by a Gaussian beam. The positive mask, (*a*), was cut from metal foil and the negative, (*b*), produced by evaporating metal through it onto a flat glass plate.

(*a*)

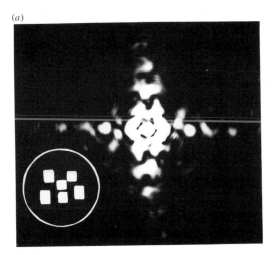

(*b*)

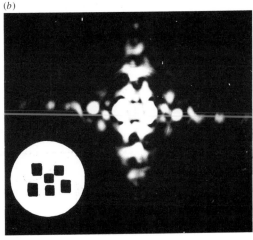

must be zero everywhere except for a small central region. The moduli of ψ_1 and ψ_2 in the same region must therefore be equal, their phases differing by π. The intensity functions are the same.

Experimental confirmation of Babinet's theorem is not easy, mainly because of the strength of the central peak. If the unobstructed beam is very large, the peak is extremely strong and usually dominates the diffraction pattern of the mask which is mainly transparent. To get a convincing experimental confirmation of the theorem, the following rules should be followed.

(1) The edges of the unobstructed beam should be blurred. This suppresses the outer parts of its transform. In fact, using a Gaussian beam gives the most concentrated central peak with the weakest wings.
(2) The positive and negative masks should each be about 50% transmitting, to give the strongest diffraction patterns possible.
(3) The masks should contain fine detail so as to give rise to a strong diffraction pattern well outside the central peak.

8.3 Interference

We have so far considered only the effect of modifying a single wavefront; we shall now consider the effects occurring when two or more wavefronts interact. These effects are called *interference*. In this chapter, we shall concern ourselves mainly with wavefronts from identical objects.

For identical apertures we can make use of the principle of convolution (§4.6). For example, two similar parallel apertures can be considered as the *convolution* of one aperture with a pair of δ-functions, one at the origin of each. The interference pattern is therefore the product of the diffraction pattern of one aperture and that of the pair of δ-functions (§4.4.5). We can therefore divide such an interference problem into two parts – the derivation of the Fourier transform of the single aperture and that of the set of δ-functions. The transform of the single aperture is called the *diffraction function* and that of the set of δ-functions is called the *interference function*; the complete diffraction pattern is the product of the two. This is shown for two circular holes in Fig. 8.11.

8.3.1 Interference pattern of two circular holes

We can regard a pair of circular holes, with separation a, as the result of convoluting a single hole with a pair of δ-functions. Now from

§4.4.5 the transform of the two δ-functions is given by

$$\psi(u,v) = 2\cos(ua/2).$$ (8.37)

Thus the diffraction pattern of the two holes is the diffraction pattern of one of them multiplied by a cosinusoidal function, varying in a direction parallel to the separation a, Fig. 8.12.

The zeros of the function (8.37) occur at values of θ given by

$$ua/2 = (m + \tfrac{1}{2})\pi,$$ (8.38)

where m is an integer. Since $u = k\sin\theta = 2\pi\sin\theta/\lambda$, this simplifies to

$$a\sin\theta = (m + \tfrac{1}{2})\lambda.$$ (8.39)

It will be realized that what we have achieved is a rather round-about method of deriving an expression for Young's fringes. There are, however, several reasons for using this approach: first, we have derived the full expression for the profile of the fringes, not just the spacing; secondly, we have demonstrated that the convolution method gives the correct result for a simple example; and, thirdly, we have

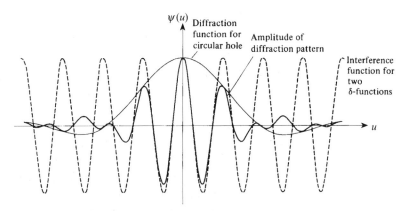

Figure 8.11 Product of diffraction function and interference function gives amplitude of complete diffraction pattern of two circular holes.

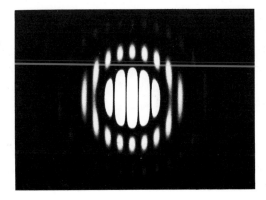

Figure 8.12 Fraunhofer diffraction pattern of two circular holes.

prepared the ground for more complicated systems, such as that which follows.

8.3.2 Interference pattern of two parallel apertures of arbitrary shape

We can regard a pair of similar parallel apertures (Fig. 8.13) as the convolution of a single aperture with two δ-functions. The diffraction pattern is therefore the product of the diffraction pattern of a single aperture and the interference function, which is a set of sinusoidal fringes. This is illustrated in Fig. 8.14, and is one of the most important results in diffraction theory. The argument is obviously applicable to a pair of apertures of any shape.

8.3.3 Interference pattern of a regular array of identical apertures

An array of apertures can be regarded as the convolution of a set of δ-functions with one aperture. From §4.4.5 we find that, if the δ-functions form a regular one-dimensional lattice with spacing d, the transform is

$$\psi(u, v) = \sum_{n=0}^{N-1} \exp(-iund), \tag{8.40}$$

Figure 8.13 Pair of parallel apertures.

Figure 8.14 (*a*) Diffraction pattern of one of the apertures in Fig. 8.13. (*b*) Complete diffraction pattern of the mask in Fig. 8.13.

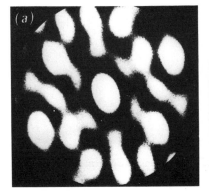 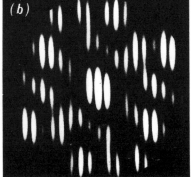

where N is the number of pinholes. When $N \to \infty$, the sum is (§4.4.5):

$$\psi(u, v) = \sum_{m=-\infty}^{\infty} \delta(u - 2\pi m/d). \qquad (8.41)$$

The index m is called the *order of diffraction*. When N is finite, the sum of the geometrical series (8.40) is

$$\psi(u, v) = \frac{1 - \exp(-iuNd)}{1 - \exp(-iud)}. \qquad (8.42)$$

The intensity is given by

$$I(u, v) = |\psi(u, v)|^2 = \frac{\sin^2(uNd/2)}{\sin^2(ud/2)}. \qquad (8.43)$$

This expression, which is plotted in Fig. 8.15 for $N = 6$, has some interesting properties. It is zero whenever the numerator is zero except when the denominator is also zero; then it is N^2. As the number of apertures increases, the number of zeros increases and the pattern becomes more detailed. The peaks of intensity N^2 – called the *principal maxima* – become outstanding compared to the smaller subsidiary maxima, of which there are $N - 2$ between the principal maxima. In fact, these principal maxima approximate to the δ-functions of (8.41), namely $\delta(u - 2m\pi/d)$.

The conditions for the production of principal maxima are that $ud/2 = m\pi$. Since $u = 2\pi \sin \theta/\lambda$, we have

$$d \sin \theta = m\lambda, \qquad (8.44)$$

the well-known equation for the diffraction grating.

Figure 8.15
Form of function
$\sin^2(uNd/2)/$
$\sin^2(ud/2)$ for $N = 6$.

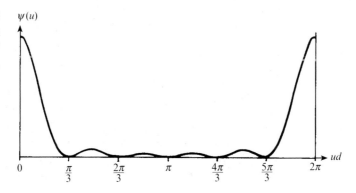

8.3.4 Diffraction gratings

The diffraction grating is a one-dimensional periodic array of similar opertures. If the grating is used in transmission, the apertures are narrow slits; if in reflexion they are narrow mirrors. Because they are important interferometric instruments, gratings will be discussed in depth in §9.2; here we shall only briefly outline the Fraunhofer diffraction theory of the basic grating, because it provides a useful basis for understanding other systems, such as the hologram and image formation. If each slit has transmission profile $b(x)$, and the line spacing is d, the transmission function is

$$f(x) = b(x) \otimes \sum_{n=-N/2}^{N/2} \delta(x - nd) \qquad (8.45)$$

where a total number N slits has been assumed. When $N \to \infty$, the transform of the $\sum$ is given by (8.41) and

$$\psi(u) = B(u) \sum_{m=-\infty}^{\infty} \delta(u - 2\pi m/d). \qquad (8.46)$$

Now we should recall the general definition of u (8.15); for light incident at angle θ_0 to the axis, and diffracted to angle θ, we have

$$u = \frac{2\pi}{\lambda}(\sin \theta - \sin \theta_0). \qquad (8.47)$$

Since $\sin \theta$ and $\sin \theta_0$ lie between -1 and 1, the maximum *observable* value of u is $4\pi/\lambda$. So that, although (8.46) is defined for all m, the diffraction condition is in general

$$m\lambda = d(\sin \theta - \sin \theta_0) \qquad (8.48)$$

where for a given θ_0, $(-1 - \sin \theta_0)d/\lambda \le m \le (1 - \sin \theta_0)d/\lambda$.

The amplitudes of the various orders of diffraction are given by the transform of the individual aperture, $B(u)$. A common example is a square-wave grating (Ronchi ruling) where $b(x) = \text{rect}(2x/d)$. Without repeating the details, one immediately sees from Fig. 4.3 that the even orders of diffraction are missing and the odd orders have steadily decreasing intensity. Although Fig. 4.3 does not show the zero order, it must be added here because $b(x)$ is positive definite and $\psi(0)$ is its integral (see §8.2.8), which is non-zero. The existence of a strong zero order for a positive-definite function has important consequences which will be discussed in §§9.2.4 and 12.2.5.

A finite grating (all *real* gratings are, of course, finite) is given by summing (8.45) to finite N. This is conveniently expressed by multiplying the infinite sum by a 'window function' of length Nd which 'transmits' only N δ-functions. Then the grating is represented by

$$f(x) = b(x) \otimes \left[\sum_{n=-\infty}^{\infty} \delta(x - nd) \cdot \text{rect}(x/Nd) \right]. \tag{8.49}$$

Notice the order of the operations; it is important to carry out the product first, and the convolution afterwards, in order to represent a finite number of complete slits. The reverse order might give incomplete slits at the end. The difference here is minor, but it is not difficult to construct examples for which the order of convolution and product is very important. Convolution and multiplication are *not* commutative.

The diffraction pattern of (8.49)

$$\psi(u) = B(u) \cdot \left[\sum \delta \left(u - \frac{2\pi m}{d} \right) \otimes \text{sinc} \left(\frac{uNd}{2} \right) \right] \tag{8.50}$$

has the following characteristics. There are well-defined orders of diffraction (for large N) as defined in (8.48) but each one has a $\text{sinc}(uNd/2)$ profile. This has width (to the first zero) $\Delta u = 2\pi Nd$, which is $(1/N)$ of the distance between the orders.

8.3.5 Interference pattern of a lattice of pinholes

We can now extend our results to an array of pinholes, periodic in x and y, which we may call a two-dimensional lattice. We can approach this through a set of four pinholes, at positions $\pm(x_1, y_1), \pm(x_2, y_2)$, see Fig. 8.16($a$). We have to evaluate the expression

$$\begin{aligned} \psi(u, v) &= \sum \exp[-\text{i}(ux + vy)] \\ &= 2[\cos(ux_1 + vy_1) + \cos(ux_2 + vy_2)] \\ &= 4\cos \left(u\frac{x_1 + x_2}{2} + v\frac{y_1 + y_2}{2} \right) \cos \left(u\frac{x_1 - x_2}{2} + v\frac{y_1 - y_2}{2} \right). \end{aligned} \tag{8.51}$$

As in §8.3.1, we see that this function has maxima at values of u and v given by the equations

$$u(x_1 + x_2) + v(y_1 + y_2) = 2m_1\pi,$$
$$u(x_1 - x_2) + v(y_1 - y_2) = 2m_2\pi, \tag{8.52}$$

where m_1 and m_2 are integers. The interference pattern is therefore the product of two sets of linear fringes, each set being perpendicular to the separation of the pairs of holes (Fig. 8.12). Such fringes are called *crossed fringes* and are shown in Fig. 8.16(*b*).

By reasoning analogous to that of §8.3.3, and which will be formalized in §8.4, we can see that as the lattice of pinholes, with these four points providing the unit cell, increases in extent, the conditions for constructive interference become more precisely defined, and in the limit the interference pattern becomes a collection of points, also arranged on a lattice (Fig. 8.16(*c*)). This is called the *reciprocal lattice* of the original lattice, because *u* and *v* are reciprocally related to the separations of the pairs of holes in Fig. 8.16(*a*).

8.3.6 Reciprocal lattice in two dimensions

The concept of the reciprocal lattice in two dimensions is derived formally as follows (this will serve as a basis for the equivalent three-dimensional derivation in §8.4). Suppose that the positions of the pinholes in an infinite periodic lattice can be defined in terms of two given lattice vectors **a** and **b** as

$$f(x, y) = \sum_{h,k=-\infty}^{\infty} \delta(\mathbf{r} - h\mathbf{a} - k\mathbf{b}). \qquad (8.53)$$

This puts a δ-function at every point of a periodic lattice whose unit cell is the parallelogram whose sides are **a** and **b**.†

The Fourier transform of (8.53) is, writing **u** for the vector (u, v):

$$\psi(u, v) = \sum_{h,k=-\infty}^{\infty} \exp[-\mathrm{i}\mathbf{u} \cdot (h\mathbf{a} + k\mathbf{b})]. \qquad (8.54)$$

This expression can be simplified if we define two new vectors $\mathbf{a}^\star$ and

---

† For a given lattice, there are many different ways of choosing **a** and **b**, but there are usually one or two which are obviously simplest.

Figure 8.16 (*a*) Two pairs of pinholes; (*b*) diffraction pattern of (*a*), showing crossed fringes; (*c*) reciprocal lattice – the diffraction pattern of an extended lattice of pinholes based on (*a*) as unit cell.

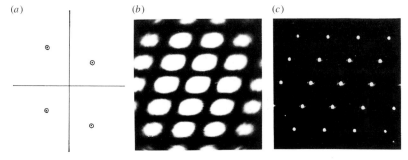

$\mathbf{b}^\star$ in the plane of $\mathbf{u}$, such that

$$\mathbf{a}\cdot\mathbf{a}^\star = 1, \quad \mathbf{b}\cdot\mathbf{b}^\star = 1 \tag{8.55}$$

$$\mathbf{a}^\star\cdot\mathbf{b} = \mathbf{b}^\star\cdot\mathbf{a} = 0 \quad (\text{i.e. } \mathbf{a}^\star \perp \mathbf{b} \text{ and } \mathbf{b}^\star \perp \mathbf{a}). \tag{8.56}$$

The vectors $\mathbf{a}^\star$ and $\mathbf{b}^\star$ are not parallel, and so $\mathbf{u}$ can be expressed as a linear combination of them:

$$\mathbf{u} = (h^\star\mathbf{a}^\star + k^\star\mathbf{b}^\star)2\pi \tag{8.57}$$

where $h^\star$ and $k^\star$ are, for the present, any numbers. Then (8.54) becomes

$$\begin{aligned} \psi(u,v) &= \sum_{h,k=-\infty}^{\infty} \exp[-2\pi\mathrm{i}(hh^\star\mathbf{a}^\star\cdot\mathbf{a} + kk^\star\mathbf{b}^\star\cdot\mathbf{b})] \\ &= \sum \exp[-2\pi\mathrm{i}(hh^\star + kk^\star)]. \end{aligned} \tag{8.58}$$

This sum, for general $h^\star$, $k^\star$ is usually small, being an infinite sum of complex numbers having unit modulus, which mainly cancel. However, if $h^\star$ and $k^\star$ are integers, every term is unity, and $\psi(u,v)$ is infinite. Thus $\psi(u,v)$ is an *array of δ-functions* on the lattice defined by the lattice vectors $\mathbf{a}^\star$, $\mathbf{b}^\star$. This array is the reciprocal lattice.

The vectors $\mathbf{a}^\star$, $\mathbf{b}^\star$ are easy to identify (Fig. 8.17). If the angle between $\mathbf{a}$ and $\mathbf{b}$ is γ, then the relationships $\mathbf{a}^\star\cdot\mathbf{a} = 1$ and $\mathbf{a}^\star\cdot\mathbf{b} = 0$ define $\mathbf{a}^\star$ as the vector normal to $\mathbf{b}$ having length $(a\sin\gamma)^{-1}$. Similarly $\mathbf{b}^\star$ is normal to $\mathbf{a}$ and has length $(b\sin\gamma)^{-1}$. The vectors $\mathbf{a}^\star$ and $\mathbf{b}^\star$ are called *reciprocal lattice vectors*. The diffraction pattern of a large two-dimensional periodic array of δ-functions is shown in Fig. 8.16(c). The name 'reciprocal lattice' arises because its dimensions are reciprocally

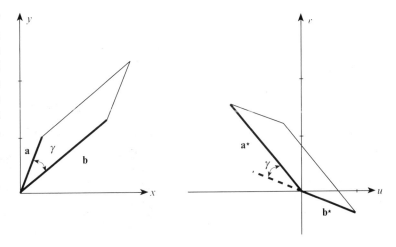

Figure 8.17 Relationship between real-space lattice vectors and reciprocal lattice vectors, showing the two-dimensional unit cells in each space.

related to those of the direct lattice; if we reduce **a** and **b** by a given factor, the reciprocal lattice expands by the same factor.

This concept of the reciprocal lattice becomes more important in connexion with three-dimensional interference which we shall discuss in §8.4.

8.3.7 Interference pattern of a lattice of parallel apertures

If we have an extended lattice of similar apertures, Fig. 8.18(*a*), we may consider it as the convolution of a single aperture with the lattice having translations **a** and **b**. Then the diffraction pattern (Fig. 8.18(*b*)) is the product of the diffraction pattern of the lattice and that of the single aperture. In other words, the reciprocal lattice pattern is multiplied by the diffraction pattern of the unit. When the unit is a square hole, for example, then the influence of the diffraction pattern is easily seen; if the pattern is more complicated, as in Fig. 8.19(*a*) which shows a set of holes representing a lattice of molecules, the result is less clear (Fig. 8.19(*b*)) but the correspondence is quite definite. We may look upon the diffraction pattern in another way. A single unit of Fig. 8.19(*a*) gives a particular diffraction pattern (Fig. 8.19(*c*)); the effect of putting the units on a lattice is, apart from making the pattern stronger, to make the diffraction pattern observable only at the reciprocal lattice points. This process is called *sampling*; it is important in dealing with diffraction by crystals, and has many applications in image processing and communication theory.

If we regard the set of apertures as a two-dimensional diffraction grating, the reciprocal lattice represents its set of orders. Each reciprocal lattice point is an order of diffraction (§8.3.3), specified now by

Figure 8.18
(*a*) Lattice of parallel apertures.
(*b*) Diffraction pattern of (*a*)

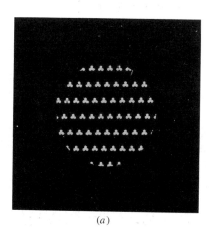

(*a*)

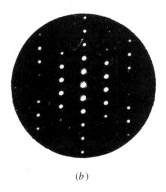

(*b*)

two integers, $h^\star$ and $k^\star$, instead of one. In three dimensions (§8.4) we shall see that three integers are needed.

8.3.8 Diffraction by a random of array parallel apertures

Suppose that the diffracting object consists of a collection of parallel apertures arranged randomly. We can regard the collection as the convolution of the single aperture with a set of δ-functions representing the aperture positions.

We therefore need to determine the diffraction pattern of a set of N randomly-arranged δ-functions. This problem is expressed mathematically as

$$\psi(u,v) = \sum_{n=1}^{N} \delta(x - x_n)\,\delta(y - y_n)\exp[-i(ux + vy)]$$

$$= \sum \exp[-i(ux_n + vy_n)], \tag{8.59}$$

where the nth aperture has random origin at (x_n, y_n) (Fig. 8.20). This

Figure 8.19 (*a*) Set of holes representing a lattice of chemical molecules. (*b*) Diffraction pattern of (*a*). (*c*) Diffraction pattern of a unit of (*a*).

(*a*)

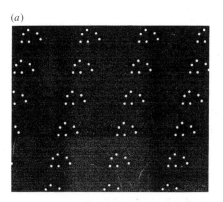

(*b*)

(*c*)

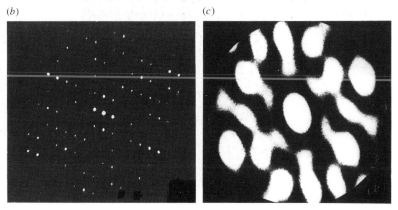

sum cannot be evaluated in general. But the intensity of the transform

$$I(u,v) = |\psi(u,v)|^2 \tag{8.60}$$

can be evaluated by writing the square of the sum (8.59) as a double sum:

$$
\begin{aligned}
|\psi(u,v)|^2 &= |\sum_{n=1}^{N} \exp[-i(ux_n + vy_n)]|^2 \\
&= \sum_{n=1}^{N}\sum_{m=1}^{N} \exp\{-i[u(x_n - x_m) + v(y_n - y_m)]\}. \tag{8.61}
\end{aligned}
$$

Now since x_n and x_m are random variables, $(x_n - x_m)$ is also random and so the various terms usually make randomly positive or negative contributions to the sum. There are two exceptions to this statement. Firstly, the terms with $n = m$ in the double sum all contribute a value $e^{i0} = 1$, and there are N of them, so that the expected value of the double sum (8.61) is N. Secondly, when $u = v = 0$, *all* the terms in the sum contribute 1, and the value of (8.61) is N^2, so that we can write the statistical expectation:

$$I(u,v) = N + N^2\bar{\delta}(u,v), \tag{8.62}$$

where $\bar{\delta}(u,v)$ has the value of unity at $(u,v) = (0,0)$ and is zero elsewhere.† The function (8.62) represents a bright spot of intensity N^2 at the origin and a uniform background of intensity N.

Of course a truly random distribution does not exist in practice and the above description must really be modified. If the N points are all within a finite region (say a square of side D) the terms in the double sum (8.61) will all have positive values even if u and v deviate from zero by as much as $\pi/2D$. So the spot at the origin has

† Kronecker delta.

Figure 8.20 Random set of similar apertures, showing the origin (x_n, y_n) of an individual aperture.

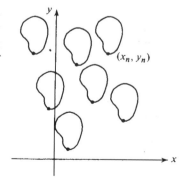

a finite size, of this order of magnitude. In addition, the randomness of the distribution might be restricted to avoid the overlapping of neighbouring apertures. This can be shown to result in a weak structure appearing in the background term.

Returning to the diffraction pattern of the random array of apertures, we now recall that the object was expressed as the convolution of a single aperture with the random array of δ-functions. Its diffraction pattern is then the product of the diffraction pattern of a single object and the function (8.62). At all points except the origin and its immediate vicinity the result is an intensity just N times the intensity of the single aperture's diffraction pattern. Only at the origin itself there appears a bright spot, with intensity N^2 times that of the zero order of the single aperture diffraction pattern. The result is illustrated by Fig. 8.21. If the number of apertures becomes *very* large, the bright spot is the only observable feature. A practical application of this analysis to astronomical imaging will be discussed in §12.8.

8.4 Three-dimensional interference

Fraunhofer diffraction by three-dimensional obstacles has several applications, the most important being to crystal diffraction. It is not just a straightforward extension from one and two dimensions to three, because the theory developed so far has essentially described the diffraction pattern as the solution of a boundary-value problem,

Figure 8.21 (*a*) Mask of random parallel apertures; (*b*) diffraction pattern of one unit of (*a*); (*c*) complete diffraction pattern of (*a*); the centre inset is an under-exposed part of the diffraction pattern showing the strong spot at the centre.

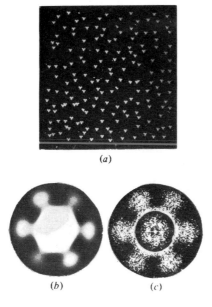

(*a*)

(*b*)   (*c*)

in which the incident wave on the surface (mask) was allowed to develop according to Huygens's principle, as expressed in the Kirchhoff–Huygens theory. If the 'mask' is three-dimensional, it may be that the boundary conditions are over-defined, and a diffraction pattern does not exist. In fact we shall see that the Fourier transform alone does not describe the diffraction pattern, but another condition, described by the construction of an 'observation sphere' has also to be satisfied, and tells us *which parts* of the transform contribute to the pattern.

8.4.1 Crystals and convolutions

Crystals are three-dimensional gratings and diffract waves of suitable wavelengths: neutrons, electrons, atoms and X-rays. The general principles of diffraction by any of these waves are the same; just the relevant parameters must be used. The discussion here will centre around X-rays. The theory was originally developed by M. von Laue for the *weak scattering* case, which means that the probability of a wave being scattered *twice* within the crystal is negligible.

A crystal is a collection of atoms. For simplicity let us consider a crystal composed of identical atoms only. From the point of view of X-ray diffraction, since X-rays are scattered only by electrons, a crystal can be considered as a set of atomic positions (δ-functions) convoluted with the electron density function for one atom. The atomic positions repeat on a *lattice*;† i.e. a small group of atoms, called the *unit cell*, is repeated regularly in three dimensions. Therefore we can regard the crystal as composed of the unit cell convoluted with the lattice positions. These ideas are illustrated in two dimensions in Fig. 8.22. This would lead to an infinite crystal. We therefore limit its extent by multiplying the convolution by the shape function, the external boundary.

From the convolution theorem, therefore, we see that the transform of the electron density can be expressed as the transform of the shape function convoluted with a product of three transforms: that of the atom, that of the set of δ-functions representing the atomic positions in the unit cell, and that of the crystal lattice.

This is a *complete* outline of the theory of X-ray diffraction. All that remains is to fill in the details. Unfortunately, this would require several textbooks since each aspect is complicated. In the following section we shall discuss diffraction by the crystal lattice, and in §8.6.2

---

† It should be noted that the term *lattice* is not synonymous with the *structure*; it is merely a name for the framework upon which the structure is built.

we shall touch on the problems involved in determining the atomic positions within the unit cell.

8.4.2 Diffraction by a three-dimensional lattice

We are concerned with the diffraction pattern produced by a three-dimensional lattice of δ-functions. Suppose we have an incident wave with wave-vector $\mathbf{k}_0$, and that it is diffracted to a direction with vector $\mathbf{k}$. In order to conserve energy, the incident and diffracted waves must have the same frequency,

$$\omega_0 = ck, \tag{8.63}$$

and therefore the moduli of $\mathbf{k}$ and $\mathbf{k}_0$ must be equal;

$$|\mathbf{k}| = |\mathbf{k}_0|. \tag{8.64}$$

Alternatively, we can say that the waves must have the same time-variation, $\exp(-i\omega_0 t)$, since this must pass unchanged through the calculation of diffraction by a *stationary* lattice. (Diffraction by a moving lattice is different and is dealt with in §8.5.) The condition (8.64) can be represented geometrically by saying that $\mathbf{k}_0$ and $\mathbf{k}$ must be radius vectors of the same sphere, which is called the *Ewald sphere, reflecting sphere* or *sphere of observation* (Fig. 8.23). An order of diffraction satisfying this condition is called a Bragg reflexion, after

Figure 8.22
Two-dimensional
representation of a
crystal structure.

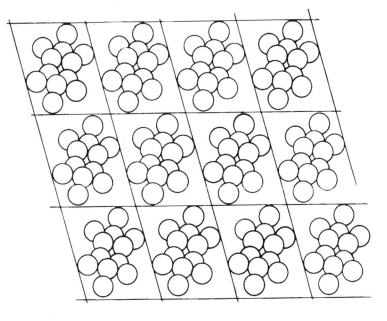

W. L. Bragg who, in 1912, introduced the idea of reflexion of X-rays by lattice planes.

Let us proceed with the calculation of the amplitude of the wave diffracted in the direction of $\mathbf{k}$. The δ-function of the lattice point $\mathbf{r}'$ acts as a secondary source of strength proportional to the value of the incident wave at the point,

$$\exp[i(\mathbf{k}_0 \cdot \mathbf{r}')]. \tag{8.65}$$

We shall assume the constant of proportionality to be unity. This source scatters a wave in the direction $\mathbf{k}$ which can be written as

$$\psi(\mathbf{k}) = \exp\{i[(\mathbf{k} \cdot \mathbf{r}) + \phi]\}, \tag{8.66}$$

where ϕ is, as yet, an arbitrary phase. But this wave originates from $\mathbf{r}'$ as (8.65), so that

$$\exp[i(\mathbf{k} \cdot \mathbf{r}' + \phi)] = \exp[i(\mathbf{k}_0 \cdot \mathbf{r}')]. \tag{8.67}$$

Thus

$$\phi = (\mathbf{k}_0 - \mathbf{k}) \cdot \mathbf{r}' \tag{8.68}$$

and

$$\psi(\mathbf{k}) = \exp\{i[(\mathbf{k} \cdot \mathbf{r}) + (\mathbf{k}_0 - \mathbf{k}) \cdot \mathbf{r}']\}. \tag{8.69}$$

The total diffracted beam with wave-vector $\mathbf{k}$ is therefore given by summing (8.69) over all positions $\mathbf{r}'$ of the lattice of δ-functions with unit-cell vectors $\mathbf{a}$, $\mathbf{b}$, $\mathbf{c}$. Following §8.3.6,

$$f(\mathbf{r}') = \sum_{h,k,\ell=-\infty}^{\infty} \delta(\mathbf{r} - h\mathbf{a} - k\mathbf{b} - \ell\mathbf{c}), \quad (h, k, \ell \text{ integers}) \tag{8.70}$$

which reduces to the summation

$$\Psi(\mathbf{k}) = \exp(i\mathbf{k} \cdot \mathbf{r}) \sum_{h,k,\ell=-\infty}^{\infty} \exp\{i[(\mathbf{k} - \mathbf{k}_0) \cdot (h\mathbf{a} + k\mathbf{b} + \ell\mathbf{c})]\}. \tag{8.71}$$

Figure 8.23 Sphere of observation.

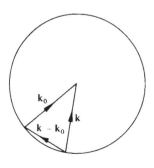

In the same way as we saw in §8.3.6, the summation is clearly zero unless the phases of all the terms are multiples of 2π:

$$(\mathbf{k} - \mathbf{k}_0) \cdot (h\mathbf{a} + k\mathbf{b} + l\mathbf{c}) = 2\pi s \quad (s \text{ is an integer}). \qquad (8.72)$$

One trivial solution to this equation is

$$\mathbf{k} - \mathbf{k}_0 = 0 \quad (s = 0), \qquad (8.73)$$

which also satisfies (8.64). But there is also a host of other solutions.

8.4.3 Reciprocal lattice in three dimensions

These other solutions to (8.72) can be derived by means of a reciprocal lattice (§8.3.5). The vectors $\mathbf{k} - \mathbf{k}_0$ between points in the reciprocal lattice are also solutions of (8.72). In three dimensions, we define reciprocal lattice vectors $\mathbf{a}^*$, $\mathbf{b}^*$ and $\mathbf{c}^*$ in terms of the real lattice vectors by means of the equations

$$\left. \begin{array}{l} \mathbf{a}^* = V^{-1} \mathbf{b} \times \mathbf{c}, \\ \mathbf{b}^* = V^{-1} \mathbf{c} \times \mathbf{a}, \\ \mathbf{c}^* = V^{-1} \mathbf{a} \times \mathbf{b}, \end{array} \right\} \qquad (8.74)$$

where V is the volume of the unit cell in real space:

$$V = \mathbf{a} \cdot \mathbf{b} \times \mathbf{c}. \qquad (8.75)$$

It now follows, as in §8.3.6, that if $(\mathbf{k} - \mathbf{k}_0)/2\pi$ can be written as the sum of integral multiples of $\mathbf{a}^*$, $\mathbf{b}^*$ and $\mathbf{c}^*$,

$$(\mathbf{k} - \mathbf{k}_0)/2\pi = h^*\mathbf{a}^* + k^*\mathbf{b}^* + l^*\mathbf{c}^*, \quad (h^*, k^*, l^* \text{ are integers}), \quad (8.76)$$

Figure 8.24
Reciprocal lattice
with unit-cell vectors.

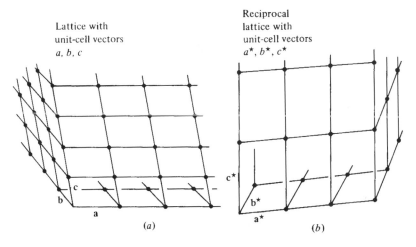

Lattice with
unit-cell vectors
a, b, c

Reciprocal
lattice with
unit-cell vectors
a^*, b^*, c^*

(a)

(b)

the summation (8.71) diverges; otherwise it is zero. This defines the three-dimensional reciprocal lattice of δ-functions at the points (8.76) (Fig. 8.24).

The observed diffraction pattern consists of those beams which satisfy both (8.64) and (8.76). The two conditions are represented geometrically by the observation sphere and the reciprocal lattice respectively. One therefore draws the observation sphere and the reciprocal lattice superimposed and looks for intersections (Fig. 8.25). The observation sphere passes through the origin of reciprocal space (because $\mathbf{k} - \mathbf{k}_0 = 0$ is a point on it) and its centre is defined by the direction of the vector $\mathbf{k}_0$. Mathematically, the exact intersection of a sphere and a set of discrete points is negligibly probable; but because neither an exactly parallel beam nor a purely monochromatic source of X-rays exists, diffraction by a crystal does in fact occur. (One important point is the trivial solution (8.73) which ensures that at least one 'diffracted' beam – the undeflected one – exists to carry away the incident energy.) By controlling the direction of the incident beam $\mathbf{k}_0$ and moving the crystal and recording screen in appropriate ways, it is possible to produce a section of the reciprocal lattice with, say, one of the indices h^*, k^*, ℓ^* constant. Such a photograph is shown in Fig. 8.26.

8.4.4 Diffraction by a complete crystal

We can see from Fig. 8.26 that the intensities vary in an irregular way; some orders of diffraction are strong and some weak. This variation

Figure 8.25 Two-dimensional representation of intersection of sphere of observation with reciprocal lattice, showing directions of incident beam $\mathbf{k}_0$ and of three possible diffracted beams $\mathbf{k}_1$, $\mathbf{k}_2$ and $\mathbf{k}_3$.

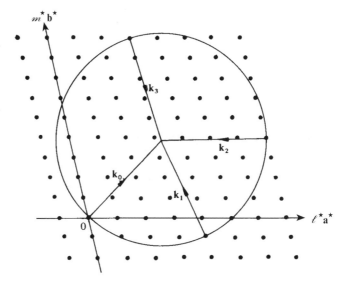

arises as it did in Fig. 8.18 as a result of multiplying the reciprocal lattice by the transform of the atomic positions within the unit cell, and a major part of crystallography consists of interpreting these variations (§8.6.2). In the case of X-ray and neutron diffraction it is usually correct to assume that the scattering is weak, so that only single scattering events need to be taken into account. Electron diffraction is different, and corrections must be applied for multiply-scattered waves. The results are too complicated to discuss here (see Cowley, 1984) but also contribute to the differences between the intensities of the various orders. One can also see that the spots in Fig. 8.26 have finite size. This is caused by the geometry of the apparatus – finite size of X-ray focus, angular divergence of beam and so on. Even if these factors could be allowed for, however, the spots would still have a non-zero size because of the shape function of the crystal (§8.4.1), but this effect cannot be observed directly in practice.

8.4.5 The acousto-optic effect

Another three-dimensional diffraction effect which can conveniently be treated in the weak scattering limit by the concepts of the reciprocal lattice and observation sphere is the acousto-optic effect, which is basically a situation in which a one-dimensional sinusoidal refractive index modulation is impressed on an initially uniform material. A particularly simple case is that of an ultrasonic plane wave propagating in water.† Because of the finite compressibility of water, the local

† Water is a surprisingly good medium for demonstrating the effect. Crystals such as PbGeO$_4$ have been developed to use it for solid-state applications.

Figure 8.26 Precession photograph of haemoglobin. (By courtesy of M.F. Perutz.)

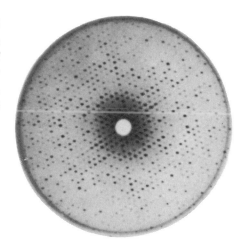

densityand hence refractive index responds to the oscillating pressure of the ultrasonic wave, creating a propagating sinusoidal modulation of amplitude A to the refractive index:

$$\mu(\mathbf{r}) = \mu_W + A\cos(\mathbf{q}\cdot\mathbf{r} - \Omega t) \qquad (8.77)$$

where μ_W is the refractive index of water at atmospheric pressure and the ultrasonic wave has frequency Ω and wave-vector $\mathbf{q}$. The velocity of sound in water is $v_s \approx 1200\,\text{m}\cdot\text{s}^{-1}$, and so for a frequency $\Omega = 2\pi \times 10\,\text{MHz}$, the wavelength $2\pi/q$ of the ultrasonic wave is about 0.1 mm, $\gg \lambda$. The water behaves as a three-dimensional phase grating with this period, and because $v_s \ll c$ we shall assume that an incident light wave sees it as a stationary modulation.

This is a well-defined problem, but one whose exact solution is elusive because the incident wave is refracted as well as diffracted by the medium. We shall only discuss here a very approximate approach that illustrates the physics of the problem; further details can be found in the books by Born and Wolf (1980), Korpel (1988) and Yariv (1991).

Diffraction of light by a weak three-dimensional sinusoidal grating can be treated by the same techniques that we used for crystal diffraction. The grating is represented by a complex transmission function

$$f(\mathbf{r}) = \exp\{i\int^{\mathbf{r}} [\mu(\mathbf{r}') - 1]\mathbf{k}_0 \cdot d\mathbf{r}'\} \qquad (8.78)$$

where the integration is carried out from the point at which the light enters the medium to $\mathbf{r}$. To simplify the problem we shall assume a sample having the form of a finite slab with thickness d normal to the incident light and replace the integral in (8.78) by an average value $[\mu(\mathbf{r}) - 1]k_0 d$; the grating is thus seen to have the form $f(\mathbf{r}) = \exp\{i[\mu(\mathbf{r}) - 1]k_0 d\}$. Now the modulation $|\mu(\mathbf{r}) - \mu_W| \ll 1$ (typically $< 10^{-6}$), so that $f(\mathbf{r})$ can be expanded for small enough d:[†]

$$\begin{aligned} f(\mathbf{r}) &= \exp\{ik_0 d\,[\mu_W - 1 + A\cos(\mathbf{q}\cdot\mathbf{r} - \Omega t)]\} \\ &\simeq \exp[ik_0 d(\mu_W - 1)][1 + iAk_0 d\cos(\mathbf{q}\cdot\mathbf{r} - \Omega t)]. \end{aligned} \qquad (8.79)$$

The Fourier transform of this function in reciprocal space $\mathbf{u}$ is

$$\begin{aligned} F(\mathbf{u}) &= \exp[ik_0 d(\mu_W - 1)][\delta(\mathbf{u}) + \tfrac{1}{2}iAk_0 d\,e^{-i\Omega t}\delta(\mathbf{u} - \mathbf{q}) \\ &\quad + \tfrac{1}{2}iAk_0 d\,e^{i\Omega t}\delta(\mathbf{u} + \mathbf{q})], \end{aligned} \qquad (8.80)$$

[†] When d is large enough for this approximation to be inapplicable, one can evaluate the integral by Bessel functions (Appendix 1), and see that higher diffraction orders will be produced. But the whole problem becomes much more complicated, as mentioned earlier.

which represents three δ-functions, a strong one at the origin and two weak ones at $\pm\mathbf{q}$.

The diffraction problem is then represented by the superposition of the sphere of observation on this Fourier transform (Fig. 8.27). It is shown for incidence normal to $\mathbf{q}$ in (Fig. 8.27(a)); it is clear that there is no diffracted beam in this case. There are only two angles of incidence at which diffraction occurs, $\pm\alpha$, shown for $+\alpha$ in (Fig. 8.27(b)). The angles α are clearly given by

$$q = \pm 2k_0 \sin\alpha. \tag{8.81}$$

Putting $q = 2\pi/\Lambda$ and $k_0 = 2\pi/\lambda$, this translates to

$$\lambda = 2\Lambda \sin\alpha, \tag{8.82}$$

which is the familiar form of the Bragg diffraction formula for lattice planes with spacing Λ (see Fig. 8.27(c)).

Generally, the size of a sample or crystal used for acousto-optic experiments may not be large compared with the acoustic wavelength. In such a case, the transmission function is multiplied by a shape function (§8.4) with value 1 within the sample, 0 outside. In its transform, each of the three δ-functions is therefore convoluted with the transform of the shape function, giving it a finite size. This means that the Bragg condition (8.82) is not exact. An extreme case occurs when the sample is thin in the direction normal to $\mathbf{q}$, so that the transform points are greatly elongated in this direction (Fig. 8.28). Diffraction can then be observed for incidence in any direction. It is easy to show that these angles satisfy the two-dimensional diffraction grating equation for oblique incidence (8.48). This is called the *Raman–Nath* scattering limit.

Because the velocity of light is much greater than that of sound, the light essentially sees the grating as stationary. However, the time-dependent amplitudes of the δ-functions $Ak_0d \exp(\pm i\Omega t)$ do have practical effects. Since the two diffracted waves are shifted in frequency by $\pm\Omega$ they can, for example, interfere to give a moving fringe pattern. The frequency shift can be simply interpreted as a Doppler effect when

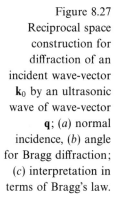

Figure 8.27 Reciprocal space construction for diffraction of an incident wave-vector $\mathbf{k}_0$ by an ultrasonic wave of wave-vector $\mathbf{q}$; (a) normal incidence, (b) angle for Bragg diffraction; (c) interpretation in terms of Bragg's law.

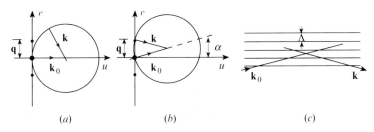

the wave is reflected by a moving grating (Problem 8.11), or as the condition for conservation of energy when a phonon (§8.5) is either absorbed or emitted by a photon.

8.5 Advanced topic: inelastic scattering of thermal neutrons by phonons

It is easy to extend our analysis of the acousto-optic effect to other systems. An important one is the thermally-excited crystal lattice. At non-zero temperature, the atoms vibrate about their equilibrium lattice positions, and it is convenient to analyze their motions into the superposition of harmonic waves, called *phonons*, each one having a wave-vector $\mathbf{q}$ and frequency Ω. In the limit of long wavelength, phonons and ultrasonic waves are the same thing, but the phonon wavelengths necessary to describe thermal motion can be as small as twice the interatomic spacing. An important feature of phonons is their dispersion relation $\Omega(\mathbf{q})$ (§2.3), which contains a lot of physics, and neutron diffraction provides one of the best methods of studying it. Because energy is exchanged between the phonon and the neutron during the diffraction, it is called *inelastic scattering*.

Neutrons produced by either by a nuclear pile or by spallation† are allowed to come to thermal equilibrium with a moderator at room temperature T. Their velocity distribution is then given by the classical

† A pulse of energetic electrons from a linear accelerator is aimed at a uranium target. There they produce gamma-rays, which in turn generate neutrons. The method has the advantage over a nuclear pile that a pulsed beam can be produced.

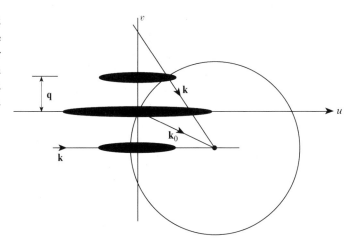

Figure 8.28 Reciprocal space construction for diffraction by an ultrasonic wave in a thin sample.

Boltzmann distribution, and their root-mean-square velocity in the x direction satisfies

$$\tfrac{1}{2}mv_x^2 = \tfrac{1}{2}k_B T. \tag{8.83}$$

This velocity gives a de Broglie wavelength λ:

$$\lambda = \frac{h}{mv_x} = \frac{h}{(mk_B T)^{\frac{1}{2}}} \tag{8.84}$$

which is of the order of $10^{-10}\,\text{m}$ – about the same as the lattice spacing. The phase velocity of the neutron waves is easily shown to be $v_x/2$ which, at room temperature, is about $800\,\text{m} \cdot \text{s}^{-1}$, comparable with the speed of sound. It is the similarity of both the wavelength and the velocity of the neutrons to those of phonons that makes them so suitable for diffraction experiments.

Neutrons are scattered weakly by the nuclei of the atoms in a crystal, and the geometry of their diffraction follows the von Laue theory that we discussed for X-rays in §8.4. We shall consider the diffraction by a single phonon of wave-vector $\mathbf{q}$ as an example. When the phonon propagates through the crystal, we have a density wave (as in the ultrasonic case) and the atoms sample it at points on the periodic lattice. Thus the three-dimensional Fourier transform is that of the atomic positions – the reciprocal lattice – convoluted with that of the density wave – the three δ-functions of (8.80).† The diffraction orders which can be observed for neutrons incident with wave-vector $\mathbf{k}_0$ and frequency ω_0 are then given by the intersections of an observation sphere with this three-dimensional transform. As pointed out in §8.4.2, the observation sphere construction represents conservation of energy during diffraction. In this case there is a Doppler shift, as in the acousto-optic effect, and $\omega_0 \rightarrow \omega_0 \pm \Omega$. This can be taken into account by constructing an observation sphere with the appropriate radius for the diffracted wave, as shown schematically in Fig. 8.29. Measurements of the direction and energy $\hbar(\omega_0 \pm \Omega)$ of the diffracted neutrons supplies both $\mathbf{q}$ and Ω of the phonon.

Of course, in reality there are many phonons propagating simultaneously in the crystal, but the observation sphere selects them one-by-one; for given angles of incidence and diffraction only one phonon is involved. We shall not go into further practical details, but show an example of typical results obtained for a magnesium crystal (Fig. 8.30). A full discussion is given in the book by Squires (1978).

---

† A one-dimensional example is treated in more detail in §9.2.3.

8.6 Advanced topic: phase retrieval

When a diffraction pattern is recorded – by photograhy, or a detector or any technique not involving interference with a reference wave† – the phase of the wave is lost. Diffraction is a powerful method of investigating the structure of matter, and it would be convenient if an inverse Fourier transform could be applied to a diffraction pattern to reveal directly the structure of the diffracting object. However, the reverse transform has to be applied to the *complex amplitude* pattern, and if the phases are lost, this is not known completely. This is called the *phase problem*, and finding a solution to it is extremely important, particularly in crystallography.

Although in principle the phase problem can have no general solution (there is an infinite number of mathematical functions which give the same diffraction pattern intensities), in practice the addition of some reasonable constraints leads to a unique solution, and techniques

† Example of techniques using interference with a reference wave to preserve phase information are holography (§12.6), aperture synthesis (§11.9.3) and the heavy-atom method in crystallography (§12.2.5).

Figure 8.29 Reciprocal space construction for inelastic diffraction by a phonon in a lattice. The phonon wave-vector is $\mathbf{q}$ and that of the incident neutron $\mathbf{k}_0$. Diffraction of the neutrons by the stationary lattice would be represented by intersections between the circle C_0 and reciprocal lattice points $h^* a^* + k^* b^*$ as in Fig. 8.25. In inelastic scattering, when the neutron loses energy to a phonon $\mathbf{q}$, energy conservation is represented by circle C_1 and this intercepts the modulated transform at points P_1 and P_2, giving diffracted beams $\mathbf{k}_1$ and $\mathbf{k}_2$. Likewise, a neutron can gain energy from a phonon in which case it is represented by the circle C_2 which intersects the a modulated-transform point at P_3 giving diffracted beam $\mathbf{k}_3$.

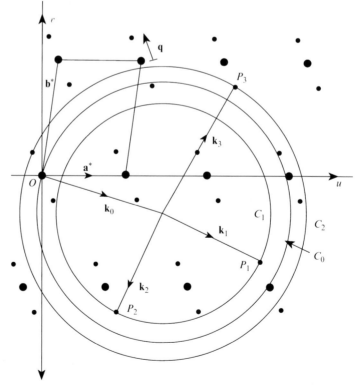

have been developed by which this solution can be found. For their pioneering work in this field, Hauptman and Karle were awarded the Nobel prize for chemistry. Today the interpretation of most crystalline X-ray diffraction patterns has become a fairly straightforward technical matter, although supplementary techniques are still needed for the most complicated crystals (§12.2.5). The most difficult part is often the preparation of the crystal itself! In this section we shall discuss the ideas behind the solutions in two fields: the first is the major field of crystallography, where the diffraction pattern of the crystal is sampled at the reciprocal lattice points (§8.4.3) only, and the second an adaptation of the same ideas to continuous optical diffraction patterns.

8.6.1 *A priori* information

As we pointed out above, some constraints are required to make the solution unique. The most important one is that the object function be *real* and *positive*. This is always true for the electron density of a crystal. In the optical examples it restricts the object to things such as opaque masks with holes in them or photographic transparencies. The second piece of *a priori* information is an estimate of some dimensional parameters, which are the dimensions of the unit cell of a crystal (known from the reciprocal lattice) and the number of atoms in it, or the overall size in the continuous optical case.

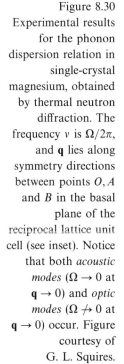

Figure 8.30 Experimental results for the phonon dispersion relation in single-crystal magnesium, obtained by thermal neutron diffraction. The frequency v is $\Omega/2\pi$, and $\mathbf{q}$ lies along symmetry directions between points O, A and B in the basal plane of the reciprocal lattice unit cell (see inset). Notice that both *acoustic modes* ($\Omega \to 0$ at $\mathbf{q} \to 0$) and *optic modes* ($\Omega \nrightarrow 0$ at $\mathbf{q} \to 0$) occur. Figure courtesy of G. L. Squires.

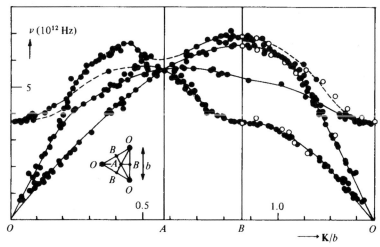

8.6.2 Direct methods in crystallography

The determination of the crystal structure from the intensities of its X-ray diffraction spots alone is called a *direct method*, and should be contrasted with other techniques which require the addition of further information, such as the heavy-atom method outlined in §12.2.5. Although such methods were first proposed around 1950, their need for quite considerable computations delayed their general acceptance till the 1970s, when powerful electronic computers became generally available. In this section we shall outline the ideas behind the direct method, with a simple example. Two useful reviews of the field are Woolfson (1971) and Hauptman (1991).

We saw in §8.4 that a crystal can be described by the convolution between the molecular electron density (or that of a group of molecules with some well-defined geometrical relationship) and the crystal lattice. The diffraction pattern of the former is then sampled at the reciprocal lattice points. The distances and angles between the diffraction spots allow the reciprocal lattice, and hence the real lattice, to be determined easily.

The amplitude of the diffraction pattern at reciprocal lattice point **h**, defined by (8.76)

$$\mathbf{h} = h^\star \mathbf{a}^\star + k^\star \mathbf{b}^\star + \ell^\star \mathbf{c}^\star \tag{8.85}$$

is related to the electron density $\rho(\mathbf{r})$ within the unit cell by

$$F(\mathbf{h}) = V^{-1} \int\!\!\int\!\!\int_{\text{cell}} \rho(\mathbf{r}) \exp(-i\mathbf{h} \cdot \mathbf{r}) \, d^3\mathbf{r}, \tag{8.86}$$

which is the three dimensional Fourier transform of $\rho(\mathbf{r})$.

Suppose the cell contains N atoms, and for simplicity let each one have electron density $Z s(\mathbf{r})$ with respect to its own origin. The difference between one atom and another is here contained in the value of Z, the atomic number. The electron density ρ can then be expressed by the convolution between $s(\mathbf{r})$ and a set of N δ-functions at the atomic positions $\mathbf{r}_j$, where the jth δ-function has strength Z_j. Equation (8.86) can then be written as a sum:

$$F(\mathbf{h}) = S(\mathbf{h}) \sum_{j=1}^{N} Z_j \exp(-i\mathbf{h} \cdot \mathbf{r}_j), \tag{8.87}$$

where $S(\mathbf{h})$ is the transform of $s(\mathbf{r})$ and is a smooth and reasonably well-known function. Finally, the intensity measured at reciprocal lattice point $\mathbf{h}$ is $|F(\mathbf{h})|^2$:

$$|F(\mathbf{h})|^2 = |S(\mathbf{h})|^2 \left| \sum_{j=1}^{N} Z_j \exp(-i\mathbf{h} \cdot \mathbf{r}_j) \right|^2 ,$$

$$= |S(\mathbf{h})|^2 \sum_{j=1}^{N} \sum_{k=1}^{N} Z_j Z_k \exp[-i\mathbf{h} \cdot (\mathbf{r}_j - \mathbf{r}_k)] . \quad (8.88)$$

In (8.88) there are four unknowns for each position j – these are Z_j and the three components of $\mathbf{r}_j$ – and there are N values of j; in all, $4N$ unknowns. Therefore, if $|F(\mathbf{h})|^2$ is measured at more than $4N$ different values of $\mathbf{h}$, *in principle* there is enough information for all the variables to be determined. Since the measurement of this number of reflexions, or even many more, is usually possible, the problem should not only be soluble but even over-determined! The question is, how can the solution be found?

The discussion here will centre around a few key points with the intention of making the ideas clear without mathematical complexity. In this vein, consider a crystal of molecular units each having N *identical* point atoms. Thus all the Z_js are equal, and will be assumed unity, and we shall put $S(\mathbf{h}) = 1$. Now consider the two functions $\rho(\mathbf{r})$ and $\rho^2(\mathbf{r})$. They are

$$\rho(\mathbf{r}) = \sum_{j=1}^{N} \delta(\mathbf{r} - \mathbf{r}_j), \quad (8.89)$$

$$\rho^2(\mathbf{r}) = \beta \sum_{j=1}^{N} \delta(\mathbf{r} - \mathbf{r}_j). \quad (8.90)$$

The former is *required* to be positive (§8.6.1) and the second is obviously so.† The Fourier transform of (8.89) is $F(\mathbf{h})$ and that of (8.90) is its auto-correlation

$$F(\mathbf{h}) \otimes F^*(-\mathbf{h}) = \sum_{\mathbf{k}} F(\mathbf{k})F^*(\mathbf{k} - \mathbf{h}) \quad (8.91)$$

But also, from (8.89) and (8.90), $\rho^2(\mathbf{r}) = \beta\rho(\mathbf{r})$ and so (8.91) becomes:

$$\beta F(\mathbf{h}) = \sum_{\mathbf{k}} F(\mathbf{k})F^*(\mathbf{k} - \mathbf{h}) . \quad (8.92)$$

This is known as *Sayre's equation* for point atoms. Let us now

---

† The factor β represents the ratio between $\delta(x)$ and its square – unknown, but definitely positive. Its value is irrelevant here. Remember that the δ-function is a mathematical abstraction representing a real atom.

separate the amplitude and phase of each $F(\mathbf{h})$ by writing $F(\mathbf{h}) = E(\mathbf{h})\exp[i\phi(\mathbf{h})]$. It then follows on multiplying (8.92) by $F^*(\mathbf{h})$ that

$$\beta E^2(\mathbf{h}) = \sum_{\mathbf{k}} E(\mathbf{h})E(\mathbf{k})E(\mathbf{k} - \mathbf{h})\exp\{i[-\phi(\mathbf{h}) + \phi(\mathbf{k}) - \phi(\mathbf{k} - \mathbf{h})]\} .$$

(8.93)

A practical method of getting a good first approximation to the solution is based on this equation, observing that $\beta E^2(\mathbf{h})$ is positive.

Measurement of the diffraction pattern gives us values for $E(\mathbf{h})$ for many values of $\mathbf{h}$. We should recall (§4.4.8) that for real ρ, $F(\mathbf{h}) = F^*(-\mathbf{h})$ so that $E(\mathbf{h}) = E(-\mathbf{h})$ and $\phi(\mathbf{h}) = -\phi(-\mathbf{h})$. Amongst the measured values, choose the three largest values of E which have reciprocal lattice vectors $\mathbf{h}$, $\mathbf{k}$ and $\mathbf{h} \pm \mathbf{k}$ (i.e., one vector is the sum or difference of the other two). Then for the value of (8.93) to be positive it is most likely that the term involving these three will make a positive contribution to the sum, since their product makes up its largest term. If so, the sum of the phases will be zero.

$$\phi(\mathbf{h}) \pm \phi(\mathbf{k}) \doteq \phi(\mathbf{h} \pm \mathbf{k}) ,$$

(8.94)

in which the sign $\doteq$ is to be read as 'expected to be equal to'. In the end, it might turn out only to be close to correct. Inspecting the various measured intensities, we continue to identify triads of $\mathbf{h}$, $\mathbf{k}$ and $\mathbf{h} \pm \mathbf{k}$ for which the product $E(\mathbf{h})E(\mathbf{k})E(\mathbf{h} \pm \mathbf{k})$ is relatively strong and use (8.94) to relate their phases. As the product gets smaller, the reliability of (8.94) gets less.

Now the actual values of the phases ϕ are determined by the origin of the unit cell, which is in principle arbitrary, although symmetry of the molecule will often dictate some preferred choice. For example, if the molecule is centro-symmetric, choosing the origin at the centre of symmetry makes all the phases zero or π. This reduces the amount of work enormously. In general, the phases of any three of the diffraction spots (in a three-dimensional example) can be assigned arbitrary values provided that they do not form a triad. Using (8.94) we then try to express all the phases in terms of the three chosen ones and the phases of a small number of additional prominent spots, which are represented by symbols. If there are several triads involving the strongest spots, their phases can be determined fairly reliably. The process works better in two and three dimensions than in one because increasing the dimensionality increases the number of possible relations.

The next stage in the determination involves reconstructing the object from the known amplitudes and the phases deduced. Since some of them have been given arbitrary symbols, it is necessary to perform a series of reconstructions, with various values assigned to the symbols. The set of phases that reconstructs an object with the least negative parts and having the closest similarity to *a priori* expectations of the structure (number of atoms, bond lengths etc) is assumed to be roughly correct. An improvement to the phases is found by retransforming this approximate structure with all the negative electron densities set to zero. The phases calculated this way are then used, together with the measured amplitudes, to return a better structure, and this process is iterated several times until the required degree of accuracy is obtained.

8.6.3 A centro-symmetrical example of the direct method

The method described above is so important in modern crystallography that we shall illustrate it with a simple one-dimensional example, as shown in Fig. 8.31. This is not as easy as it seems because the method does not work so well in one dimension; a two-dimensional one is given by Woolfson (1971) but it is too lengthy for reproduction here. We shall use a centro-symmetrical example, consisting of ten equal point 'atoms' at points x in the region $-32 < x < 32$ arranged symmetrically about $x = 0$. This means that the diffraction amplitudes are all real, so that the phases are either zero or π. We shall be able to assign the phase zero (i.e. sign '+') to one diffraction amplitude, and three more are assigned '*a*', '*b*' and '*c*'. At least one of these must be '−' since otherwise, trivially, (8.94) can be satisfied by making *all* the phases '+'. But as we shall see in §12.2.5, this corresponds to an overwhelmingly heavy atom at the origin, in which case we do not need to use the direct method. The first approximation to a solution is shown in Fig. 8.31, where the known input (*a*) should be compared with the best choice for (*a*, *b*, *c*) as (+, −, +) which is shown in (*d*). Another possibility is (*h*) which has the right number of strong peaks, although its background is considerably stronger. The results shown are obtained *before* starting the iterations.

8.6.4 Phase retrieval in optical diffraction patterns

The solution for continuous diffraction patterns is essentially similar, the difference being in the way that the initial estimate of the phases is obtained. It turns out that this is not too critical, and almost any reasonable starting point converges to the correct solution after enough iterations of the stage described in the last paragraph of

§8.6.2. In this case we are presented with the diffraction pattern intensity $|F(u,v)|^2$ of an object $f(x,y)$ which is known to be real and non-negative. We also are given the outer bound of the region of $f \geq 0$ in the (x,y) plane. Now if the outer bound is centro-symmetrical (a rectangle, for example) there will always be two possible solutions, related by symmetry about the centre; these are called *enantiomorphs* and exist in the crystallographic case too. The phase retrieval program must choose between them, and this is done by giving an initial bias. A method described by Fienup (1982) works as follows and is illustrated by Fig. 8.32.†

We represent the outer bound of the object by the function $g(x,y)$ which is zero outside the bound and has value such as $1 + \beta(x-y)$ within it, which is biased so as to prefer one enantiomorph. The transform of g is $G(u,v) = |G(u,v)| \exp[i\phi_g(u,v)]$. The first trial phase

† We are grateful to Chaim Schwartz for providing this example.

Figure 8.31 Illustrating the direct method in crystallography. (a): The original function $f(x)$ chosen, which has ten δ-functions symmetrically arranged about the origin in $-32 < x < 32$. (b): The amplitudes of the diffraction spots (for $0 < u < 32$). The strongest one is assigned phase $+$, and three more a, b and c. Phases deduced by applying (8.94) are indicated for several of the stronger spots. (c)–(j): The reconstructions of the object (in $x \geq 0$ only) obtained with the eight permutations of $+$ and $-$ for (a,b,c): (c): $+++$, (d): $+-+$, (e): $++-$, (f): $--+$ (g): $---$, (h): $-+-$ (i): $-++$, (j): $+--$.

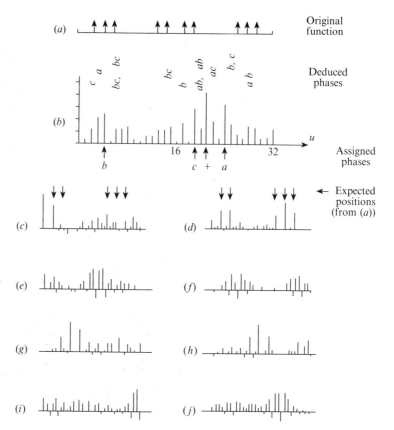

to be used on $F(u,v)$ is $\phi_g(u,v)$. So we transform the function

$$F_1(u,v) = |F(u,v)| \exp[i\phi_g(u,v)] \qquad (8.95)$$

to get the first trial function $f_1(x,y)$. All negative parts of f_1 are replaced by zero, giving $f_2(x,y)$, and this is transformed to give $F_2(u,v) = |F_2| \exp[i\phi_2(u,v)]$. Now $|F_2|$ is replaced by the known $|F|$ and the result $|F| \exp[i\phi_2(u,v)]$ replaces (8.95) in the cycle, which is then repeated until a satisfactory result is obtained. Essentially the method works because the transform of the bound g gives a typical rate of change for the phase, and this is sufficient for a starting value. The origin of the function obtained clearly coincides with that of g. Convergence is certainly not rapid; the example of Fig. 8.32 shows the result after 500 iterations, and it is still not identical with the original. Since this is a relatively simple structure, it is clear why considerably more effort must be put into obtaining the initial estimate of the phases in the crystallographic work, where the object is much more complicated, before the iteration is started.

Figure 8.32 Phase retrieval for a continuous object. (*a*) The original object $f(x,y)$; (*b*) the outer bound function $g(x,y)$ with its asymmetrical bias; (*c*) the image $f_1(x,y)$ obtained after the first cycle; (*d*) the image $f_{500}(x,y)$ obtained after the 500th cycle.

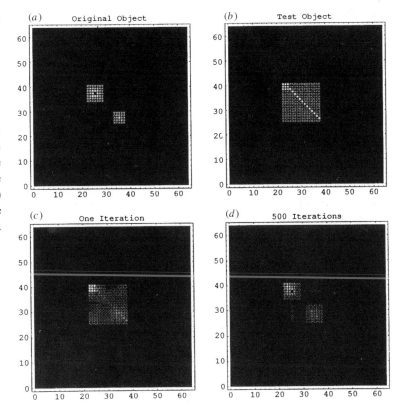

CHAPTER NINE

Interferometry

9.1 Introduction

In Chapter 8 we discussed the theory of Fraunhofer diffraction and interference emphasizing in particular the relevance of Fourier transforms. In this chapter we shall describe the applications of interference to measurement; this is called *interferometry*. Some of the most accurate dimensional measurements are made by interferometric means, particularly using waves of different types – electromagnetic, matter, acoustic etc. The variety of techniques is enormous, and we shall limit ourselves in this chapter to a discussion of several distinctly different interferometric principles, without any intention of describing the variety of instruments or methods within the classes. There are several monographs on interferometry which discuss practical aspects in greater detail, for example Tolansky (1973), Steel (1983) and Hariharan (1985).

The discovery of interference effects by Young (§1.2.4) enabled him to make the first interferometric measurement, a determination of the wavelength of light. Even this primitive system, a pair of slits illuminated by a common point source, can be surprisingly accurate, as we shall see in §9.1.1. In general interference is possible between waves of any non-zero degree of mutual coherence (§11.4), including different sources (light beats), but for the purposes of this chapter we shall simply assume that waves are either completely coherent (in which case they can interfere) or incoherent (in which case no interference effects occur between them). In the case of complete coherence, there is a fixed phase relationship between the waves, and interference effects are observed that are stationary in time, and can therefore be observed with primitive instruments such as the eye or photography. We therefore combine *coherent* waves by adding their complex amplitudes, and then calculate the intensity by taking the square modulus of the sum; we combine *incoherent* waves by calculating their individual intensities first, and then adding. The need for coherence between the interfering waves dictates at optical frequencies that they must all originate from the same source; the various ways of dividing an incident wave into separate parts and interfering them, after they have been influenced by the system to be measured, constitute different interferometers.

9.1.1 Interferometry by Young's fringes

Young's fringe experiment constitutes the basic interferometer, and it is worth dwelling briefly on some of its basic aspects. According to Huygens's principle, each slit behaves as a source of coherent waves, and the wavefronts are circular in the two-dimensional projection shown in Fig. 9.1. The maxima and minima of the interference pattern arise at points where the waves interfere constructively (amplitudes add) or destructively (amplitudes subtract). In the simplest case, where the two slits emit with the same phase, constructive interference occurs when the path difference is an integer number of wavelengths and destructive interference when the number is integer-plus-half. The loci of such points lie on a family of hyperbolae whose foci are at the two slits. If the slits are replaced by pinholes (which are more satisfactory from the analytical point of view – see footnote on p. 162 – the loci of the fringes in three dimensions are a family of hyperboloids having their foci on the pinholes. Their intersection with a plane screen at a large distance gives approximately straight fringes.

A simple way of producing two coherent sources is to use a single point source and its image in a plane mirror; this is called *Lloyd's mirror* and is shown in Fig. 9.2. If the source is nearly in the plane of the mirror the separation of the source and its image is quite small,

Figure 9.1 Set-up for Young's fringes.

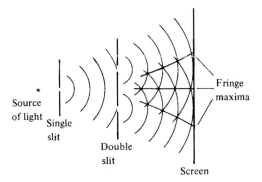

and well-separated interference fringes can be produced. Obviously, however, the zero fringe – that which is equidistant from the two-sources – cannot be produced by this method; but if one extrapolates back to the position where it should be, one finds that there is a minimum of intensity there, not a maximum. There is therefore some asymmetry between the source and its image; this can be traced to the change of phase that occurs when light is reflected from a medium of higher refractive index (§5.4.2) or from a conductor (§5.6.1).

Although one would not usually employ Young's fringes for high-resolution measurements, it is of interest to see what could be achieved. The slits are represented by a transmission function consisting of two δ-functions separated by distance a (§8.3.1) and we showed that the the amplitude of the diffraction pattern observed on a distant screen (Fraunhofer conditions) is then

$$\psi(u) = 2\cos(ua/2). \qquad (9.1)$$

Writing this in terms of the angular variable $\sin\theta$, where $u = k_0\sin\theta$

$$\psi(\sin\theta) \;=\; 2\cos(k_0 a\sin\theta/2)\,; \qquad (9.2)$$
$$I = |\psi(\sin\theta)|^2 \;=\; 4\cos^2(k_0 a\sin\theta/2)\,. \qquad (9.3)$$

If two separate wavenumbers k_1 and k_2 contribute to the source, the intensities add incoherently and so, assuming equally bright sources,

$$I = 4\left[\cos^2\left(\frac{k_1 a}{2}\sin\theta\right) + \cos^2\left(\frac{k_2 a}{2}\sin\theta\right)\right]. \qquad (9.4)$$

The two sets of $\cos^2$ fringes will be out of phase and therefore cancel one another to give a uniform intensity when

$$\frac{k_1 a}{2}\sin\theta - \frac{k_2 a}{2}\sin\theta = (2n+1)\frac{\pi}{2}. \qquad (9.5)$$

The condition for wavelength resolution is, intuitively, that at least

Figure 9.2 Set-up for Lloyd's single-mirror fringes. The phase change on reflexion is simulated by making the radii of the circles representing the reflected wave interleave those representing the direct wave.

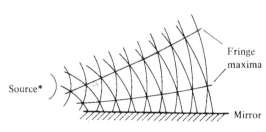

Source*

Image*

Fringe maxima

Mirror

one such cancellation will occur at an observable angle $\theta < 90°$; this occurs if

$$(k_1 - k_2) > \pi/a$$

or, for $k_1 \approx k_2$, in terms of wavelength λ

$$\frac{\lambda_2 - \lambda_1}{\lambda} > \frac{\lambda}{2a} \qquad (9.6)$$

where $\lambda = (\lambda_1 + \lambda_2)/2$. This is called the *limit of resolution*; its inverse, $\lambda/(\lambda_2 - \lambda_1)$ is called the *resolving power*. For example, if $a \sim 1\,\text{mm}$ and $\lambda \sim 0.5\,\mu\text{m}$, a resolving power of order 4000 has been achieved, which is quite good for such a primitive experiment. If the source contains more than two wavelengths, the fringe pattern gets too complicated for simple interpretation (Fig 9.3) and more sophisticated techniques are required, such as Fourier spectroscopy (§11.6), but this discussion contains the physics of all two-beam interferometry!

A great practical improvement over the two-slit system is a periodic array of slits – the diffraction grating.

9.2 Diffraction gratings

A diffraction grating is a periodic one-dimensional array of similar apertures, usually narrow slits or mirrors. We have seen in §8.3.4 that the diffraction pattern of such an array is a periodic series of δ-functions, whose strengths are determined by the exact shape and dimensions of the apertures. The positions of the δ-functions are determined only by the period of the array; from (8.47–8.48),

$$u = k_0(\sin \theta - \sin \theta_0) = u_m = 2\pi m/d \qquad (9.7)$$

Figure 9.3 Interference fringes from (*a*) a monochromatic source, (*b*) a polychromatic line source (Hg lamp), (*c*) a broad band source.

(*a*)   (*b*)   (*c*)

where m is the *order of diffraction*. Because k_0 enters the definition of u, the angle of diffraction θ depends on the wavelength $\lambda = 2\pi/k_0$; this dependence makes diffraction gratings important tools for spectroscopy. We shall discuss them in the framework of the scalar approximation, despite the real need for a vector formulation (Hutley, 1982).

9.2.1 Production of diffraction gratings

Although this book is concerned primarily with general principles, methods of production of diffraction gratings must be briefly described because some of the succeeding theory depends upon acquaintance with them.

The first serious gratings were made by scribing a series of lines on glass or metal with a fine diamond. Rowland used an accurate screw to translate the diamond laterally through a small distance between each pair of lines. Obviously much is implied in this sentence: the diamond and the flat upon which the grating is to be ruled must be carefully chosen; the screw and flat must be accurately adjusted relative to each other; the diamond point must not change during the ruling operation; and the temperature of the whole apparatus must be kept constant so that no irregular expansions occur. Thus machines for making gratings, called *ruling engines*, are extremely complicated and costly.

Recent years have seen a vast change in the methods of production of gratings. First, gratings have been ruled by cutting a very fine screw thread on a cylinder. The cylinder is then coated with a plastic which forms a surface-replica that can be removed after setting. In this way surprisingly good gratings can be produced very cheaply.

But more important is the development of *holographic diffraction gratings*. As will be pointed out in §12.6, holograms are essentially *complicated* diffraction gratings. They can also be designed to be *simple* diffraction gratings. The development of high-resolution emulsions, particularly photo-resists for the microelectronic industry, has made it possible to photograph a very fine interference pattern between plane-waves which produces, in a single exposure, a grating with many thousands of lines. For example if two coherent plane waves from a laser with wavelength of say $0.5\,\mu\text{m}$ interfere at an angle of $2\alpha = 60°$, the interference pattern has Young's fringes with a spacing $\lambda/\sin\alpha = 1\,\mu\text{m}$. Because the laser lines are very sharp the number of fringes is enormous, and a grating many centimetres long can be produced in a single exposure. This technique completely avoids the problem of errors in line position, which is very troublesome in ruled gratings.

Another advantage of the holographic grating is that the line-spacing can be arranged to be non-uniform in a planned way so as to correct for known aberrations in the associated optics or to reduce the number of accessory optical elements required. For example, a self-focusing grating can be produced by using as the source of the grating the interference pattern between two spherical waves.

Most serious diffraction gratings are reflexion gratings, being either ruled on an optically-flat reflecting surface, or being produced holographically on such a surface by etching through the developed photoresist. This preference arises because reflexion gratings are generally *phase gratings*, whose efficiency, both in theory and in practice, can be considerably larger than that of transmission gratings (§9.2.4). Gratings can also be produced on cylindrical and spherical surfaces in order to add a further dimension to the possible correction of aberrations.

9.2.2 Resolving power

One of the important functions of a diffraction grating is the measurement of the wavelengths of spectral lines; because we know the spacing of the grating we can use equation (9.7) to measure wavelengths absolutely. The first question we must ask about a grating is 'What is the smallest separation between two wavelengths that will result in two separate peaks in the spectrum?' This defines the *limit of resolution*. We shall see that the limit results from the finite length of a grating.

The problem can be considered in terms of §8.3.3, where we saw that the diffraction pattern of finite number of equally-spaced apertures has both principal and secondary maxima. In the case of N slits there are $N - 1$ zero values of the intensity between the principal maxima. If two different wavelengths are present in the light falling on a grating, the intensity functions will add together; we need to find the conditions under which the principal maxima can clearly be discerned as double. We therefore have to consider in more detail the exact shape of the interference function.

From (8.43), the intensity function (normalized to unity at $u = 0$) is

$$I(u) = \frac{\sin^2(uNd/2)}{N^2 \sin^2(ud/2)} \; . \tag{9.8}$$

This has principal orders with $I(u) = 1$ at $u = 2m\pi/d$ and zeros at

$$u = 2(m + p/N)\pi/d, \qquad (9.9)$$

where p is an integer $1 \le p \le N - 1$. Between the zeros are subsidiary orders. Now if the the intensities of the two incident wavelengths are equal, their combined intensity is the sum of two functions like (9.8), when expressed in terms of the angle θ, as shown in Fig. 9.4. A reasonable criterion for resolution was suggested by Rayleigh, who considered that the two wavelengths would just be resolved if the principal maximum of one intensity function coincided in angle with the first zero ($p = 1$) of the other. This is a useful criterion, if a little pessimistic, and will be discussed in greater depth in §12.3 where it is used in an imaging context. Using the Rayleigh criterion, we find from (9.9) that the first zero is separated from the main order by $\delta u = 2\pi/Nd$. Now from the definition of u as in (9.7), we can write

$$\frac{\delta u}{u} = \frac{\delta k}{k} = (-)\frac{\delta \lambda}{\lambda} . \qquad (9.10)$$

The *resolving power* is defined as $\lambda/\delta\lambda_{\min}$, which for order m is

$$\frac{\lambda}{\delta\lambda_{\min}} = \frac{uNd}{2\pi} = mN. \qquad (9.11)$$

This result shows that the resolving power obtainable does not depend solely upon the line spacing; if a coarse grating is made, a higher order can be used and the resolving power may be as good as that of a finer grating. If L is the total length of the grating, $d = L/N$ and the resolving power is equal to

$$\frac{\lambda}{\delta\lambda_{\min}} = mN = \frac{Nd}{\lambda}(\sin\theta - \sin\theta_0) = \frac{L}{\lambda}(\sin\theta - \sin\theta_0). \qquad (9.12)$$

Thus, for given angles of diffraction θ and incidence θ_0, the resolving power depends only on the total length of the grating. The highest resolving power is obtained when $\theta_0 \to -\pi/2$ and $\theta \to \pi/2$, whence

$$\frac{\lambda}{\delta\lambda_{\min}} \to \frac{2L}{\lambda}. \qquad (9.13)$$

Gratings should therefore be made as long as possible. In fact we might as well just make a pair of slits at distance L apart, although they would use the light very inefficiently. This situation, Young's slits, has already been discussed in §9.1.1. For example, from (9.13) with $\lambda = 0.5\,\mu\text{m}$, a grating $5\,\text{cm}$ long should give a resolution limit approaching 2×10^5 although it is difficult to use θ and θ_0 around 90°.

In practice, it is rare for a grating to have a resolving power equal to the theoretical value; errors in ruling are inevitably present and they can affect the performance considerably. With a very good grating and well-corrected optical components, resolving powers of over half the theoretical value can be obtained, but this is unusual.

9.2.3 Effects of periodic errors – ghosts

There is one type of error that often arises in ruled gratings and does not affect the resolving power but is nevertheless undesirable for other reasons; this is a periodic error in line position. It can arise from a poor screw or by a badly designed coupling between the screw and the table carrying the grating (§9.2.1), and has the effect of enhancing some of the secondary maxima.

As an example we shall analyse the situation where an error in line position is repeated every qth line; the true spacing is qd, and therefore q times as many orders will be produced. Most of them will be very weak, but some may be strong enough to be appreciable compared with the main orders. To make the problem soluble analytically we assume the line positions x_p to contain a small error which is sinusouidal in position,

$$x_p = pd + \epsilon \sin 2\pi p/q. \tag{9.14}$$

Figure 9.4 The addition of two diffraction-grating functions for different wavelengths, showing resolution of the two wavelengths according to the Rayleigh criterion.

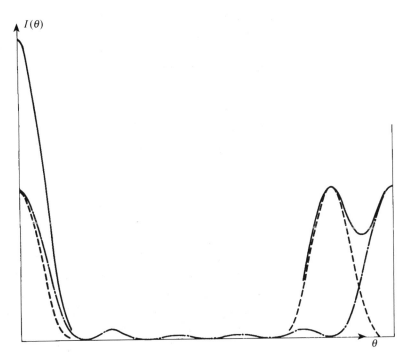

The grating is then represented by the set of δ-functions

$$f(x) = \sum_p \delta(x - pd - \epsilon \sin 2\pi p/q), \tag{9.15}$$

whose Fourier transform is

$$
\begin{aligned}
F(u) &= \sum_p \exp[-iu(pd + \epsilon \sin 2\pi p/q)] \\
&\approx \sum_p [\exp(-iupd)(1 - iu\epsilon \sin 2\pi p/q)] \tag{9.16}
\end{aligned}
$$

when $\epsilon \ll d$. On writing $\sin 2\pi p/q$ as $\frac{1}{2}i[\exp(-2i\pi p/q) - \exp(2i\pi p/q)]$, this is easily shown to be

$$
\begin{aligned}
F(u) &= \sum_m \delta\left(u - \frac{2\pi m}{d}\right) - \frac{u\epsilon}{2}\sum_m \delta\left[u - \frac{2\pi}{d}\left(m + \frac{1}{q}\right)\right] \\
&\quad + \frac{u\epsilon}{2}\sum_m \delta\left[u - \frac{2\pi}{d}\left(m - \frac{1}{q}\right)\right], \tag{9.17}
\end{aligned}
$$

where m is the order of diffraction. The summations show that, in addition to the principal maxima, there are also maxima at angles given by the two 'orders' $m + 1/q$ and $m - 1/q$. That is, each order m is flanked by two weak satellite lines, with intensities proportional to ϵ^2, at a separation of $1/q$th of the orders in reciprocal space (Fig. 9.5). These lines are called *ghosts* ; their intensities are are also proportional to $\sin^2 \theta$, implying that they increase rapidly with m. Errors in line position do not result in ghosts around the zero order.

The idea that periodic displacements in position from the points of a regular lattice leads to ghost orders of diffraction has applications in many areas of physics, of which we shall mention a few.

(1) In a compressible medium, an acoustic wave results in such a periodic displacement, and in §8.4.5 we saw an elementary analysis of the acousto-optic effect.

(2) Thermal motion of the atoms of a crystal can be analyzed as a superposition of sinusoidal displacements, *phonons*, each of which results in ghost orders. These, together, give a diffuse background and apparent broadening to to the otherwise sharp X-ray diffraction spots from the crystal; this is called the *Debye–Waller effect*.

Figure 9.5
Representation of the
orders of diffraction
from a grating with
periodic errors in line
position.

$I(m)$

$\bar{4}$ $\bar{3}$ $\bar{2}$ $\bar{1}$ 0 1 2 3 4 m

(3) In the study of alloys, *superlattices* arise when there is spatial ordering of each component within the basic lattice, sometimes in quite a complicated manner (e.g Cu–Zn). Superlattices can also be created artificially in semiconductors by deposition of atoms in a calculated sequence. The existence of a superlattice is indicated by the appearance of ghost orders of diffraction between the main X-ray diffraction spots. A particularly important example occurs in some magnetic materials (e.g. MnF_2) where the atomic spins are ordered on a lattice that is different from the crystal lattice, and ghost orders are then evident in the diffraction pattern of slow neutrons, whose spin causes them to interact with the atomic spins. The ghosts are absent in an X-ray diffraction pattern of the same crystal because X-rays are not spin-sensitive.

(4) Finally we should mention that the one-dimensional analysis carried out in this section is equivalent to the spectral analysis of a *frequency modulated* (FM) radio wave. The ghost orders of diffraction, which contain the information on the frequency and amplitude of the modulation, are called *side-bands* in telecommunications (see §10.3.4).

9.2.4 Diffraction efficiency

The discussion of gratings has so far concentrated on the interference function, the Fourier transform of the set of δ-functions representing the positions of the individual apertures. This transform has now to be multiplied by the diffraction function, which is the transform of one aperture.

Let us first consider a simple *amplitude transmission grating* for which the apertures are slits, each with width b (which must be less than their separation d). The diffraction function is then the transform of such a slit (§8.2.3),

$$\psi(u) = b \operatorname{sinc}(bu/2). \tag{9.18}$$

At the order m, $u_m = 2\pi m/d$. For the first order, as b is varied, the maximum value of $\psi(u_1)$ is easily shown to occur when $b = d/2$; thus the optimum slit width is half the spacing. But even with this value the efficiency of the grating is dismally small. The light power P_m reaching the various orders is proportional to the values of $|\psi(u_m)|^2$ namely,

$$P_0 \propto d^2/4, \quad P_{\pm 1} \propto d^2/\pi^2, \quad P_{\pm 2} = 0, \text{ etc} \quad (\text{for } b = d/2). \tag{9.19}$$

The constant of proportionality can be deduced by putting $b = d$, whereupon the grating becomes completely transparent. Then all the incident light falls in the zero order and we have powers

$$P'_0 \propto d^2, \quad P'_i = 0 \quad (i > 0) \quad (\text{for } b = d). \tag{9.20}$$

The *diffraction efficiency* is defined as the fraction of the incident light diffracted into the strongest non-zero order. Maximized by the choice of $b = d/2$, it still only reaches $P_1/P'_0 = \pi^{-2}$, about 10%. This figure can hardly be improved upon, within the limitations of real, positive transmission functions; the way to higher efficiencies is through the use of *phase gratings*.

9.2.5 Blazed gratings

The discussion in the previous section shows us how inefficient an amplitude transmission grating must necessarily be. Rayleigh originated the idea of combining the effects of refraction or reflexion with interference to make a phase grating that could concentrate most of the intensity in one particular order. The principle is illustrated by Fig. 9.6. Each element in the transmission grating shown in Fig. 9.6(a) is made in the form of a prism, of which the angle is such that the deviation produced is equal to the angle of one of the orders of diffraction; correspondingly, in Fig. 9.6(b) a reflexion grating is shown in which each element is a small mirror.

Such gratings are widely used. Instead of using any available sharp diamond edge for ruling a grating, a special edge is selected that can make optically-flat cuts at any desired angle. Gratings so made are called *blazed gratings*. It will be noted that a diffraction grating can be blazed only for one particular order and wavelength, and the high efficiency applies only to a restricted wavelength region.

The scalar-wave theory of the blazed grating is an elegant illustration

Figure 9.6 (a) Blazed transmission grating. The value of θ must satisfy the two equations $n\lambda = d \sin \theta$ and $\theta = (\mu - 1)\alpha$. (b) Blazed reflexion grating. The value of θ must satisfy the two equations $n\lambda = d \sin \theta$ and $\theta = 2\alpha$.

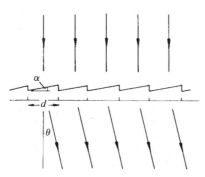
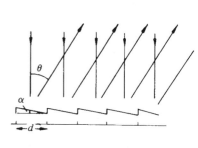

of the use of the convolution theorem. Suppose that a plane wavefront incident normally on the grating is deviated by angle β on being reflected or transmitted by an individual facet. For a reflexion grating, $\beta = 2\alpha$ is determined by the geometry only; for a transmission grating it may be wavelength dependent too, but Fig. 9.7 notwithstanding, we shall restrict our interest to the former. Following the analysis in §8.2.5, the individual facet is represented by a phase ramp of form $\exp(ik_0 \sin \beta)$. We now describe the individual facet, of width b, by the complex transmission function $g(x) = \text{rect}(x/b) \exp(ik_0 \sin \beta)$. The complete grating is therefore represented by

$$
\begin{aligned}
f(x) &= g(x) \otimes \sum \delta(x - nd) & (9.21) \\
&= [\text{rect}(x/b) \exp(ik_0 \sin \beta)] \otimes \sum \delta(x - nd). & (9.22)
\end{aligned}
$$

The Fourier transform of (9.22) is

$$
\begin{aligned}
F(u) &= [\delta(u - k_0 \sin \beta) \otimes \text{sinc}(ub/2)] \cdot \sum \delta(u - 2\pi m/d) \\
&= \text{sinc}[b(u - k_0 \sin \beta)/2] \cdot \sum \delta(u - 2\pi m/d). & (9.23)
\end{aligned}
$$

One sees in Fig. 9.8(*a*) that the maximum of the envelope function (sinc), which indicates the value of u giving the highest intensity, has moved from the origin to $k_0 \sin \beta$. If $k_0 \sin \beta = 2\pi m_0/d$, this coincides with order m_0 (usually, but not necessarily, the first order). This way, β can chosen to maximize the intensity in a specified order, for a given wavenumber k_0. The corresponding wavelength is called the *blazing wavelength*.

We can now calculate the diffraction efficiency. At the blazing wavelength the intensity I_m of the order m is given by $|F(2\pi m/d)|^2$:

$$
I_m = \text{sinc}^2 \left[\frac{b\pi}{d}(m - m_0) \right]; \qquad (9.24)
$$

in the ideal case (complete transmission) where $b = d$, $I_m = 0$ for all orders except $m = m_0$. The diffraction efficiency in the m_0th order is therefore 100%! In practice, a grating with $b = d$ is difficult to construct; there are usually some obstructed regions at the edges of the facets. Then we have $b < d$ and the orders $m \neq m_0$ have small but

Figure 9.7 Several spectral orders from a blazed transmission grating. The top and bottom of the photograph are under-exposed so that the relative intensity of the blazed order can be appreciated.

non-zero intensities, with a consequent reduction in the efficiency (see Fig. 9.7).

At a wavenumber $k_1 \neq k_0$ the phase ramp is $\exp(ik_1 \sin \beta)$ and

$$F(u) = \text{sinc}[b(u - k_1 \sin \beta)/2] \cdot \sum \delta(u - 2\pi m/d) \qquad (9.25)$$

$$I_m = \text{sinc}^2 \left[\frac{b\pi}{d} (m - m_0 \frac{k_1}{k_0}) \right] , \qquad (9.26)$$

in which we used the blazing condition to write $\sin \beta$ in terms of k_0. This is illustrated by Fig. 9.8(*b*); the diffraction efficiency is no longer 100%, but can still be quite high if $k_1 \approx k_0$. The modification for non-zero angle of incidence is simple, and will be left as a problem to the reader. It should be remarked that the blazing wavelength can be altered somewhat by changing the angle of incidence.

In detail, the structure of a diffraction grating is comparable in scale with the wavelength, and so the scalar theory of diffraction is not really adequate. In particular, polarization-dependent effects are very much in evidence. A fuller discussion is given by Hutley (1982).

9.3 Two-beam interferometry

As well as answering basic questions about the nature of light, the phenomenon of interference also opened up vast possibilities of accurate measurement. As we have seen, even Young's fringes can give quite an accurate measure of the wavelength of light, and with more carefully designed equipment optical interferometry has become one of the most accurate measurement techniques in physics.† In this section we shall describe several interferometers based on the interference of two separate waves, with examples of their applications. For

Figure 9.8 Diffracted intensity in the orders of a blazed grating; (*a*) at the wavelength for which the blazing was designed (all energy goes, theoretically, into the +1 order); (*b*) at a slightly different wavelength (the +1 order predominates, but other orders appear weakly).

† Time and frequency-counting techniques have relative accuracy about one order-of-magnitude better!

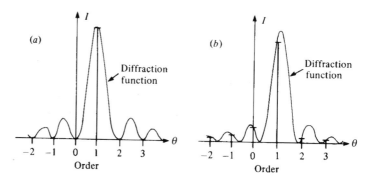

measurement purposes it is usual today to employ laser light sources, and the interferometers used are the ones appropriate to such sources; only when imaging is required as part of the interferometer (such as in interference microscopy, §12.4.6) are quasi-monochromatic sources preferred.

9.3.1 Jamin and Mach–Zehnder interferometers

These interferometers use partial reflexion at a beam-splitter to create two separate coherent light waves, and recombine them at a second beam-splitter. The optical path lengths can be compared, and small differences detected. The Jamin form is very stable to mechanical disturbances, but is less flexible in that the beams are relatively close to one another (Fig. 9.9). In the Mach–Zehnder inerferometer a greater separation is possible (Fig. 9.10). Both interferometers can be adjusted, using a white light source, to have zero path difference under specified conditions. A typical application of either is to the measurement of refractive index of gases. In Fig. 9.9, one sees that the two beams go through closed glass tubes of length L, one of which is evacuated and the other contains gas at a known pressure. The difference in optical path $\overline{\delta l}$ then allows the refractive index of the gas to be measured:

$$\overline{\delta l} = (\mu - 1)L. \tag{9.27}$$

In most measurements of this type one does not actually measure a fringe shift, but uses a compensator (usually a parallel glass plate of known thickness and refractive index inserted at a variable angle to one of the beams) to bring the fringes back to the initial (null) position. Using electronic detection, the null position can be sensed to about 10^{-3} fringes. Thus, from (9.27), for $L = 20\,\mathrm{cm}$ and $\lambda = 0.5\,\mu\mathrm{m}$, an accuracy of measurement $\delta\mu = \pm 10^{-3}\lambda/L = \pm 2.5 \times 10^{-9}$ is achievable.

9.3.2 Michelson interferometer

The Michelson interferometer produces beams that are not only widely-separated but also propagate in directions at right angles. These two features make it a very versatile instrument and, with its modifications, it is the best known of all the interferometers. It should not be confused with the Michelson stellar interferometer, which is described in §11.9.1.

The principle is illustrated in essence in Fig. 9.11. Light enters from the left and is partly reflected and partly transmitted by the semi-silvered mirror S; the two beams are reflected from the mirrors M_1 and M_2, and the resultant interference fringes are observed by the eye at A or B. Because the rays reflected from M_1 have to pass through three thicknesses of the mirror S, whereas those reflected from M_2 have to pass through only one, an extra plate at P is inserted to give equality between the two paths. This plate is needed because glass is dispersive and so only by having the same amount of the same glass in both beams can the optical paths be made equal at *all* wavelengths. This compensating plate must therefore be of the same thickness as S and placed at the same angle.

With the Michelson interferometer many different sorts of fringes can be obtained – straight, curved, or completely circular, in monochromatic or white light. These can all be understood in terms of a single theory if we regard the problem as a three-dimensional one, the dif-

Figure 9.9 Jamin interferometer used for the measurement of refractive index of a gas. At R the glass plates are fully reflecting, and at S they are about 50% reflecting. The interference is visible at outputs A and B.

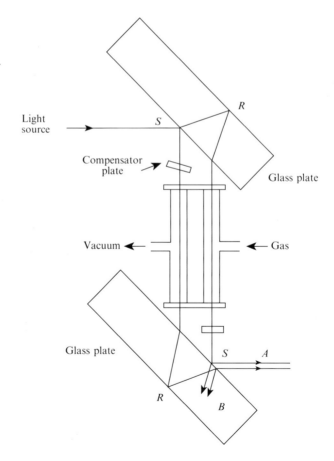

ferent sorts of fringes resulting from looking at the same pattern from different directions.

The source must be a broad one, but we can simplify the understanding of the interferometer by considering one 'ray' at a time,

Figure 9.10
Mach–Zehnder
interferometer. M_1
and M_2 are mirrors
and S_1 and S_2 are
50% reflecting
beam-splitters.

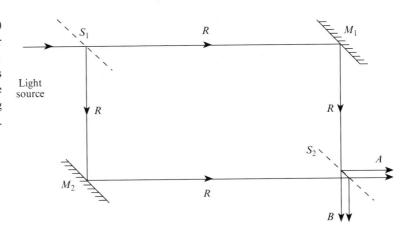

Figure 9.11
Michelson
interferometer. P is
the compensating
plate. The output can
be observed at A
or B.

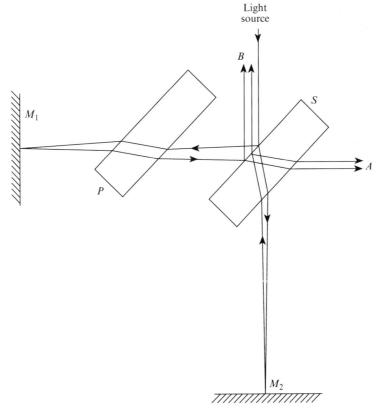

coming from one point on the source. If we ignore the finite thickness of the components, we see from Fig. 9.12 that O has an image in S at O_S and an image at O_2 in M_2; O_S has an image O_{S1} in M_1, and O_2 has an image O_{2S} in S. The images O_{S1} and O_{2S} are the two virtual sources that give rise to interference.

It can easily be seen that O_{S1} and O_{2S} can be brought as closely together as we require. Small adjustments in M_1 and M_2 can change their relative positions, and it is even possible to have one behind the other. The different sorts of fringes arise from the relative positions of O_{S1} and O_{2S}, and the scales of the fringes depend upon their separation.

If O_{S1} and O_{2S} are side-by-side, the situation is just like Young's experiment, and we get a set of straight† fringes, normal to the vector $O_{S1}O_{2S}$. The closer the points, the wider the fringe separation. A more usual situation is for O_{2S} to be behind or in front of O_{S1}. The directions of constructive and destructive interference then lie on cones around the line $O_{S1}O_{2S}$. From Fig. 9.13(a) we see that, on a screen at distance $L \gg O_{2S}O_{S1} \equiv s$ the wave amplitude at P, the point corresponding to an angle of observation θ, is

$$\psi_P \sim \exp(ik_0\overline{O_{2S}P}) + \exp(ik_0\overline{O_{S1}P}) \tag{9.28}$$

$$\sim 2\exp\left[ik_0\frac{\overline{O_{2S}P} + \overline{O_{S1}P}}{2}\right]\cos\left[k_0\frac{\overline{O_{2S}P} - \overline{O_{S1}P}}{2}\right]. \tag{9.29}$$

† The fringes are actually hyperbolic, but are indistinguishable from straight fringes in most practical situations (§9.1.1)

Figure 9.12 Principle of Michelson interferometer.

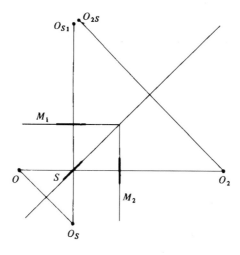

For this large L, $\frac{1}{2}(\overline{O_{2S}P} - \overline{O_{S1}P}) \simeq \frac{1}{2}s\cos\theta$, so the observed intensity can be written

$$|\psi_P|^2 \sim 4\cos^2\left[\frac{k_0 s \cos\theta}{2}\right] = 2[1 + \cos(k_0 s \cos\theta)]. \qquad (9.30)$$

This defines a set of circular fringes (of constant θ) in the plane containing P. The exact value of s determines whether the centre ($\theta = 0$) will be bright or dark. The scale of the fringe pattern is determined by s. If, for example, s is such that $k_0 s = m_0 \pi$ so that the centre is bright, the mth bright fringe is at angle θ_m such that $k_0 s \cos\theta_m = (m_0 - m)\pi$. For small angles θ, this gives

$$k_0 s \theta_m^2 \simeq 2m\pi. \qquad (9.31)$$

Thus the angular radii of the circular fringes, as in Fig. 9.13(*b*), are proportional to $\sqrt{m}$ on a scale depending on $k_0 s$.

As in §8.4, it is instructive to consider this problem in terms of the three-dimensional Fourier transform of the two points; this is a set of planar sinusoidal fringes, represented in Fig. 9.14. The different fringes observed are different aspects of this Fourier transform.

To understand this statement we make use once again of the concept of the sphere of observation (§8.4.2). Now, however, we are dealing with coherent sources and not scatterers so that their phase difference is always zero and does not depend upon an incident beam $\mathbf{k}_0$. Thus expression (8.65) must be replaced by unity; this can be done by putting $\mathbf{k}_0 = 0$ which results in the observation sphere having radius $2\pi/\lambda$ and being *centred on the origin* of reciprocal space. The sphere of observation therefore penetrates the Fourier transform, and has its centre on a maximum (Fig. 9.14).

As the points O_{S1} and O_{2S} become closer, the scale of the fringes becomes larger, and as the disposition of the points changes the transform rotates into different orientations. Fig. 9.15 shows how different types of fringes arise, when the intersection is projected onto the observation screen. If white light is used, the sphere must be

Figure 9.13 (*a*) Path difference at P between waves at angle θ to the axis, when O_{2S} lies behind O_{S1}. (*b*) Circular fringes seen in a Michelson interferometer.

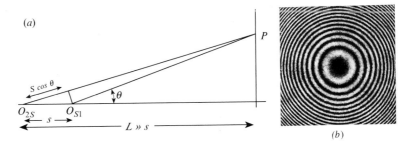

(*a*)

(*b*)

considered to have finite thickness; as we can see from Fig. 9.15(d), the centre fringe only is sharp and the others are coloured and soon merge together. Such coloured fringes are the best way of identifying the zero-order interference.

9.3.3 Localization of fringes

So far we have considered fringes as orginating from a single point source O. If the source is extended and incoherent, each point on it creates its own fringe pattern in a different place; the various fringe patterns superimpose and generally cancel one another. However, there is a region where all the fringe patterns coincide, and so even

Figure 9.14
Schematic
representation of the
Fourier transform of
two points – plane
sinusoidal fringes –
cutting the sphere of
observation.

Figure 9.15 Different
types of fringes from
Michelson
interferometer:
(a) and (b) show how
'straight' (see
footnote on p. 236)
and circular fringes
are produced, and
obviously
intermediate types of
fringes are possible;
(c) shows how the
fringes become finer
as O_{S1} and O_{2S} move
further apart;
(d) shows how
broadened fringes are
produced if a range
of wavelengths (e.g.
white light) is used.

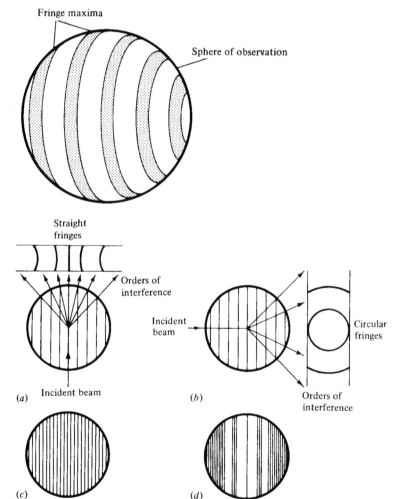

with a broad incoherent source a fringe pattern can be seen *localized* in this region; in the Michelson interferometer the fringes are localized in the region of the mirrors M_1 and M_2 (which the observer sees as coincident).

The following method of understanding the localization of fringes can be applied to any interferometer but will be illustrated by reference to the Michelson (Fig. 9.16). We consider a ray that travels from the point O through the interferometer and reaches the observer's eye. It will have divided on the way, and as we saw in §9.3.2, appears to have originated at two apparent sources O_{S1} and O_{2S}. Both rays must enter the eye for the interference pattern to be visible. In a projection on the plane defined by the eye, O_{S1} and O_{2S}, the two rays intersect at X. We shall assume that the zero-order fringe is visible which requires the interferometer to be adjusted to bring O_{S1} and O_{2S} side-by side, so that the path lengths $O_{S1}X$ and $O_{2S}X$ are equal. The zero order of diffraction then lies on the plane $\mathscr{Z}_O$ bisecting $O_{S1}O_{2S}$ (§9.1.1), which passes through X. Now consider a second point Q on the source. This has a specific geometrical relation to O and the ray through O which is preserved in the reflexions, so that its images Q_{S1} and Q_{2S} satisfy

$$O_{S1}Q_{S1} = O_{2S}Q_{2S}, \qquad (9.32)$$

and the angles $\quad \angle O_{S1}Q_{S1}X = \angle O_{2S}Q_{2S}X. \qquad (9.33)$

From the congruent triangles $O_{S1}Q_{S1}X$ and $O_{2S}Q_{2S}X$ it therefore follows that the zero-order plane $\mathscr{Z}_Q$ of the interference pattern between Q_{S1} and Q_{2S} also passes through X where it intersects $\mathscr{Z}_O$ along a line perpendicular to the plane of the diagram; thus *all* the zero-order fringes from points on the extended object intersect along this line. The pattern is therefore localized around X, the point of intersection between the projections of the rays on the plane containing O_{S1} and O_{2S}.

The higher-order fringes can also be located geometrically. When $O_{S1}O_{2S} \ll O_{S1}X$, the first-order surface $\mathscr{F}_O$ is approximately a plane at angle $\beta = \lambda/O_{S1}O_{2S}$ to $\mathscr{Z}_O$. One can see that when the angle is small the first-order planes derived from O and Q intersect along a line parallel to the zero order, in the plane through X normal to the ray (an *exact* geometrical construction which is independent of Q does not exist).

In the Michelson interferometer, X lies in the plane of the mirrors. The fringes are therefore *localized in the plane of mirrors*. This is an important aspect of the Michelson interferometer and its derivatives when used with extended incoherent sources; the observer must focus on the mirrors in order to see the fringe pattern.

9.3.4 The Michelson–Morley experiment

One of the most important experiments leading to the modern era of physics was carried out by Michelson and Morley around 1887, employing the superb accuracy that Michelson's genius enabled him to extract from his interferometer. He was concerned by the fact that in order to explain the aberration of light – the apparent change of the direction of light from a star that occurs because the Earth is in motion around the Sun – Fresnel had had to assume that the 'aether' (the assumed medium in which electromagnetic waves propagate) must be at rest as an opaque body moves through it. He therefore set himself the task of measuring the velocity of the Earth with respect to the aether.

Starting with the assumption that the Earth's velocity relative to the aether was of the same order of magnitude as its orbital velocity, Michelson showed that his interferometer could make the measurement with reasonable certainty. The difficulty was that the effect to be

Figure 9.16 Construction to show localization of Michelson fringes in the plane of the mirrors. The diagram is drawn when one mirror is tilted from the nominal position so that X can be located.

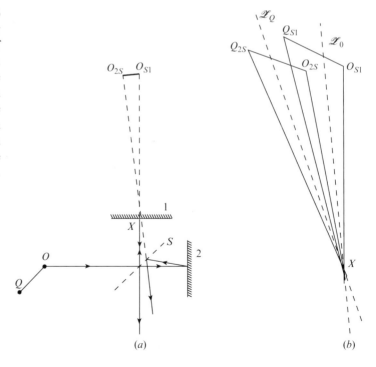

(a)   (b)

measured is a second-order one. The velocity of light could be found only by measuring the time taken for a light signal to return to its starting point, and the difference between the time for a journey *up and down* the path of the Earth and that *across* the path, to take the two extremes, is a second-order quantity derived as follows.

According to classical physics, the time t_1 for the up-and-down journey of a path L is

$$t_1 = \frac{L}{c+v} + \frac{L}{c-v} \qquad (9.34)$$

where v is the velocity of the Earth. For the transverse passage the light would effectively have to travel a longer path $L' = 2L[1 + (v^2/c^2)]^{\frac{1}{2}}$, the time taken being $t_2 = l'/c$. Expanded to second order in v/c we have

$$t_1 \approx \frac{2L}{c}\left(1 + \frac{v^2}{c^2}\right); \qquad t_2 \approx \frac{2L}{c}\left(1 + \frac{1}{2}\frac{v^2}{c^2}\right). \qquad (9.35)$$

The time difference $t_1 - t_2 = (L/c)(v^2/c^2)$, which corresponds to a path difference of Lv^2/c^2. If v is small compared with c, it would appear that the measurement of this quantity would not be possible.

But *was* it too small? The orbital velocity of the Earth is about 10^{-4} of the velocity of light. If L is 100 cm the path difference is about 10^{-6} cm, or about $\lambda/50$; this was too small for measurement using visual techniques, but large enough to suggest that with some modification a measurable effect might be expected.

The chief factor in producing a measurable path difference was an increase in the path L; the interferometer was mounted on a stone slab of diagonal about 2 m (Fig. 9.17) and the light reflected so that it traversed this diagonal several times, giving a total distance L of 11 m. Since there was no *a priori* knowledge of what might be the direction of the path of the Earth, the whole apparatus was caused to rotate and the maximum difference in path should be twenty-two times that previously calculated — just under half a fringe. Michelson and Morley were confident that they could measure this to an accuracy of about 5%.

This experiment is described in some detail because it is one of the most important experiments in optics. It illustrates the importance of developing techniques to measure very small quantities; the complete account of the care taken in avoiding spurious effects is well worth reading in the original (Michelson, 1927). The result was most surprising and disappointing; no certain shift greater than 0.01λ was found (Fig. 9.18). The broken line in Fig. 9.18 shows $\frac{1}{8}$ of the displacement expected from the orbital velocity of the Earth. It appeared the velocity of the Earth was zero!

There was just the possibility that the orbital velocity of the Earth at the time of the experiment happened to cancel out the drift velocity of the Solar System. This could not happen at all seasons of the year and therefore more measurements were made at intervals of several months. The result was always zero. This result was one of the mysteries of nineteenth-century physics. It was perplexing and disappointing to Michelson and Morley, whose skill and patience seemed to have been completely wasted. But now we know that it was not so; in 1905 Einstein came forward with a new physical principle, relativity, the main assumption of which is that the velocity of light is invariant whatever the velocity of the observer, which was completely in accordance with the experiments. Thus out of an apparently abortive experiment, a new physical principle arose and a new branch of physics had its beginning.

Figure 9.17 Interferometer used in the Michelson–Morley experiment (From Michelson, 1927.)

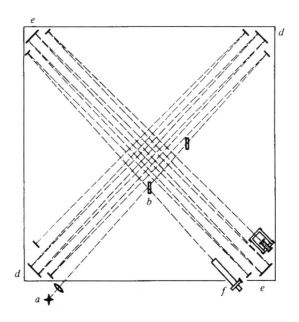

Figure 9.18 Typical diurnal variation of the fringe shift. (From Michelson, 1927.)

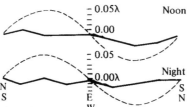

9.4 The Sagnac common-path interferometer

In this instrument, the two interfering beams traverse identical, or almost identical paths, but propagate in opposite directions. As a result, with almost no trouble it can be set up to give white-light fringes, since the two path lengths are automatically equal. Small differences can be introduced by making offsets from the ideal counter-propagating condition. Fig. 9.19 shows the simplest form, but variants using more mirrors are possible. There are as usual two output waves; one at A, which is easily accessible and one at B which returns in the direction of the source. There is no need for the compensating plate of the Michelson interferometer because both beams pass through the beam-splitter plate an equal number of times. If the amplitude reflexion coefficient of the splitter is $\mathscr{R}$ and its transmission coefficient $\mathscr{T}$, clearly one of the waves at A has amplitude $\mathscr{T}^2$ and the other $\mathscr{R}\overline{\mathscr{R}}$. The waves therefore interfere destructively when there is zero path distance at this exit (see §5.6.2), but cancel exactly only if $\mathscr{T}^2 = \mathscr{R}^2$; this requires a carefully constructed beam-splitter. At B, we have two waves with amplitudes $\mathscr{T}\overline{\mathscr{R}}$ which therefore interfere constructively at zero path difference, with unit contrast for all values of $\mathscr{T}$ and $\mathscr{R}$. Exit B is therefore often preferred for quantitative use, despite the need for additional optics (not shown in the figure) to separate the incident and outgoing waves.

If light passes through the interferometer at an angle to that shown in Fig. 9.19, a path difference is introduced because the two counter-propagating beams do not coincide exactly, and have slightly different path lengths in the beam-splitter plate. But the two emerging rays are always parallel, so if an extended incoherent source is used it follows from §9.3.3 that the fringe pattern is *localized at infinity*.

9.4.1 Velocity of light in a moving medium

The Sagnac interferometer is important for two reasons. The first is historical; it provided the first way of measuring relativistic effects in light propagation. In 1859, Fizeau constructed an interferometer of this type to measure the velocity of light in moving water (Fig. 9.20). The water flows as shown in the figure, and so one of the light beams propagates parallel to the flow velocity, and the second one anti-parallel to it. The difference in velocity between the two waves could thus be measured. Of course, the classical theory of motion in a moving frame of reference that gave light velocities $c_+ = c/\mu + v$ and $c_- = c/\mu - v$ for the two cases, did not explain the results, and Fizeau found it necessary to employ an 'aether drag' coefficient $(1 - \mu^{-2})$ to explain them (this term had previously been introduced to explain anomalous results of stellar aberration due to the motion of the Earth, §9.3.4). Einstein's theory of relativity explains the results correctly by showing c_+ and c_- to be:

$$c_\pm = \frac{c/\mu \pm v}{1 \pm v/\mu c}.$$
(9.36)

It is interesting that, unlike the Michelson–Morley experiment, the effect to be measured is a first-order correction in v, and can therefore be observed fairly easily.

9.4.2 Optical gyroscopes

An important modern application of the Sagnac interferometer is to make an optical gyroscope. Suppose that the whole interferometer rotates in its plane at an angular velocity Ω. A phase difference is produced between the two counter-propagating beams; this is called the *Sagnac effect*. Because it involves light propagating in a non-inertial (rotating) system it should properly be treated by general

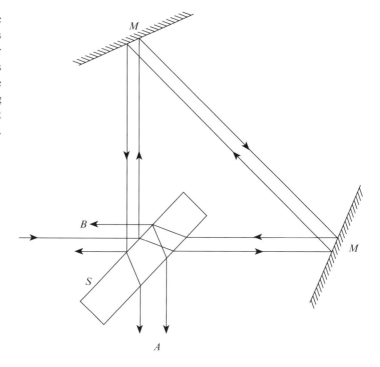

Figure 9.19 Sagnac interferometer. S is the beam-splitter plate. Because of its thickness, the counterpropagating beams do not coincide exactly.

relativity†, but it turns out that special relativity gives the right answer, as follows.

Consider a circular interferometer‡ of radius R, with path length L between the first and second passage of light through the beam-splitter (for a single turn, $L = 2\pi R$, but several turns may be involved). The velocity is $v = R\Omega$ and the light travels in a medium of refractive index μ. Using the velocity-addition formula for an inertial frame, in one sense (clockwise, the same as Ω) the velocity of light is c_+ and in the other sense (counter-clockwise) it is c_- (9.36). These velocities represent the two speeds of light as measured in the laboratory frame. During the time t_+ that the clockwise light takes to traverse the length L of the interferometer, the beam-splitter moves a distance $R\Omega t_+$, so that $c_+ t_+ = L + R\Omega t_+$. Likewise, for the anti-clockwise sense, $c_- t_- = L - R\Omega t_-$. Combining these two, we find

$$\Delta t = t_+ - t_- = L \left[\frac{1}{c_+ - R\Omega} - \frac{1}{c_- + R\Omega} \right]. \tag{9.37}$$

† Chow *et al.* (1985); special and general relativity give the same result since no gravitational field is involved.
‡ We represent the triangle of Fig. 9.19 by a circle, so that the velocity $R\Omega$ of the medium is a constant, and get an approximate result.

Figure 9.20 Fizeau's experiment to determine the velocity of light in moving water (schematically).

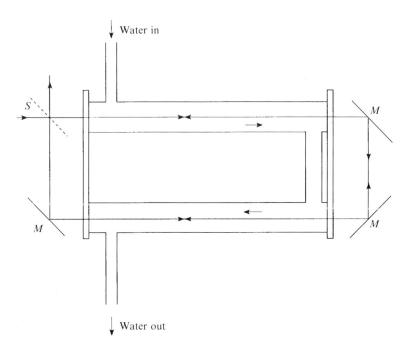

Water in

Water out

Substituting for c_+ and c_-, (9.36) we find

$$\Delta t = \frac{2LR\Omega}{c^2 - R^2\Omega^2} \simeq \frac{2LR\Omega}{c^2}. \tag{9.38}$$

Note that μ does not appear in this equation.

This time difference can be translated to a phase difference or a frequency difference, depending on the method of sensing it. In an interferometric optical gyroscope, a light wave of frequency ω enters the interferometer, and the phase difference $\omega\Delta t$ is measured. The effect is very small, and it is usual to use a coil of optical fibre to increase it by making L very long; the fact that the fibre is made of glass does not affect the result because (9.38) does not involve μ. R is simply made as large as practically convenient (it determines the size of the instrument). For example, if $L = 100$ m, $R = 0.1$ m, $\Omega = 1$ rad·s^{-1} and $\lambda = 0.5\,\mu$m the phase shift is

$$\Delta\phi = \omega\Delta t = \frac{2\pi c}{\lambda} \cdot \frac{2LR\Omega}{c^2} \simeq 0.8\,\text{rad}. \tag{9.39}$$

It is quite practical to measure such a phase shift, but a useful gyroscope must be accurate to about 10^{-3} times the rate of rotation of the Earth, $15°$hr^{-1}, so that phase shifts of order 10^{-7}rad would have to be measured. Because of attenuation, the longest practical value for L is a few kilometres.

We shall not go into the various sensitive detection techniques that have been developed nor into the problems that have had to be solved in order to develop this effect into a successful technology. Many of the problems centre around non-reciprocal processes in the fibre materials, which affect the propagation of the clockwise and counter-clockwise wave differently and thus mimic the Sagnac effect. For example, if the two beams have different intensities (beam-splitter $\mathcal{R} \neq 50\%$) and there is a non-linear refractive index component (§13.5) the optical paths of the two beams will differ.

The ring laser gyroscope (Fig. 9.21) represents another way of using the Sagnac effect. A closed loop resonator is constructed with (at least) three mirrors defining a cycle of length L assumed, for simplicity, to be filled with a lasing material of refractive index μ. The laser operates (see §14.5.1) at a resonance frequency for which the cycle length is a whole number m of wavelengths λ/μ. When the resonator rotates, the light travelling clockwise satisfies

$$L + tR\Omega = L(1 + \frac{\mu}{c}R\Omega) = m\lambda_+/\mu, \tag{9.40}$$

whereas the anti-clockwise wave satisfies

$$L - tR\Omega = L(1 - \frac{\mu}{c}R\Omega) = m\lambda_-/\mu. \qquad (9.41)$$

The very small wavelength difference is

$$\Delta\lambda = \lambda_+ - \lambda_- = \frac{2LR\mu^2\Omega}{cm}, \qquad (9.42)$$

assuming that the same mode number m is optimum in both senses. Translating this small $\Delta\lambda$ into a frequency difference between the exiting waves we have, using $\lambda = \mu L/m$ from (9.40–9.41),

$$\Delta f = \frac{\Delta\lambda\, c}{\mu\lambda^2} = \frac{2LR\mu\Omega}{m\lambda^2} = \frac{2R\Omega}{\lambda}. \qquad (9.43)$$

Measuring Δf allows Ω to be determined. This approach has, of course, assumed a 'circular triangle', but gives the result approximately. An estimate of its value, for $R = 0.1\,\text{m}$, $\Omega = 1\,\text{rad} \cdot \text{s}^{-1}$ and $\lambda = 0.5\,\mu\text{m}$ is $0.4\,\text{MHz}$. In principle measurement of such a frequency (or less when Ω is much smaller) should not be difficult, but early efforts to implement laser ring gyroscopes were bedevilled by frequency locking; as $\Delta f \rightarrow 0$, scattering in the optics caused the two counter-propagating modes to become mixed, therefore stimulating emission at the same frequency. Further details can be found in Kim and Shaw (1986), Lefevre (1993).

Figure 9.21 Ring laser gyroscope.

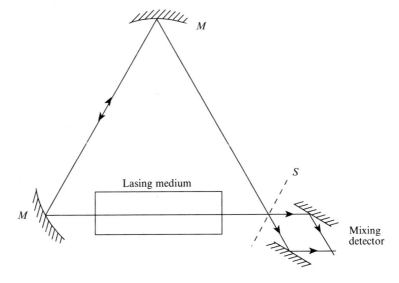

9.5 Interference by multiple reflexions

Two-beam interference provided both an initial verification of the wave theory of light and a method of measuring wavelengths with an accuracy of a few per cent. Because, as we showed in §9.1.1, the intensity distribution in two-beam interference is sinusoidal, the positions of the maxima and minima cannot be located with any great accuracy. Multiple-beam interference overcomes this difficulty; the conditions for reinforcement of several beams are much more precise than those for the reinforcement of two, and very sharp maxima can be obtained.

We have already discussed one way of using multiple-beam interferometry, the diffraction grating, in which a set of accurately constructed periodic apertures gave us a set of interfering waves with regularly incremented phases. Another way of producing such a set of waves uses multiple reflexions from a plane parallel transparent plate, or between parallel mirrors. As the wave is reflected backwards and forwards, a constant addition to its phase is made in each cycle, and if a little of the wave is extracted at each reflexion, the result is a set of waves with progressively increasing phase. We shall therefore consider the problem of multiple reflexions between two parallel surfaces each having amplitude reflexion coefficient $\mathscr{R}$ and transmission coefficient $\mathscr{T}$; if we assume that no energy is lost, then

$$\mathscr{R}^2 + \mathscr{T}^2 = 1. \tag{9.44}$$

Let us first calculate the phase difference introduced in one cycle. Consider the wave transmitted by a plate of thickness d and refractive index μ, when the wave travels at angle θ to the normal *within the medium* (Fig. 9.22). Just before the direct ray OAX leaves the surface, its wavefront is AD. After reflexion at B, the first reflected ray is BCY; AD is also a wavefront of this ray. The path difference between the two is $\overline{ABD} = \mu(AB + BD)$. By constructing A', which is the reflexion of A in the lower surface, it is clear that

$$AB + BD = A'D = 2d\cos\theta \tag{9.45}$$

so that the phase difference between the interfering wavefronts AD is

$$2\pi\overline{ABD}/\lambda = k_0\overline{ABD} = 2k\mu d\cos\theta \equiv g. \tag{9.46}$$

It is important to emphasize that the path difference is *not* twice the projected thickness AB of the plate. Moreover, g *decreases* (like $\cos\theta$) as the angle of incidence increases. Both these points are somewhat counter-initiutive.

Now let us look at the amplitudes of the multiply-reflected waves (Fig. 9.23). We should remember (§5.6.2) that $\mathcal{R}$ is defined for reflexion from one side of each reflector – let's say the inside. So reflexion from the outside will have coefficient $\overline{\mathcal{R}} = -\mathcal{R}$ (§5.6.2). The amplitudes of the waves are as shown in the figure. An exiting wave, either in reflexion or transmission, will combine the waves having these amplitudes with phase increments g at each stage. The situation is very similar to the diffraction grating except that the waves have steadily decreasing amplitudes.

Let us consider the transmitted light. The series is

$$\psi(g) = \mathcal{T}^2 \sum_{p=0}^{\infty} \mathcal{R}^{2p} \exp(ipg). \tag{9.47}$$

This function can be evaluated by two methods:

Figure 9.22 Path differences for rays reflected from top and bottom of plane film.

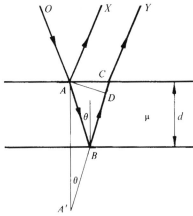

Figure 9.23 Multiple internal reflexions in a transparent plate. $\mathcal{R}$ and $\mathcal{T}$ refer to amplitudes.

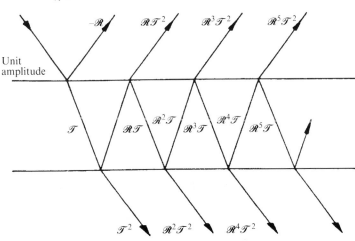

– as a Fourier series with coefficients $a_p = \mathscr{R}^{2p}$;
– as a geometric series with factor $\mathscr{R}^2 \exp(ig)$.

First we treat (9.47) as a Fourier series. It represents (§4.2.3) a periodic function with period $\Delta g = 2\pi$. The function within each period is the Fourier transform of the coefficients $\mathscr{R}^{2p}$, where p is considered as a continuous variable. Writing

$$\mathscr{R}^{2p} = \exp(2p \ln \mathscr{R}) \tag{9.48}$$

we see that we need the transform of

$$f(p) = \exp(-\alpha p) \ (p \geq 0); \quad f(p) = 0 \ (p < 0) \tag{9.49}$$

which is

$$F(g) = \int_0^\infty \exp[-(\alpha + ig)p]dp = (\alpha + ig)^{-1} \tag{9.50}$$

since p and g are conjugate variables [the exponent in (9.47) is $\exp(ipg)$]. Substituting $-2 \ln \mathscr{R}$ for α we have the required result:

$$F(g) = (-2 \ln \mathscr{R} + ig)^{-1}; \quad |F(g)|^2 = [4(\ln \mathscr{R})^2 + g^2]^{-1}. \tag{9.51}$$

The function (9.51) is called a *Lorentzian*. Its intensity $|F(g)|^2$ is illustrated in Figs. 9.24 (broken line) and 11.3; superficially it looks like a Gaussian curve, but it decays more slowly in the wings. Now, the Fourier series (9.47) is equal to $F(g)$ convoluted with $\sum \delta(g - 2m\pi)$; the convolution has intensity $|\psi(g)|^2$ and is shown in Fig. 9.24 (full line). In §4.4.6 we defined the 'half-peak width' w; for the Lorentzian we find that half the maximum value occurs when

$$g_H = \pm 2 |\ln \mathscr{R}| \tag{9.52}$$

so the half-peak width is given by

$$w = 2g_H = 4 |\ln \mathscr{R}| \approx 4(1 - \mathscr{R}), \tag{9.53}$$

the approximation being valid when $\mathscr{R} \approx 1$. In this case the overlap of neighbouring peaks is negligible, and so this result applies also to $|\psi(g)|^2$.

Second, we can evaluate (9.47) as a geometric series. This is the conventional method of attacking the problem. We write

$$\psi(g) = \mathscr{T}^2 \sum_{p=0}^\infty [\mathscr{R}^2 \exp(ig)]^p, \tag{9.54}$$

which is a geometrical series having sum

$$\psi(g) = \mathscr{T}^2 / [1 - \mathscr{R}^2 \exp(ig)]. \tag{9.55}$$

The intensity

$$
\begin{aligned}
I(g) &= |\psi(g)|^2 = \mathcal{T}^4/(1 + \mathcal{R}^4 - 2\mathcal{R}^2 \cos g) \\
&= \frac{\mathcal{T}^4}{(1 - \mathcal{R}^2)^2 + 4\mathcal{R}^2 \sin^2(g/2)} \\
&= \left(\frac{\mathcal{T}^2}{1 - \mathcal{R}^2}\right)^2 \cdot \frac{1}{1 + F \sin^2(g/2)},
\end{aligned}
\tag{9.56}
$$

where $F \equiv 4\mathcal{R}^2/(1 - \mathcal{R}^2)^2$ is called the *finesse*.

Expression (9.56) has some interesting features. We notice that the function has periodic maxima of value $[\mathcal{T}^2/(1 - \mathcal{R}^2)]^2$ at $g = m\pi$. If there is no absorption, $[\mathcal{T}^2/(1 - \mathcal{R}^2)]^2 = 1$, and so we reach the apparently paradoxical conclusion that, *even if the transmission coefficient $\mathcal{T}$ is almost zero*, at $g = m\pi$ all the light is transmitted! Of course it is not really a paradox; the strong transmitted wave results from constructive interference between many multiply-reflected weak waves. When F is large ($\mathcal{R} \simeq 1$) these maxima are very narrow; between them the function has value of order $1/F \ll 1$ (Fig. 9.25). The visibility of the fringes, which we shall define in §11.4.2, is then $F/(2 + F)$.

Calculation of the width of the peaks from (9.56) The transmission has fallen to half its peak value when $F \sin^2(g/2) = 1$. Thus

$$
g_H = 2 \sin^{-1}(F^{-\frac{1}{2}}) \approx 2F^{-\frac{1}{2}}.
\tag{9.57}
$$

If $\mathcal{R} \to 1$, the width $w = 2g_H \simeq 4(1 - \mathcal{R})$ as we had in (9.53). Table 9.1 gives the values of w corresponding to different values of $\mathcal{R}$. For comparison it should be noted that the corresponding value for Young's fringes is 0.50, so that until we attain a value of $\mathcal{R}$ greater than 0.6 no improvement in sharpness is obtained. (For values of $\mathcal{R}$ less than about 0.5, I does not even fall to values less than $\frac{1}{2}I_{\max}$.)

Figure 9.24 Intensity of the Lorentzian function (broken line); intensity when the amplitude of the Lorentzian is convoluted with a periodic array of δ-functions (full line).

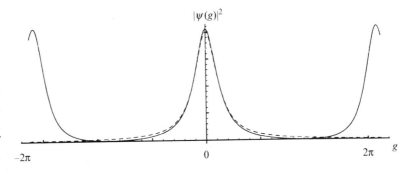

Table 9.1 *Values of half-peak width w for different values of $\mathscr{R}$*

$\mathscr{R}$	0.5	0.6	0.7	0.8	0.9	0.95	0.98
w	0.54	0.36	0.24	0.15	0.07	0.03	0.01

The reflected wave behaves in a complementary manner. We leave it to the reader to show that there is almost complete reflexion for all values of g, except for narrow dark lines at $g = m\pi$.

Multiple reflexion interference effects can be seen when any of the parameters in (9.46), $g = 2k_0 \mu d \cos \theta$ changes, namely k_0 (or λ), μ, d or θ. Fig 9.26 shows convergent light transmitted by a mica sheet, coated to make $\mathscr{R} \to 1$. Thus θ is a function of position, and sharp fringes occur wherever $g = n\pi$. In addition, d occasionally changes at molecular steps on the surface. Finally, because mica is birefringent, there are two values of μ and the fringes appear double.

9.5.1 The Fabry–Perot interferometer or étalon

An important practical application of interference by multiple reflexions is the Fabry–Perot interferometer. Its basic construction is simplicity itself: it consists of two flat glass plates, arranged in a mechanical support with spacers between them so that they are parallel to one another. The two inner surfaces are coated so as to have a high reflexion coefficient, but to transmit a small amount of light (Fig. 9.27) This simple description glosses over several important qualifications which can make the Fabry–Perot an expensive instrument. First, the

Figure 9.25 Form of $I(g)$ from (9.56) for different values of $\mathscr{R}$.

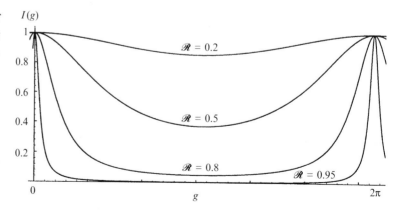

plates must be flat to a very high degree of accuracy (better than $\lambda/50$). Second, the inner surfaces (reflective) must be very accurately parallel to one another. Third, the distance d, which can be very large (centimetres) must not change with time, due to temperature or other fluctuations. On the other hand, the outer surfaces of the plates are largely irrelevant. They must be optically flat, but play no part in the analysis. It is convenient if they are not quite parallel to the inner surfaces, or are anti-reflexion coated, so as not to give rise to reflexions which could confuse the interference pattern.

If an extended monochromatic source of light is observed through the interferometer, sharp bright rings are seen at angles θ given by $2k_0\mu d\cos\theta = 2m\pi$, or $\mu d\cos\theta = m\lambda/2$. In this case, μ is the refrac-

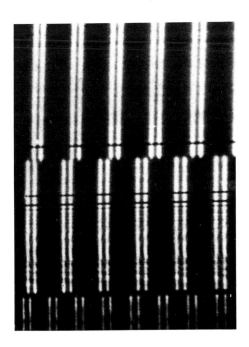

Figure 9.26 Multiple reflexion fringes in birefringent mica, from Tolansky (1973).

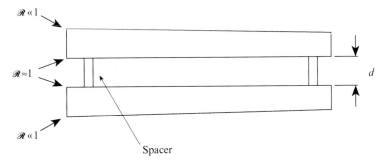

Figure 9.27 Schematic construction of a Fabry–Perot étalon. The outer surfaces are not quite parallel to the inner ones.

$\mathcal{R} \ll 1$

$\mathcal{R} \approx 1$

$\mathcal{R} \ll 1$

d

Spacer

tive index of the air (or other medium) between the plates. Just like the Michelson interferometer fringes, these rings have radii basically depending on the square roots of the natural numbers (see equation (9.31)). But the rings are now very sharp. If we consider g_H to indicate how closely fringes of different wavelength can be situated before they merge,† the resolving power is given for large F by (9.57)

$$\frac{g}{\delta g} = \frac{g}{g_H} = \pi m F^{\frac{1}{2}}. \tag{9.58}$$

This is equal to $\lambda/\delta\lambda$, $d/\delta d$ or $\mu/\delta\mu$, depending on what is being measured. For example, if the intensity reflexion coefficient is $\mathscr{R}^2 = 0.95$ and $d = 2.5\,\mathrm{cm}$, we have for $\lambda = 0.5\,\mu\mathrm{m}$, $m = 2d/\lambda = 10^5$ at the centre of the ring pattern and $F = 1600$. Then $g/\delta g = 2.5 \times 10^7$, which is considerably better than the other interferometers considered so far. The reason is clear; not only is the order m large, but also the number of interfering waves N is large, and the resolving power $g/\delta g = mN$ (9.11). The effective number of reflexions contributing to the interference is $2\pi F^{\frac{1}{2}}$. However, in order to get such resolution, the phase after more than $2\pi F^{\frac{1}{2}}$ reflexions must be accurate to better than $\lambda/4$, so the plates must be flat and *parallel* over the area used to $\lambda/8\pi F^{\frac{1}{2}}$, in this example to $\lambda/1000$, which is about 5 Å!

The Fabry–Perot interferometer is widely used for studying the fine structure of spectral lines. If a source emits several lines, the overlapping ring patterns may be confusing, and it is usual to separate out the line of interest by first passing the light through a 'lesser' spectrometer, such as a prism or diffraction grating (Fig. 9.28). The limited spectral region separated out by this first spectrometer must be such that all the rings observed as one group (satellite of the same spectral line for example) do indeed have the same order. This means that the region must be smaller than $\delta\lambda = \lambda/2m$, which is called the *free spectral range* of the interferometer.

9.5.2 Multiple reflexions in an amplifying medium

A subject that has become of great importance with the advent of the laser is the effect of an amplifying medium on the behaviour of a multiple reflexion interferometer, since this is the basis of laser resonators (§3.10 and §14.5.1). Suppose that the laser medium amplifies

---

† One can not use the Rayleigh criterion here to determine the resolution, because (9.56) has no zeros. The Sparrow criterion (§12.3) would suggest using $0.6g_H$ instead of g_H.

the wave field by factor G during a single round trip. Then (9.47) becomes:

$$\psi(g) = \mathcal{T}^2 \sum_{p=0}^{\infty} (\mathcal{R}^2 G)^p \exp(ipg). \tag{9.59}$$

The result obtained would be similar to (9.56) if $\mathcal{R}^2 G$ were less than unity. For large enough G, the value becomes unity and the sum is:

$$
\begin{aligned}
\psi(g) &= \mathcal{T}^2 \sum_{p=0}^{\infty} \exp(ipg) \\
&= \mathcal{T}^2 \sum_{q=0}^{\infty} \delta(g - 2\pi q).
\end{aligned}
\tag{9.60}
$$

The spectrum is a series of ideally sharp lines; for this function $g_H = 0$. This is the basic reason that laser lines are so sharp. Now if one asks what happens if $G\mathcal{R}^2$ becomes greater than unity, one is asking a question which mathematics cannot answer. In practice, in a continuous laser the amplification factor G eventually settles down, at high enough intensity, to a value which is equal to $\mathcal{R}^{-2}$, so that stability is achieved at that intensity. In a pulsed laser the amplification is large as the pulse starts, and gradually gets smaller as the population inversion is wiped out. The number of terms in the series for which $G\mathcal{R}^2$ is greater than unity remains finite; towards the end of the pulse G falls below unity and the series terminates.

One would expect from the above arguments that a continuous laser would emit a number of perfectly sharp lines (longitudinal modes) separated by $\delta k = \pi/\mu d$ (from (9.60)) indicating a wavelength separation $\delta\lambda = \lambda^2/2\mu d$. As will be discussed in §14.6, the lines are not ideally sharp because of noise (spontaneous emission) and thermal fluctuations; in addition, the number of lines emitted is rather small

Figure 9.28 Use of Fabry–Perot étalon for high-resolution spectroscopy.

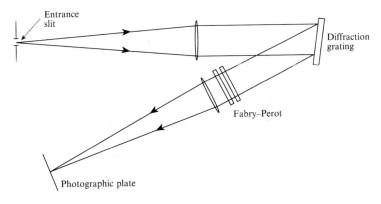

(sometimes only one) centred around the wavelength for which the gain G is maximum. A pulsed laser clearly does not produce ideally sharp lines; if the wave train continues only for a time T, the linewidth $\delta\lambda$ must be greater than λ^2/cT.

9.5.3 The confocal resonator: transverse modes

One widely-used example of a periodic system of the type discussed in §3.10 is the confocal resonator, for which $L = R_1 = R_2$ (Fig. 9.29). This is just stable in the geometrical sense, and the foci of the two mirrors coincide (hence the name). If such a resonator contains an amplifying (lasing) medium, we have a system which closely approximates that discussed in the previous section, in that geometrical optics do not predict any light leakage, even for mirrors of finite size. We shall not discuss the properties of this resonator in detail, but use it to illustrate the idea of *transverse mode patterns*.

Consider the amplitude $a(x, y)$ of light travelling to the right in the common focal plane (Fig. 9.29). Since this is the light amplitude in the focal plane, the amplitude at the other focal plane of the mirror M_2 (which is coincident with it) must be its Fourier transform $A(u, v)$ (§8.2), but the light is travelling to the left. However, this system is symmetrical about the focal plane and so the direction of travel of the waves is unimportant, and a stable mode of operation is seen when the Fraunhofer diffraction pattern is identical to the original function $a(x, y)$ in amplitude and phase at each point.

Few continuous functions have this property: that the function and its Fourier transform are identical in form. We have already met two such functions which transform into themselves: the infinite periodic array of δ-functions, and the Gaussian. The former is of infinite extent and is not continuous. The latter, the Gaussian function, is in fact the first of a set of functions all of which have the required property; the complete set is the set of Gauss–Hermite polynomials that should be

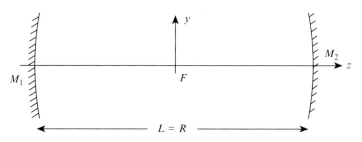

Figure 9.29 Confocal resonator.

familiar to any student of quantum mechanics as the wave functions of a harmonic oscillator.† They are expressed as

$$a_n(x) = H_n(x/\sigma)\exp(-x^2/2\sigma^2), \tag{9.61}$$

where the functions $H_n(x)$ obey the recurrence relation:

$$2xH_n = H_{n+1} + 2nH_{n-1}; \quad H_0 = 1 \tag{9.62}$$

which gives $H_1 = 2x$, $H_2 = 4x^2 - 2$, etc. In Fig. 9.30(a) we show two examples of these functions.

In two dimensions any product of the form

$$a_{lm}(x, y) = H_l(x/\sigma)H_m(y/\sigma)\exp[-(x^2 + y^2)/2\sigma^2] \tag{9.63}$$

† The wave function in p-space is the Fourier transform of that in q-space, and the Hamiltonian of the harmonic oscillator, which can be written $\mathcal{H} = \frac{1}{2}(p^2 + q^2)$, is invariant on interchanging p and q.

Figure 9.30
(a) Examples of Gauss–Hermite functions;
(b) photographs of laser modes.

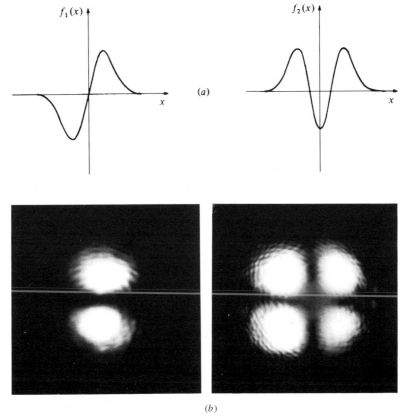

$f_1(x)$ x (a) $f_2(x)$ x

(b)

satisfies our requirements. They can be seen as the intensity distribution across the output beam of a slightly misaligned continuous-wave laser; two examples are shown in Fig. 9.30(*b*). The various functions are known as *transverse modes* and are referred to by the number pair (l, m). They should be compared with the similar modes in optical fibres (§10.3.1), although their origin there is rather different.

The requirement of identity of scale allows us to calculate the focal size at the common focus when the focal length is F. For the mode $(0, 0)$, which is a Gaussian spot,

$$a(x, y) = \exp[-(x^2 + y^2)/2\sigma^2], \tag{9.64}$$

the diffraction pattern is the transform of (9.64):

$$A(u, v) = \exp[-(u^2 + v^2)\sigma^2/2]. \tag{9.65}$$

The physical size of the pattern is given (for small angles) by (8.9):

$$p_x = uF/k_0, \quad p_y = vF/k_0 \tag{9.66}$$

and we required p_x and p_y to be identical coordinates with x and y. Thus for $A(p_x k_0/F, p_y k_0/F)$ and $a(x, y)$ to be the same function:

$$\exp[-(x^2 + y^2)/2\sigma^2] = \exp[-k_0^2(x^2 + y^2)\sigma^2/2F^2] \tag{9.67}$$

giving $\sigma^2 = F/k_0$. Thus the radius of the spot, of order $\sigma/2^{\frac{1}{2}}$, is $(F/2k)^{\frac{1}{2}} = (F\lambda/4\pi)^{\frac{1}{2}}$. This, of course, is the spot size at the mid-point of the laser tube. When the beam reaches the mirrors, it will have the form of the Fresnel diffraction pattern of this spot at distance F, which is easily seen to have about twice the above radius.

9.6 Advanced topic: Berry's phase in interferometry

Two waves with orthogonal polarizations can not interfere directly. It is therefore important to ensure that the initial polarization of a wave entering an interferometer is preserved or restored before the interference pattern is observed. From a practical point of view, polarization changes within an interferometer often occur when light is reflected at oblique incidence from a mirror or beam-splitter, and also when the light path does not lie in a single plane. Even if interference occurs, the phase may be different from a naive expectation, because of phase changes on reflexion.

If the light path is not planar, further phase changes can arise for topological reasons, depending on whether the light path describes a right-handed or left-handed helical route; such phases are called *geometrical* or *Berry phases* (Berry, 1984, 1987). We shall illustrate this phenomenon with a simple example.

The discussion is considerably simplified if we build our interferometer with polarization-preserving mirrors and beam-splitters. These are not conventional laboratory components; in general, reflexion at a mirror reverses the sense of a circularly-polarized wave, but this can be avoided if reflexion occurs from a dielectric surface at an angle above the Brewster angle. If this property is required at an angle of incidence around 45°, it is necessary to use internal reflexion; but in order to avoid other phase changes, reflexion must not be critical. These conditions are satisfied by a prism of, say, MgF$_2$, ($\mu < \sqrt{2}$) as shown in Fig. 9.31(c). We emphasize that the above mirrors have been introduced in order to simplify the following discussion; they have no effect on the value of the Berry phase.

With such polarization-preserving reflectors we can construct a three-dimensional Mach–Zehnder interferometer in which the light beams traverse the edges of a cube. The edges of the cube are exactly equal, so we expect, *a priori*, that there will be constructive interference at exit *A* and destructive interference at *B*.† Two possible

† The asymmetry arises because of the beam-splitter: see §5.6.2. *B* will not concern us here.

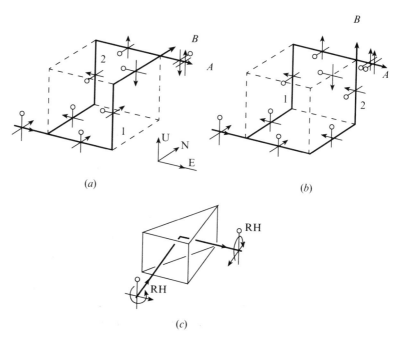

Figure 9.31 'Cubical Mach–Zehnder' interferometers, each with zero path difference and two routes between the beam-splitters; in (*a*) the routes labelled 1 and 2 have LH and RH helicities respectively, whereas in (*b*) they both have RH helicity. (*c*) shows the relationship between the field vectors on reflexion at the special mirrors in this example.

constructions for the interferometer using different cube edges, each having path difference zero, are shown in Fig. 9.31. In each of the interferometers, we follow through the orientation of the **E** vector of the incident wave after successive reflexions; this is done for initially $\parallel$ ($\uparrow$) and $\perp$ ($\updownarrow$) fields. What appears in Fig. 9.31 is that in version (*a*) there is *destructive* interference at *A* (the exiting fields have opposite orientations) while in (*b*) there is *constructive* interference.

The difference can be traced to the sense of rotation of the light beams. Designating the three directions by North, East and Up as shown, in (*a*) beam 1 goes EUNE while beam 2 goes ENUE; the first is a rudimentary left-handed helix, while the second is right-handed. In (*b*), both beams are ENUE, which is right-handed. It is the different sequences of the finite rotations which introduce the π phase difference between the two cases. This phase change is topological in origin, and depends only on the geometry of the system; it is an example of a wide range of such phase changes, in classical and quantum physics, which have been derived on general grounds by Berry (1984). They are identically zero in two-dimensional arrangements.

For any chosen three-dimensional interferometer the geometrical phase can be deduced by plotting the propagation vectors **k** of the waves as they traverse the two arms of the interferometer as two loci on the surface of a sphere, as in Fig. 9.32(*a*). Before the initial beam-splitter the **k**-vectors of the two waves coincide, and likewise after the second one. Thus the two loci have common end-points. If we define Ω as the solid angle subtended at the centre of the sphere by the segment of the sphere enclosed between the two loci, we shall show that the topological phase difference between the two is $\pi - \Omega/2$.

Figure 9.32
(*a*) Construction of a general **k**-route on the surface of a sphere; (*b*) construction of locus ENUE appropriate to the cube interferometer in Fig. 9.31(*a*). $\Omega/2$ is the spherical angle enclosed by the three axes.

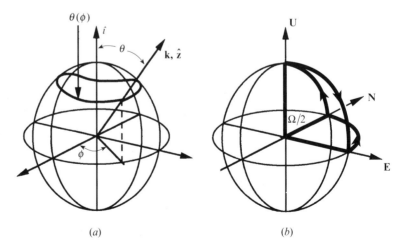

(*a*)   (*b*)

This construction makes it easy to calculate the topological phase for any interferometer (Problem 9.10). Clearly, if the interferometer lies in a plane, $\Omega = 0$.

This construction can be derived in the case of electromagnetic waves as follows (Lipson, 1990). We shall assume that the changes in $\mathbf{k}$ are continuous (so that the loci on the sphere are uniquely defined, although in the example we quoted there are actually discontinuous changes in $\mathbf{k}$ at the mirrors!) We postulate an observer of the electromagnetic wave who travels slowly along each of the routes, measuring the vector field in his own frame of reference, which changes continuously so that the local z-axis always coincides with $\mathbf{k}$. In order to maintain this situation, the observer has to rotate his frame, and we define $\boldsymbol{\alpha}(t)$ as his angular velocity with respect to the laboratory frame.

In the frame rotating at $\boldsymbol{\alpha}$, we relate the time-derivative of a general vector $\mathbf{V}$ to that in the inertial frame by

$$\left(\frac{\partial \mathbf{V}}{\partial t}\right)_{\text{rotating frame}} = \left(\frac{\partial \mathbf{V}}{\partial t}\right)_{\text{inertial frame}} - \boldsymbol{\alpha} \times \mathbf{V}. \qquad (9.68)$$

We can apply this to Maxwell's equations (see footnote on p. 000 in §9.4.2) ((5.9)–(5.10)) which yield, in the observer's space,

$$\nabla \times \mathbf{H} = \epsilon_0 \left(\frac{\partial \mathbf{E}}{\partial t} - \boldsymbol{\alpha} \times \mathbf{E}\right) ; \qquad (9.69)$$

$$\nabla \times \mathbf{E} = -\mu_0 \left(\frac{\partial \mathbf{H}}{\partial t} - \boldsymbol{\alpha} \times \mathbf{H}\right). \qquad (9.70)$$

The other two Maxwell equations ((5.7)–(5.8)) are unchanged. The wave equation which replaces (5.12) then follows as

$$\frac{\partial^2 \mathbf{E}}{\partial t^2} + \boldsymbol{\alpha} \times \frac{\partial \mathbf{E}}{\partial t} = c^2 \nabla^2 \mathbf{E}. \qquad (9.71)$$

This equation is analogous to that obtained in classical mechanics for Foucault's pendulum swinging on a rotating Earth, in which the second term arises from the *Coriolis force*.

In contrast to the usual wave equation (5.12), a linearly-polarized plane-wave is not a solution of (9.71); however, left- and right-handed circularly-polarized waves are solutions. Substituting waves

$$\mathbf{E}_{\pm} = E_0(1, \pm \mathrm{i}, 0) \exp[\mathrm{i}(\omega t - kz)], \qquad (9.72)$$

we immediately find the dispersion relation

$$c^2 k^2 = \omega^2 \pm \alpha_z \omega, \quad \Rightarrow \quad \omega_{\pm} \approx ck \pm \alpha_z/2 \qquad (9.73)$$

when $\alpha \ll \omega$. The phase difference between the waves along the two routes now results from the slight difference between the velocities that arises if α_z is positive for one and negative for the other, as a result of their opposite helicities.

The phase difference between beginning and end of a route is

$$\int_0^{z,t} (k \, \mathrm{d}z - \omega \, \mathrm{d}t) \pm \frac{1}{2} \int_0^t \alpha_z \, \mathrm{d}t = \Delta\Phi_0 \pm \frac{1}{2} \int_0^t \alpha_z \, \mathrm{d}t \equiv \Delta\Phi_0 + \gamma, \quad (9.74)$$

$\Delta\Phi_0$ indicating the usual (kinetic) phase difference expected from the optical path length of the route. The extra term γ can be easily interpreted by the construction on the sphere. The direction of $\mathbf{k}$ has angular position (θ, ϕ), Fig. 9.32(*a*). Then the locus is the curve $\theta(\phi)$ and $\boldsymbol{\alpha}$ is the sum of the two orthogonal components

$$\boldsymbol{\alpha}(t) = \frac{\mathrm{d}\phi}{\mathrm{d}t}\hat{\imath} + \frac{\mathrm{d}\theta}{\mathrm{d}t}\hat{\jmath}, \qquad (9.75)$$

where $\hat{\imath}$ is the unit vector to the pole and $\hat{\jmath}$ is the common normal to $\hat{\imath}$ and $\mathbf{k}$. Thus the projection of $\boldsymbol{\alpha}$ on the z-axis in the rotating coordinates, defined as being parallel to $\mathbf{k}$, is

$$\alpha_z = \frac{\mathrm{d}\phi}{\mathrm{d}t}(\hat{\imath} \cdot \hat{z}) = \frac{\mathrm{d}\phi}{\mathrm{d}t}\cos\theta. \qquad (9.76)$$

Integrating this along the route gives

$$\gamma = \frac{1}{2} \int_0^t \alpha_z \, \mathrm{d}t = \frac{1}{2} \int_0^t \frac{\mathrm{d}\phi}{\mathrm{d}t}\cos\theta \, \mathrm{d}t = \frac{1}{2} \int \cos\theta(\phi)\mathrm{d}\phi. \qquad (9.77)$$

For a closed loop, the solid angle subtended at the centre of the sphere is $\Omega = \oint [1 - \cos\theta(\phi)] \, \mathrm{d}\phi = 2(\pi - \gamma)$. Thus, when we have drawn on the sphere the $\mathbf{k}$-routes corresponding to the two arms of the interferometer, $\gamma = \pi - \Omega/2$, where Ω is the solid angle subtended by the enclosed segment. We recall that this γ corresponds to one of the circularly polarized waves. The wave with opposite sense gives $-\gamma$, and the *difference* between the two, namely 2γ, is directly measurable since $\Delta\Phi_0$ in (9.74) cancels out. Experiments on variations of the cube interferometer carried out by Chiao *et al.* (1988) and on propagation in helically-coiled fibres (Tomita and Chiao, 1986) confirm this result. In the cube interferometer we described in Fig. 9.31, $\gamma = \pi/2$, and it is the phase difference $2\gamma = \pi$ which results in the destructive interference (Fig. 9.32(*b*)).

CHAPTER TEN

Optical waveguides and modulated media

10.1 Electromagnetic waves in restricted systems

This chapter will deal with two examples of electromagnetic wave propagation in systems where the scalar-wave approximation is inadequate, essentially because of the small dimensions of the constituent parts. The first is the optical waveguide, already familiar in everyday life as the optical fibre, which has caused a revolution in the communications industry. The second example is the dielectric multilayer system which, in its simplest form (the quarter-wave anti-reflexion coating) has been with us for more than a century, and can today be used to make optical filters of any degree of complexity that are common elements in the laboratory.

10.2 Optical waveguides

Transmission of light along a rod of transparent material by means of repeated total internal reflexion at its walls must have been observed countless times before it was put to practical use. In this section we shall describe the geometrical and physical optical approaches to this phenomenon, and derive some of the basic results for planar and cylindrical guides, the latter of which is a model for the optical fibre. Optical fibres have many uses, two of which will be described briefly at the end of the section; the first is for transmitting images, either faithfully or in coded form, without the use of lenses; the second is for optical communication.

10.2.1 Geometrical theory of wave guiding

The principle of the optical fibre can be illustrated by a two-dimensional model (corresponding really to a wide strip rather than a fibre) shown in Fig. 10.1. The strip has thickness $2a$ and refractive index μ_2, and is immersed in a medium of lower refractive index μ_1. A plane-wave incident inside the strip at angle $\hat{\imath}$ to the x-axis is reflected completely at the wall (§5.5.1) if $\hat{\imath}$ is greater than the critical angle $\hat{\imath}_c = \sin^{-1}(\mu_1/\mu_2)$. Since the two sides of the strip are parallel, the wave is then reflected to and fro at the same angle repeatedly, ideally with no losses (Fig. 10.1(a)). According to geometrical optics, any ray with $\hat{\imath} < \hat{\imath}_c$ can propagate in this way. However, physical optics requires us to look at the sum of all the waves travelling in the same direction, and to ensure that they interfere constructively. If we do this we naively calculate the phase difference between adjacent waves travelling parallel to one another to be (Fig. 10.1(b)):

$$\Delta\phi = \mu_2 k_0 (BC - EC) = \mu_2 k_0 \, B'E = 4\mu_2 k_0 a \cos \hat{\imath}, \qquad (10.1)$$

where k_0 is 2π divided by the wavelength in free space. The requirement for constructive interference is then (as in §9.5)

$$\Delta\phi = 4\mu_2 k_0 a \cos \hat{\imath} = 2n\pi. \qquad (10.2)$$

Each integer value of n defines an allowed mode of propagation. There will always be at least one solution to equation (10.2) given by $n = 0$, $\hat{\imath} = \pi/2$. As can be seen from Fig. 10.2(a) the number of additional solutions having $\hat{\imath} > \hat{\imath}_c$ is the integer part of

$$\frac{2\mu_2 k_0 a}{\pi} \cos \hat{\imath}_c = \frac{2\mu_2 k_0 a}{\pi}(1 - \mu_1^2/\mu_2^2)^{\frac{1}{2}}. \qquad (10.3)$$

Now unfortunately the calculation is not quite as simple as this, because we have neglected to take into account the phase change $\alpha(\hat{\imath})$ which occurs on reflexion at angles exceeding the critical (§5.5.2). We should then write instead of (10.2)

$$\Delta\phi = 4\mu_2 k_0 a \cos \hat{\imath} + 2\alpha(\hat{\imath}) = 2n\pi. \qquad (10.4)$$

The solution $n = 1$, $\hat{\imath} = \pi/2$ is now the first solution because $\alpha(\pi/2) = \pi$. However, since $\alpha(\hat{\imath}_c) = 0$, there will usually be one mode less than

Figure 10.1
Geometrical optics of
light propagation
along a fibre in one
dimension. In (b), *BE*
is the wavefront
common to rays *AB*
and *B′E*.

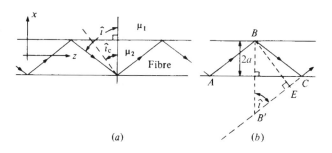

(a)   (b)

suggested by (10.3), as shown in Fig. 10.2(*b*). The modes for the two principal polarizations will not be identical because of the difference between $\alpha_\parallel$ and $\alpha_\perp$.

The modes described above, with $\hat{\imath} > \hat{\imath}_c$, are theoretically loss-less modes, and can propagate along an ideal fibre as far as the absorption coefficient of the medium will permit. In addition there are lossy modes with $\hat{\imath} < \hat{\imath}_c$, which die away after a certain number of reflexions, and are important only for very short fibres.

The above geometrical approach does in fact give a fairly complete picture of propagation in the slab, including in particular the

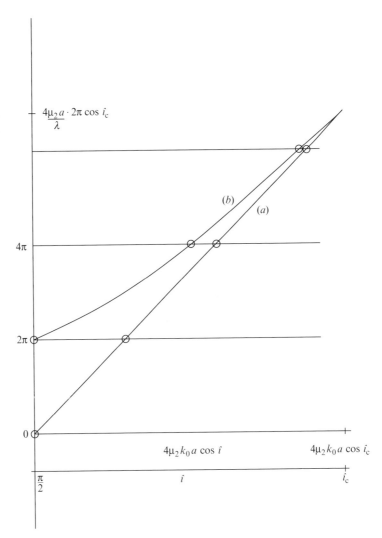

Figure 10.2 Graphical determination of the number of modes propagating in a slab; (10.2) is represented by line (*a*), (10.4) by line (*b*).

features of propagation modes and differences between the $\perp$ and $\parallel$ polarizations. In addition, it is easy to see that the wave entering the slab (and fibres in general) must do so at an angle sufficiently close to the axis that critical reflexion occurs. This means a restriction on possible angles of incidence in the exterior medium at the end of the slab (Fig. 10.3) which seriously influences the efficiency with which incoherent light can be fed into it, particularly when $\mu_2 - \mu_1 \ll \mu_1$.

However, the geometrical approach becomes clumsy to use quantitatively in any extension of this simple model. Two cases are of great importance. First, there is the optical fibre, which has a cylindrical cross-section, and supports some modes in which the light rays spiral around the axis and are not confined to a single plane (*skew rays*). Secondly, both slab waveguides and optical fibres can have continuously varying refractive index, in which case there is no well-defined plane at which critical reflexion occurs, but the angle $\hat{\imath}$ changes gradually along z, as in the situation of a mirage, §5.5.5. These are called *graded index* systems and are very important practically.

These problems are better treated by a more general approach in which Maxwell's equations are solved from scratch in the required environment. The method is much more fruitful, and has the enormous advantage of highlighting the similarity of the electromagnetic wave equation to Schrödinger's equation for matter waves (§2.3.2). Solutions to this latter equation, with which the reader may well be familiar from studies of quantum mechanics (see, e.g., Cohen-Tannoudji *et al.*, 1977; Gasiorowicz, 1974) help us both to solve particular problems easily and to develop a taste for possible useful configurations. In fact, many quantum-mechanical concepts such as tunnelling and the band theory of solids have found direct application to optical waveguides in analagous situations.

10.2.2 Wave equation for a planar waveguide

Continuing our two-dimensional planar model for a waveguide we shall now construct and solve the wave equation for the same system

Figure 10.3 Cone of angles allowing light to enter propagating modes.

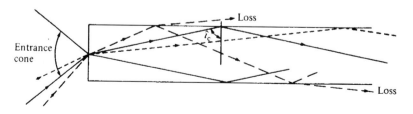

as was shown in Fig. 10.1. Specifically, we have propagation in the z-direction, while the refractive index $\mu(x)$ varies in the x-direction. There is no functional dependence on y in this model, but we have $\perp$ polarization ($E = E_y$ only) or $\parallel$ polarization (E in (x, z) plane) as two independent possibilities (§5.4.2.).

The wave equation begins with (5.11):

$$-\frac{\epsilon}{c^2}\frac{\partial^2 \mathbf{E}}{\partial t^2} = \nabla \times (\nabla \times \mathbf{E}) = \nabla(\nabla \cdot \mathbf{E}) - \nabla^2 \mathbf{E}. \tag{10.5}$$

As we saw in Chapter 6, Gauss's law, $\nabla \cdot \mathbf{D} = 0$, does not imply $\nabla \cdot \mathbf{E} = 0$ unless ϵ is a homogeneous (i.e. spatially uniform) scalar number. In the present situation this is not so. Recalling that $\epsilon(x) = \mu^2(x)$,

$$
\begin{aligned}
0 = \nabla \cdot \mathbf{D} = \epsilon_0 \nabla \cdot (\epsilon \mathbf{E}) &= \epsilon_0(\epsilon \nabla \cdot \mathbf{E} + \nabla \epsilon \cdot \mathbf{E}) \\
&= \epsilon_0 \epsilon \nabla \cdot \mathbf{E} + \epsilon_0 E_x \frac{\partial \epsilon}{\partial x}.
\end{aligned} \tag{10.6}
$$

Now the wave equation (10.5) becomes:

$$\frac{\epsilon}{c^2}\frac{\partial^2 \mathbf{E}}{\partial t^2} = \nabla^2 \mathbf{E} - \nabla(\nabla \cdot \mathbf{E}) = \nabla^2 \mathbf{E} + \nabla(E_x \frac{1}{\epsilon}\frac{\partial \epsilon}{\partial x}), \tag{10.7}$$

which reduces to the usual Maxwell wave equation (5.12) if the term

$$\nabla\left(E_x \frac{1}{\epsilon}\frac{\partial \epsilon}{\partial x}\right) \tag{10.8}$$

is small enough with respect to $\nabla^2 \mathbf{E}$ to be neglected. For the $\perp$ mode, $E_x = 0$ and so (10.8) is identically zero. But for the $\parallel$ mode $E_x \neq 0$, although in many examples $\partial \epsilon / \partial x = 0$ except for a limited number of discontinuities. In what follows, we shall assume that the term (10.8) is negligible, in which case there is no difference between the equations for $\perp$ and $\parallel$ (although the *boundary conditions* they satisfy are not identical); this is called the *weak guiding approximation*.

For the wave $E = E(x)\exp[i(k_z z - \omega t)]$ propagating in the z-direction, we can substitute into (10.7) $\partial/\partial z = ik_z$, $\partial/\partial t = -i\omega$ and $\partial/\partial y \equiv 0$ and get

$$\frac{\partial^2 E}{\partial x^2} - k_z^2 E = \frac{-1}{c^2}\epsilon\omega^2 E = \frac{-\mu^2(x)}{c^2}\omega^2 E \tag{10.9}$$

$$\frac{\partial^2 E}{\partial x^2} = [k_z^2 - \mu^2(x)k_0^2]E. \tag{10.10}$$

For analogy's sake we shall write Schrödinger's time-independent wave equation in the same way so that the similarity can be seen:

$$\frac{\partial^2 \psi}{\partial x^2} = \frac{2m}{\hbar^2}[-\mathscr{E} + \mathscr{V}(x)]\psi \tag{10.11}$$

One sees immediately that there will be corresponding solutions for refractive index profile $-\mu^2(x)$ and potential well $\mathscr{V}(x)$. Then $-k_z^2$ corresponds to the energy eigenvalue $\mathscr{E}$. It is also clear that a propagating mode, for which k_z is real, corresponds to a bound state in quantum mechanics, for which $\mathscr{E} < \mathscr{V}(\infty)$.

The specific form of $\mu(x)$ which represents the optical slab waveguide shown in Fig. 10.4 is:

$$\mu(x) = \mu_2 \quad (|x| < a, \text{ called the 'core'}) \tag{10.12}$$

$$\mu(x) = \mu_1 < \mu_2 \quad (|x| \geq a, \text{ called the 'cladding'}) \tag{10.13}$$

and is equivalent to a square-well potential. The guided wave solutions of this problem therefore (using the quantum mechanics analogy of a particle in a potential well to guide us directly to the solution) lie in the region $\mu_1^2 k_0^2 < k_z^2 < \mu_2^2 k_0^2$. If we therefore define

$$\mu_2^2 k_0^2 - k_z^2 \equiv \alpha^2 \qquad k_z^2 - \mu_1^2 k_0^2 \equiv \beta^2$$

$$\alpha^2 + \beta^2 = (\mu_2^2 - \mu_1^2)k_0^2 \equiv V^2 \tag{10.14}$$

we have, in the core region $|x| < a$,

$$\frac{\partial^2 E}{\partial x^2} = -\alpha^2 E$$

$$\Rightarrow E = E_{2s} \cos \alpha x + E_{2a} \sin \alpha x \tag{10.15}$$

where the suffixes 's' and 'a' refer to 'symmetrical' and 'antisymmetrical' modes. In the cladding region, $|x| > a$, we have likewise:

$$\frac{\partial^2 E}{\partial x^2} = \beta^2 E$$

$$\Rightarrow E = E_{1l}e^{\beta x} + E_{1r}e^{-\beta x} \quad (\beta \geq 0) \tag{10.16}$$

where suffixes l and r indicate 'left' and 'right'. Within the core of the slab, the function is wave-like (oscillatory); in the cladding it is evanescent, and for E to remain finite it must decay like $\exp(-\beta|x|)$ at large $|x|$. The core and cladding solutions join together in a continuous fashion, which will shortly be detailed. The complete function is therefore 'trapped' or 'localized' in a region centred on the

Figure 10.4 Refractive index profile $\mu(x)$ for the slab, and the equivalent Schrödinger potential.

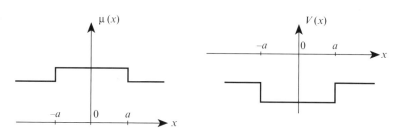

core. This is the essence of the guided wave.† The inherent symmetry of the system about the plane $x = 0$ has suggested using symmetrical and antisymmetrical solutions in (10.15). Considering just the region $x > 0$ ($x < 0$ follows by symmetry or antisymmetry), clearly only the solutions (10.16) with $E_{1l} = 0$ are acceptable. At $x = a$ the field components E_y, E_z, H_y and H_z parallel to the interface (see §5.4.1) must be continuous. For the $\perp$ mode, $E_z = 0$ and continuity of E_y then requires, for the cosine solution to (10.15),

$$E_{2s} \cos \alpha a = E_{1r} e^{-\beta a}. \tag{10.17}$$

The field H_z can be calculated from Maxwell's equation (5.4) for the $\perp$ case $\mathbf{E} = (0, E_y, 0)$:

$$-\frac{\partial \mathbf{B}}{\partial t} = \frac{i\omega}{c^2 \epsilon_0} \mathbf{H}$$

$$= \nabla \times \mathbf{E} = \left(\frac{\partial E_y}{\partial z}, 0, -\frac{\partial E_y}{\partial x} \right), \tag{10.18}$$

whence continuity of H_z implies continuity of $\partial E_y / \partial x$. Thus the analogy with Schrödinger's equation is complete. For the cosine solution to (10.15) this gives

$$\alpha E_{2s} \sin \alpha a = \beta E_{1r} e^{-\beta a}. \tag{10.19}$$

Dividing (10.19) by (10.17):

$$\alpha a \tan \alpha a = \beta a. \tag{10.20}$$

Similarily, choosing the sine solution to (10.15) gives

$$-\alpha a \cot \alpha a = \beta a. \tag{10.21}$$

We can repeat the calculation for the $\parallel$ polarization for which $\mathbf{H} = (0, H_y, 0)$. The argument analagous to (10.17)–(10.19) then lead to the equations, equivalent to (10.20)–(10.21),

$$\frac{\alpha a}{\mu_1^2} \tan \alpha a = \frac{\beta a}{\mu_2^2}; \tag{10.22}$$

$$-\frac{\alpha a}{\mu_1^2} \cot \alpha a = \frac{\beta a}{\mu_2^2}. \tag{10.23}$$

In practice the difference between μ_1 and μ_2 is often extremely small, in which case there is negligible difference between the two types of solution.

We shall limit our attention to the $\perp$ mode, for which (10.20)

† We could extend the analogy by allowing β to be imaginary, and k_z complex. Then we should arrive at the lossy modes. This is left as an exercise to the reader.

and (10.21) are eigenvalue equations whose solutions define particular values of α and β. These must satisfy $\alpha^2 + \beta^2 = V^2$, a constant, from (10.14). Only certain values of k_z result, and these correspond to *propagation modes* of the slab. The equations can not be solved analytically, but the solutions can be found graphically by plotting βa as a function of αa according to (10.20) and (10.21) and finding their intersections with the circle representing (10.14), as shown in Fig. 10.5. The circle has radius Va, and as a increases, one finds more and more intersections with the curves. Only the quadrant $\alpha, \beta > 0$ is relevant since β was defined as positive, and the figure is symmetrical about the β-axis.

The most interesting feature that appears from Fig. 10.5 is that the number of propagating modes is finite. There is always at least one mode (even as $V \to 0$); in general, the number of modes is $1 + n$, where n is the integer part of $2aV/\pi$ (cf. §10.2.1).

Typical forms of $E(x)$ are shown in Fig. 10.6. Alternate solutions are symmetric [cosine-like, with $E_{2a} = 0$ in (10.15)] and antisymmetric [sine-like, with $E_{2s} = 0$]. The lowest mode ($n = 0$) has a single peak in the centre of the slab; higher modes have more and more peaks.

10.2.3 Dispersion

Another important feature of the propagation is the *inter-mode dispersion relation*. The significance of dispersion with regard to the transmission of information was discussed in §§2.3 and 2.8. We have considered the frequency $\omega = k_0 c$ as a constant in the analysis so far. In order to create a dispersion curve, we now have to look at the

Figure 10.5 Graphical construction to find the modes in a waveguide slab. The curves labelled 's' and 'a' represent symmetric and antisymmetric modes respectively and the circle, radius aV, is shown for $a = 1, 1.7, 4$.

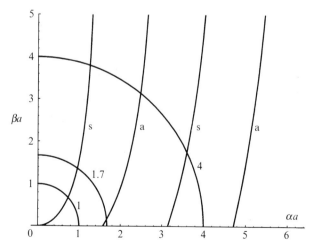

dependance of k_z (the propagation wave-vector along the slab) on k_0. It is easiest to do this by reading off from the intersections in Fig. 10.5 the value of β as a function of V. Then we calculate from (10.14)

$$k_z = (\beta^2 + \mu_1^2 k_0^2)^{\frac{1}{2}}. \tag{10.24}$$

The result is shown schematically in Fig. 10.7. It will be seen that for any given mode n:

(1) propagation starts when $\beta = 0$, i.e. at $k_0 = n\pi/2a(\mu_2^2 - \mu_1^2)^{\frac{1}{2}}$;

(2) when β is small, $k_z \approx \mu_1 k_0$; the wave propagates as if in the cladding medium;

(3) when β is large $\beta \approx V$, and so $k_z = \mu_2 k_0$; propagation is then dominated by the core.

If one looks at the distribution of energy (E^2 from Fig. 10.6) it is clear that the propagation velocity is dominated by the medium in which most of the energy is located.

Figure 10.6 $E(x)$ for the modes $n = 0, 1$, and 3 in a slab.

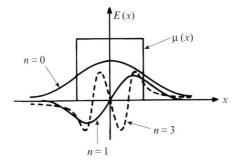

Figure 10.7 Schematic diagram of the dispersion curve for a slab waveguide. The curved region is exaggerated for clarity.

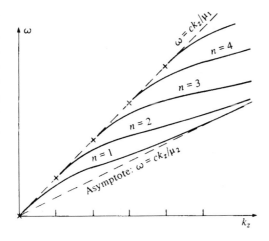

10.2.4 Single-mode waveguide

The slab with only a single propagating mode ($n = 0$) is particularly important for communication purposes, and is called a *single-mode waveguide*. The reason for its importance is that in a multi-mode guide the wave and group velocities, ck_0/k_z and cdk_0/dk_z which can be found from Fig. 10.7, differ from mode to mode and as a result information sent along a fibre in wave-group form will be distorted when several modes propagate simultaneously. Use of a single-mode waveguide avoids this cause of distortion, although pulse spreading due to the non-linear form of $k_z(k_0)$ within the single mode still occurs.

In order to make a single-mode guide, we require $a < \pi/2V$. This implies:

$$a < 0.25\lambda/(\mu_2^2 - \mu_1^2)^{\frac{1}{2}}, \tag{10.25}$$

where λ is the free-space wavelength. Although a symmetrical waveguide always has at least one mode, an asymmetrical one may have no modes at all if its width is very small.

10.3 Optical fibres

The discussion has so far centred around the one-dimensional waveguide. Although this configuration has many applications, by far the most common waveguiding system is an optical fibre. The basic geometry is a cylindrical core of glass with refractive index μ_2 embedded in a cladding medium of index μ_1. Ideally the cladding is infinite in extent, but in practice it is cylindrical, coaxial with the core, and has a large enough diameter to contain the evanescent waves (10.16) out to many times their decay distance β^{-1}.

We shall not repeat the slab calculation for the cylindrical case except to the extent that new features emerge. The equation that is to be solved is (10.7) with $\epsilon \equiv \mu^2(r)$, in the weak guiding approximation. Since the boundary (i.e. the interface between the two media) has axial symmetry, it is convenient to rewrite the equation for scalar E in cylindrical polar coordinates (r, θ, z):

$$\frac{\partial^2 E}{\partial x^2} + \frac{\partial^2 E}{\partial y^2} - [k_z^2 - \mu^2(x, y)k_0^2]E = 0 \tag{10.26}$$

becomes

$$\frac{\partial^2 E}{\partial r^2} + \frac{1}{r}\frac{\partial E}{\partial r} + \frac{1}{r^2}\frac{\partial^2 E}{\partial \theta^2} - [k_z^2 - \mu^2(r)k_0^2]E = 0. \tag{10.27}$$

Because of the axial symmetry, it is possible to write $E(r, \theta)$ as the product of two functions, $R(r)\Theta(\theta)$; (10.27) is then:

$$\frac{\Theta\, \mathrm{d}^2 R}{\mathrm{d}r^2} + \frac{\Theta}{r}\frac{\mathrm{d}R}{\mathrm{d}r} + \frac{R}{r^2}\frac{\mathrm{d}^2\Theta}{\mathrm{d}\theta^2} - [k_z^2 - \mu^2(r)k_0^2]\, R\Theta = 0. \qquad (10.28)$$

Dividing by $R\Theta$, and multiplying by r^2,

$$\frac{r^2}{R}\frac{\mathrm{d}^2 R}{\mathrm{d}r^2} + \frac{r}{R}\frac{\mathrm{d}R}{\mathrm{d}r} + \frac{1}{\Theta}\frac{\mathrm{d}^2\Theta}{\mathrm{d}\theta^2} - r^2[k_z^2 - \mu^2(r)k_0^2] = 0 \qquad (10.29)$$

which contains terms that are either functions of r or of θ, but not both. Thus the equation breaks up into two, one in r, the other in θ, each of which must be independent and equal to a constant, which we denote by l^2. They are

$$\frac{1}{\Theta}\frac{\mathrm{d}^2\Theta}{\mathrm{d}\theta^2} = \text{constant} \equiv -l^2 \qquad (10.30)$$

$$\text{and}\quad \frac{r^2}{R}\frac{\mathrm{d}^2 R}{\mathrm{d}r^2} + \frac{r}{R}\frac{\mathrm{d}R}{\mathrm{d}r} - r^2[k_z^2 - \mu^2(r)k_0^2] = l^2. \qquad (10.31)$$

The sum of these two equations is (10.29). Equation (10.30) introduces a new feature that did not appear in the planar waveguide and is related to the skew rays of §10.2.1. Any solution of it must satisfy $\Theta(2\pi) = \Theta(0)$, i.e.:

$$\Theta(\theta) = A \cos l\theta + B \sin l\theta \qquad (10.32)$$

where l is a non-negative integer and A and B are arbitrary constants. Looking at this part of the solution alone, we see that the light intensity E^2 is modulated angularly with an even number of peaks in the full circle. These are called *azimuthal modes* (Fig. 10.8). We have already met such modes in the confocal resonator in §9.5.3.

10.3.1 Step-index fibres

The radial equation (10.31) is no easier to solve than was (10.10) and gives radial modes which are oscillatory in the core and evanescent in the cladding. Note that the l^2 term can be included as if it were an additional dielectric constant $-l^2/k_0^2 r^2$; because r^{-2} diverges at $r = 0$ the field of modes with $l \neq 0$ must vanish there. In the simplest case, a step-index fibre, $\mu(r) = \mu_2$ $(r < a)$ and $\mu(r) = \mu_1$ $(r > a)$. The analysis is similar to that for the slab, with the cosine and sine functions replaced by Bessel functions J_0 and J_1 (see Appendix 1). Some typical mode structures are shown in Fig. 10.8. Only one mode can propagate if $a < 0.383\lambda/(\mu_2^2 - \mu_1^2)^{\frac{1}{2}}$, cf (10.25), and *single-mode fibres* with this property are very important for communication purposes because of their relatively small dispersion. If μ_2 and μ_1 are very close, the maximum core diameter $2a$ of a single-mode fibre can be considerably larger than λ. For example, using $\mu_2 = 1.535$ and $\mu_1 = 1.530$, a fibre with $2a < 6.2\lambda$ will support only a single mode.

10.3.2 Graded-index fibres

At this point we recall from §2.8 that one of the results of non-linear dispersion on the propagation of a wave-packet is to cause a progressive increase in its width. This eventually poses a limit to the repetition rate at which packets can be propagated without their merging. It turns out that the step-index fibre, even in the single-mode variety, has sufficient dispersion within a single mode to stimulate a search for fibres with lower dispersion for long-distance communication purposes. This has led to the development of *graded-index fibres*, in which the refractive index $\mu(r)$ is a continuous function of the radius. A commonly found profile for $\mu^2(r)$ is parabolic. This has smaller dispersion than the step-index fibre, although there is no proof that this profile has minimum dispersion; indeed a slightly lower power than parabolic produces some improvement. However, the parabolic profile $\mu^2(r) = A - br^2$ reminds us of the harmonic

Figure 10.8 Intensity distribution photographed for several modes in a circular fibre.

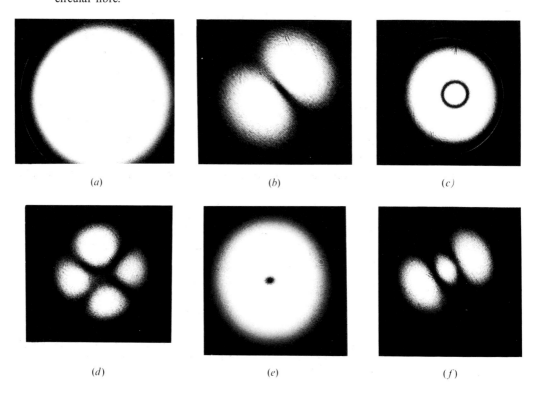

(a)   (b)   (c)

(d)   (e)   (f)

oscillator potential in quantum mechanics and since Schrödinger's equation has a simple solution for this model we shall pursue it briefly. In practice, the parabolic profile must be limited to the central region only,† and a fibre is constructed as in Fig. 10.9, which represents a cladded parabolic-index fibre, whose parameters can be adjusted so that it has a single mode only, still preserving minimal dispersion. A full discussion is given by Ghatak and Thyagarajan (1980).

The equation (10.29) can be taken as the starting point by substituting $\mu^2(r) = \mu_2^2 - b^2 r^2$. Clearly all that has been said already about azimuthal modes applies in this case too, since the form of $\mu^2(r)$ was not involved in (10.30). The radial equation (10.31) can be written

$$\frac{d^2 R}{dr^2} + \frac{1}{r}\frac{dR}{dr} + R\left(U - \alpha^2 r^2 - \frac{l^2}{r^2}\right) = 0 \tag{10.33}$$

where $U = \mu_2^2 k_0^2 - k_z^2$ and $\alpha = k_0 b$. This makes it analogous to the two-dimensional harmonic oscillator equation in quantum mechanics, where the total energy is U and the potential energy is $\alpha^2 r^2$. It suggests to us a solution of the form

$$R = e^{-\alpha r^2/2} f(r) \tag{10.34}$$

where $f(r) = \sum_{j=n_0}^{n} a_j r^j$ is a finite polynomial series.§

Substitution of (10.34) into (10.33) and comparison of coefficients of r^n, together with the requirement for n to be a finite integer (i.e. $a_k = 0$ for all $k > n$) leads us directly to the following conclusions.

(a) There are independent symmetric and antisymmetric solutions for even and odd l respectively. For $l \neq 0$, $R(0)$ must be zero.

(b) The value of $n_0 = l$. This means that the radial function can have at most $n - l + 1$ peaks, since it is a polynomial with this number of terms multiplied by the Gaussian. A typical mode pattern would look like Fig. 10.10.

(c) The allowed values $U = \mu_2^2 k_0^2 - k_z^2 = 2\alpha(n + 1)$. The value 2α is analogous to $h\nu$ in the quantum harmonic oscillator; the 'ground state energy' ($n = 0$) is $h\nu$, and not $\frac{1}{2}h\nu$, because it is a two-dimensional system. So the dispersion equation for the lowest mode (using $k_0 = \omega/c$ and $\alpha = k_0 b$) is:

$$\mu_2^2 k_0^2 - 2b k_0 = k_z^2. \tag{10.35}$$

---

† Otherwise the refractive index value outside a certain radius becomes less than unity!
§ This is the classical Sommerfeld solution of the quantum mechanical harmonic oscillator equation.

This lowest mode has $l = n = 0$, so that the electic field has amplitude

$$E(r) = R(r) = e^{-k_0 b r^2 / 2};\qquad(10.36)$$

it is a simple Gaussian profile. The radius $r = (k_0 b)^{-\frac{1}{2}}$ is just the inflexion point on this profile, so that $(k_0 b)^{-\frac{1}{2}}$ is essentially the radius of the equivalent core (compare Fig. 10.10 with Fig. 10.6). But do not take this result too seriously; it only applies to an infinite parabolic profile, and is altered significantly by the presence of a cladding medium, which must necessarily exist since μ can never fall below about 1.5.

One can see qualitatively the origin of the lesser dispersion in a graded-index fibre from Fig. 10.11. In the figure, the mode with the shorter path length is confined to the region where $\mu(x)$ is largest, while the mode with the longer path length enters regions of smaller refractive index. The two parameters – refractive index and path length – partially compensate, and give a dispersion which is less than in a step-index fibre. The dispersion of the glass itself must also be taken into account; in the normal dispersion region (§13.3.2), $\partial^2 \omega / \partial k^2$ for glass conveniently has the opposite sign to that for the fibre dispersion, and further compensation is possible.

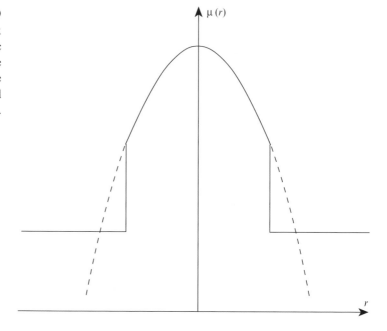

Figure 10.9
Refractive index
profile of a parabolic
graded-index fibre
(full line), and the
mathematical model
(broken line).

10.3.3 Production of fibres

A few words about how fibres are produced will take the above discussion out of the realm of pure theory. Fibres are made commercially in many kilometre lengths from specially prepared low absorption glasses. First, a short cylindrical glass rod (preform), several centimetres in diameter, is prepared with a central core of higher refractive index than the outer region. This is heated and drawn through a small orifice with diameter equal to the outer diameter required of the fibre. The inner structure all scales down in proportion. A typical absorption spectrum for a fibre glass is shown in Fig. 10.12. Notice that the units on the abscissa are dB/km, where one dB is a loss of intensity of transmitted light by $10^{0.1}$ (B is the 'Bel' which corresponds to one decade loss).

A similar method of production is used for graded-index fibres. In this case, the original glass preform is constructed from axial layers of glass of differing refractive indices, often deposited chemically by vapour deposition from sources of gradually varying composition.

Light losses from fibres come from several sources. Absorption in the glass, as mentioned above, is one; tremendous development

Figure 10.10 Typical modes in a parabolic index fibre.

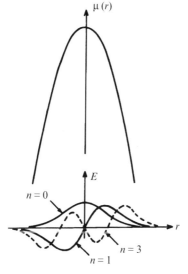

Figure 10.11 Ray equivalents of two modes in a parabolic graded-index fibre.

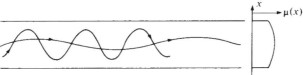

activity has resulted in this factor becoming negligible at some wave-lengths, even for hundreds of kilometres. Rayleigh scattering (§13.2) is important at short wavelengths and arises because glass is not a crystalline material and therefore has unavoidable statistical density fluctuations. Losses because of incomplete total internal reflexion in a step-index fibre are effectively avoided by the interface between μ_2 and μ_1 being buried well inside the cladding, so that it cannot be damaged or dirtied; *a fortiore* in the graded-index system where the interface is undefined. With practical dimensions, loss via the evanescent wave in the cladding can also be made negligible, although when the fibre is bent losses from this source can become noticeable.†

10.3.4 Communication through optical fibres

Optical fibres have now become the standard transmission medium for telephones and data. A typical system starts with a light-emitting-diode or semiconductor laser, emitting at the wavelength where the fibre absorption and dispersion are least (1.3–1.5 μm), whose output intensity is modulated according to the signals to be transmitted. The light is focused into a fibre. At the far end, the light is reconverted to an electronic signal by a photodetector. The maximum distance for transmission is limited by losses in the fibre, which we discussed above. It may therefore be necessary to amplify the signal periodically on very long routes. This can be done by terminating the fibre as with a photodetector, amplifying the signal electronically, removing noise that does not correspond to the transmission code and then retransmitting. Recently, schemes for amplifying the light signals *within the fibre* by

† The problem of a slab waveguide bent to a radius can be solved analytically, and is found to be analogous to a quantum-mechanical system—the tunnel diode!

Figure 10.12 Typical absorption coefficient for a fibre glass, as a function of wavelength. The limits for Rayleigh scattering and far-infra-red OH⁻ band absorption are shown as broken lines.

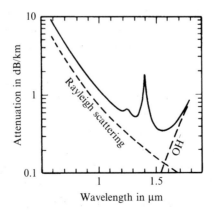

stimulated emission have been devised; essentially, part of the fibre suitably doped and pumped, is made into an optical amplifier based on the laser principle (see §14.4.3).

The great attraction of optical fibres for data transmission lies in the potential number of 'telephone conversations' which can be transmitted simultaneously on a single fibre. If we suppose that the light transmitted has frequency ω, and that a single conversation covers a band of frequencies of width ω_1 then, by mixing various conversations each with a different intermediate frequency, in principle ω/ω_1 conversations can all be used to modulate the light wave simultaneously. If $\omega = 10^{15}\,\mathrm{s}^{-1}$ and $\omega_1 = 10^5\,\mathrm{s}^{-1}$, this number is 10^{10}. Technology is nowhere near capable of using this enormous potential, but the attraction remains. This estimate has in fact neglected several points of fundamental importance, one of which is dispersion (§2.8), which reduces the potential severely although it still leaves an impressive residue. At the present, some 25 000 'conversations' can be transmitted simultaneously on a glass fibre pair compared with twenty-four on conventional cable. A popular account of present-day achievements is given by Desurvire (1992).

10.3.5 Imaging applications

For image transmission, the mode structure of light transmission in a fibre is unimportant; we are only concerned that the light be transmitted from one end to the other. A bundle of fibres is arranged in an organized array and the end is cut across cleanly. At the other end the fibres are arranged in the same way. What happens in between is unimportant. An image projected on one end is then seen at the other end. This type of device is invaluable as a method of transmitting images from inaccessible regions; one important medical application is to the internal examination of patients. The resolution of the image is just determined by the diameter of each fibre, which is typically 20–50 µm. By changing the ordering of the fibres at the far end, an image can be coded, for example changed from a circular field of view to a slit-like field. Or the reordering may be simply a rotation; inverting an image by means of a fibre bundle is cheaper, is less bulky and introduces less aberration than would a lens system although its resolution is limited (Fig. 10.13).

The subject of fibre optics is comprehensively covered by review articles and books at all levels, for example Gloge (1979), Marcuse (1989, 1991a,b), Agrawal (1989) and Keiser (1991).

10.4 Propagation of waves in a modulated medium

Another practically important electromagnetic wave propagation problem arises when the refractive index of the medium is modulated periodically. Continuing the vein of our analogy to solved quantum-mechanical problems, we immediately seek guidance fom the behaviour of an electron in a periodic crystal potential. The solutions that emerge indeed have many similarities to the well-known Bloch waves, and show band gaps that are entirely analogous to those found in the electronic structure of crystals. Although the treatment here is limited to one dimension, the subject of electromagnetic wave propagation in periodic structures has recently been applied to three-dimensional periodic media (Yablonovitch, 1993); and the results derived here have been extended to such systems. The analogy to quantum mechanics remains as the basic theme.

10.4.1 General method for multilayers

The refractive system that interests us, called a *dielectric multilayer system*, consists of a series of layers of transparent media having various refractive indices, deposited on top of each other on a substrate. A light wave is incident from free space in a direction at angle $\hat{\imath}$ to the normal, and we calculate the way in which it is reflected and transmitted. In this section, we shall set up the general framework for solving such problems and then consider a particular case, that of the periodic stack. The narrow-band interference filter will be shown as an example of its application.

Let us first consider the case of normal incidence. Waves entering the layered system along the x-axis (Fig. 10.14) are partially reflected

Figure 10.13
Inversion of an
image using a
coherent fibre bundle
10 mm long.

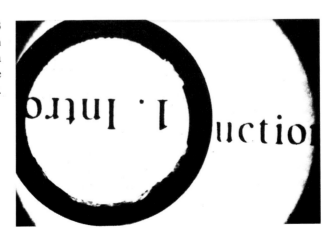

at the various interfaces. In any layer n there are in general two
waves, one travelling in each of the $+x$ and $-x$ directions. We shall
assume E to be polarized in the y-direction. We label their complex
amplitudes E_n^+ and E_n^- at the right-hand side (interface with layer
$n+1$). The phase difference for either of them is $g_n = k_n d$, where
k_n is the wavenumber in the medium, i.e. $k_n = k_0 \mu_n$. Thus the wave
amplitudes at the left-hand side of the layer are:

$$E_{nl}^+ = E_n^+ e^{+ig_n}, \tag{10.37}$$

$$E_{nl}^- = E_n^- e^{-ig_n}. \tag{10.38}$$

Now the total electric field must be continuous at the interface, so
that

$$E_{nl}^+ + E_{nl}^- = E_{n-1}^+ + E_{n-1}^-. \tag{10.39}$$

Substituting:

$$E_n^+ e^{+ig_n} + E_n^- e^{-ig_n} = E_{n-1}^+ + E_{n-1}^-. \tag{10.40}$$

Likewise, we deal with the magnetic fields, which are in the z-direction.
The amplitude of the wave propagating in the $+x$-direction is $H^+ = \mu Z_0^{-1} E^+$, and that in the $-x$-direction is $H^- = -\mu Z_0^{-1} E^-$. Thus, like
(10.40):

$$H_n^+ e^{+ig_n} + H_n^- e^{-ig_n} = H_{n-1}^+ + H_{n-1}^-. \tag{10.41}$$

Rewrite (10.40) and (10.41) in terms of the total fields

$$E_n = E_n^+ + E_n^- \tag{10.42}$$

$$\text{and} \quad Z_0 H_n = u_n(E_n^+ - E_n^-) \equiv \mu_n(E_n^+ - E_n^-). \tag{10.43}$$

Here, μ_n has been replaced by u_n; the change in notation corresponds
to that in §5.4.2, and is discussed further in §10.4.2.† We have two new
equations

$$E_{n-1} = E_n \cos g_n + \frac{iZ_0 H_n}{u_n} \sin g_n \tag{10.44}$$

$$Z_0 H_{n-1} = iu_n E_n \sin g_n + Z_0 H_n \cos g_n \tag{10.45}$$

† It is common in many books to put $Z_0 = 1$ at this stage because its value always
cancels out.

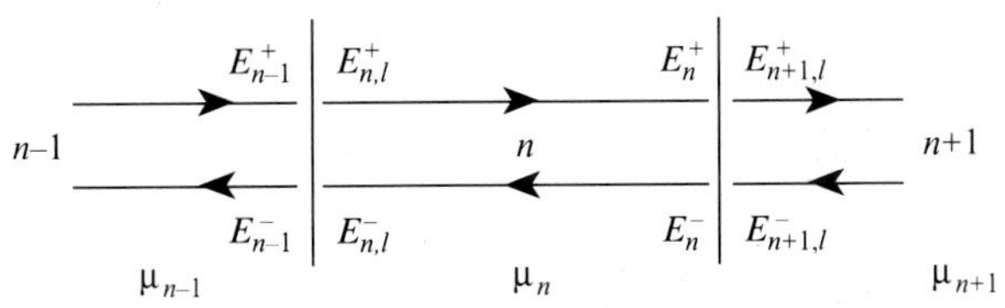

Figure 10.14
Parameters for the
multilayer
calculation.

which can be written in matrix form:

$$\begin{pmatrix} E_{n-1} \\ Z_0 H_{n-1} \end{pmatrix} = \begin{pmatrix} \cos g_n & iu_n^{-1}\sin g_n \\ iu_n \sin g_n & \cos g_n \end{pmatrix} \begin{pmatrix} E_n \\ Z_0 H_n \end{pmatrix} \equiv \mathsf{M}_n \begin{pmatrix} E_n \\ Z_0 H_n \end{pmatrix}.$$

(10.46)

The behaviour of the complete system, defined by the set of parameters (g_n, u_n) for each layer, can now be found by matrix multiplication. As a result, we have the relationship between $(E_0, Z_0 H_0)$, which contains both the incident wave and the reflected wave, and $(E_N, Z_0 H_N)$ representing the transmitted wave:

$$(E_0, Z_0 H_0) = \prod_{n=1}^{N} \mathsf{M}_n \cdot (E_N, Z_0 H_N).$$

(10.47)

Now the reflexion and transmission coefficients $\mathscr{R}$ and $\mathscr{T}$ follow. Since $E_0^- = \mathscr{R}E_0^+$, $Z_0 H_0^+ = u_0 E_0^+$ and $Z_0 H_0^- = -u_0 E_0^-$,

$$(E_0, Z_0 H_0) = [(1 + \mathscr{R}), u_0(1 - \mathscr{R})]$$

(10.48)

where the incident field is assumed to have unit value. Obviously $E_N^- = 0$, since there is no reflected wave in the $-x$ direction in the last layer, so that

$$(E_N, Z_0 H_N) = (\mathscr{T}, u_N \mathscr{T}).$$

(10.49)

The resulting matrix equation

$$(1 + \mathscr{R}, u_0(1 - \mathscr{R})) = \prod_n \mathsf{M}_n \cdot (\mathscr{T}, u_N \mathscr{T})$$

(10.50)

is easily solved by equating coefficients as will be seen in the example that follows. Notice, for future use, that the determinant $\det\{\mathsf{M}\}=1$. This implies conservation of energy; as a result, $\det\{\prod_n \mathsf{M}_n\}=1$.

10.4.2 Oblique incidence

The case of oblique incidence is easily dealt with, and will remind the reader why u was introduced. Suppose that the incident wave in free space is at an angle $\hat{\imath}$ to the x-axis. Then its angle $\hat{r}_n$ in the nth layer is given by Snell's law

$$\sin \hat{\imath} = \mu_n \sin \hat{r}_n.$$

(10.51)

The phase difference g_n now contains the x-component of $\mathbf{k}$ in the medium, i.e. $k_0 \mu_n \cos \hat{r}_n$:†

$$g_n = k_0 \mu_n d_n \cos \hat{r}_n . \tag{10.52}$$

As we saw in §5.4.2 it is now possible to express the boundary conditions for $\parallel$ and $\perp$ fields by introducing effective indices of refraction

$$u_n = \mu_n \sec \hat{r}_n \quad (\parallel \text{ polarization}) \tag{10.53}$$

$$u_n = \mu_n \cos \hat{r}_n \quad (\perp \text{ polarization}) \tag{10.54}$$

into the Fresnel coefficients for normal incidence, giving ((5.43)–(5.44)). The same argument applies here, where $\hat{r}_n$ is the angle of refraction within each medium.

10.4.3 Single-layer anti-reflexion coating: lens blooming

Although the reflexion coefficient at a single interface between air and a transparent medium, of the order of 4% as calculated in §5.4.2, might seem small, it represents a serious loss of light in optical systems such as compound lenses and instruments which contain many surfaces. The reflexion coefficient can be greatly reduced by coating the surfaces with one or more thin layers of materials with different refractive indices. Such *anti-reflexion coatings* are the most widespread use of dielectric multilayers. We shall illustrate the basic idea with the simplest case, a single-layer coating, which by suitable design can reduce the reflexion coefficient to zero at a specific wavelength, and is quite effective at neighbouring ones.

For one layer with parameters (g, u) between air (u_0) and a substrate with index u_s, we have (10.46) and (10.50),

$$\begin{pmatrix} 1 + \mathcal{R} \\ u_0(1 - \mathcal{R}) \end{pmatrix} = \begin{pmatrix} c & iu^{-1}s \\ ius & c \end{pmatrix} \begin{pmatrix} \mathcal{T} \\ u_s\mathcal{T} \end{pmatrix} \tag{10.55}$$

where $c \equiv \cos g$ and $s \equiv \sin g$. We shall use u instead of the refractive index μ, so that the results will be generally applicable to any angle of incidence and polarization. We now require $\mathcal{R} = 0$, for which (10.55) gives the pair of equations

$$1 = (c + ius_s/u_0)\mathcal{T}, \tag{10.56}$$

$$u_0 = (ius + cu_s)\mathcal{T}. \tag{10.57}$$

---

† It is a common error to assume, by erroneous intuition, that oblique incidence makes the layers 'seem thicker', and to write $g = k_0 \mu d / \cos \hat{r}$. A physical explanation, which underlies what appears here, is given in §9.5 for the Fabry–Perot interferometer.

Eliminating $\mathcal{T}$ leads to the complex equation:

$$u_0 = [u_s + ics(u - u_s^2/u)]/(c^2 + s^2 u_s^2/u^2). \qquad (10.58)$$

Clearly, the imaginary part must be zero. Since $u \neq u_s$ (otherwise the deposited layer would simply be part of the substrate) c or s must be zero. If $s = 0$, it follows that $c = \pm 1$, but the equation can not be satisfied because $u_0 \neq u_s$. Therefore $c = 0$, $s = \pm 1$ and

$$u_0 u_s = u^2. \qquad (10.59)$$

The value $s = \pm 1$ implies $g = (\text{odd})\pi/2$; the optical thickness of the layer is an odd number of quarter-wavelengths. Usually, a single quarter-wave is chosen. As an example, at normal incidence the refractive index needed to provide an anti-reflexion coating at an interface with air ($u_0 = 1$) is the square root of that of the substrate. On glass with $u_s = \mu_s \simeq 1.6$, the most common coating material is MgF$_2$, which can easily be deposited by evaporation and has a refractive index of 1.38, which is approximately correct. The reflexion coefficient as a function of wavelength for this case is shown in Fig. 10.15. Glass coated this way shows a slight purplish reflexion, because the layer reflects rather more in the blue and red; hence the common term *blooming* for this technique.

The use of more than one layer allows broader-band coatings to be designed, with better overall quality, but we shall not discuss them here (see Liddell, 1981; Macleod, 1989).

It is easy to see the physical basis of the single-layer anti-reflexion coating. By choosing a layer with u equal to the geometric mean of the air and the medium u_s, we have created two interfaces with equal reflexion coefficients (equation (5.34)). Separating them by an optical distance of $\lambda/4$ ensures that the waves reflected from the two will be in antiphase and therefore interfere destructively.

Figure 10.15 Reflexion coefficient for a quarter-wave anti-reflexion coating on glass, calculated for minimum reflectance with normal incidence at $\lambda = 5500$ Å : (*a*) ideal, $u = \sqrt{1.60}$; (*b*) MgF$_2$, $u = 1.38$.

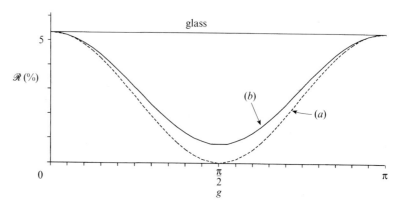

10.4.4 Periodic multilayers: selective mirrors

We shall only solve one multilayer problem in any detail, although the reader will realise that (10.47) and a small computer in fact allow calculation of the properties of *any* combination of layers which are non-absorbing, if the various (g_n, u_n) are given.†

In order to create a highly-reflective multilayer, we want to do the opposite to what we did with the anti-reflecting layer. So we need *constructive* interference between the partially-reflected waves. This can be achieved by making interfaces which have alternately positive and negative reflexion coefficients of equal value, and separating them by half-wavelength path differences (i.e., quarter-wavelength-thick layers again) as in Fig. 10.16. Let us try this idea out.

We construct a periodic system from two types of layer, which we call H for 'high' refractive index and L for 'low'. Their effective refractive indices are u_H and u_L respectively, and we shall let their optical thicknesses be equal, $g_H = g_L = g$. The system contains q pairs of these layers on a substrate with refractive index u_s. Equation (10.47) then becomes (with $c \equiv \cos g$ and $s \equiv \sin g$):

$$
\begin{pmatrix} E_0 \\ Z_0 H_0 \end{pmatrix} = \left[\begin{pmatrix} c & iu_L^{-1}s \\ iu_L s & c \end{pmatrix} \begin{pmatrix} c & iu_H^{-1}s \\ iu_H s & c \end{pmatrix} \right]^q \begin{pmatrix} E_N \\ Z_0 H_N \end{pmatrix}
$$

$$
= \begin{pmatrix} c^2 - u_H u_L^{-1} s^2 & ics(u_H^{-1} + u_L^{-1}) \\ ics(u_H + u_L) & c^2 - u_L u_H^{-1} s^2 \end{pmatrix}^q \begin{pmatrix} E_N \\ Z_0 H_N \end{pmatrix}
$$

(10.60)

$$
\equiv (M_p)^q \begin{pmatrix} E_N \\ Z_0 H_N \end{pmatrix}.
$$

(10.61)

It is easiest to evaluate $(M_p)^q$ algebraically if it is first diagonalized by

† A more interesting and much more difficult problem is the inverse: given the required properties, to calculate the values of (g_n, u_n) needed. See for example Liddell (1981).

Figure 10.16 An arbitrary multiply-reflected wave in a multilayer. Notice that any wave from *A* has the same phase at *B* if every layer has optical thickness $\lambda/4$. Phase changes on reflexion have been taken into account.

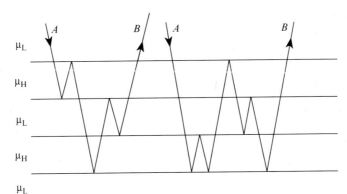

rotating the vectors $(E, Z_0 H)$, because if

$$M = \begin{pmatrix} \lambda_1 & 0 \\ 0 & \lambda_2 \end{pmatrix},$$

$$\text{then} \quad M^q = \begin{pmatrix} \lambda_1^q & 0 \\ 0 & \lambda_2^q \end{pmatrix}. \tag{10.62}$$

The values of λ are given by

$$\det\{M_p - \lambda I\} = 0$$
$$= \det\{M_p\} - \lambda \text{trace}\{M_p\} + \lambda^2$$
$$= 1 - \lambda \text{trace}\{M_p\} + \lambda^2. \tag{10.63}$$

Writing 2ξ for the trace (sum of diagonal components):

$$\lambda = \xi \pm \sqrt{\xi^2 - 1}. \tag{10.64}$$

Now ξ has the value

$$\xi = c^2 - \frac{1}{2}\left(\frac{u_H}{u_L} + \frac{u_L}{u_H}\right)s^2. \tag{10.65}$$

Note that $(u_H/u_L + u_L/u_H) \geq 2$ for any u_H, u_L. Then ξ can easily be seen to have the following characteristics (remember that $c^2 + s^2 = 1$):

- its maximum value, obtained when $c = \pm 1$, $s = 0$, is 1.
- its minimum value, when $c = 0$, $s = \pm 1$ is $-\frac{1}{2}(u_H/u_L + u_L/u_H)$ which is always less than -1. Therefore there exist regions of c and s for which λ is real (around $g = 0, \pi...$) and complex (around $g = \pi/2, 3\pi/2...$), as shown in Fig. 10.17. In general we have

$$\lambda_1 \lambda_2 = 1. \tag{10.66}$$

The particular case where $g = (\text{odd})\pi/2$ is particulary easy to treat, since the matrix M_p is already diagonal in its unrotated form; i.e. $(E_n, Z_0 H_n)$ is an eigenvector. Here $\lambda_1 = u_H/u_L$, $\lambda_2 = u_L/u_H$ and the optical thickness of each layer is

$$ud = g/k_0 = (\text{odd})\frac{\pi}{2} \times \frac{\lambda}{2\pi} = (\text{odd})\frac{\lambda}{4}. \tag{10.67}$$

We have for (10.50):

$$\begin{pmatrix} 1 + \mathscr{R} \\ u_0(1 - \mathscr{R}) \end{pmatrix} = \begin{pmatrix} \lambda_1^q & 0 \\ 0 & \lambda_2^q \end{pmatrix} \begin{pmatrix} \mathscr{T} \\ u_s \mathscr{T} \end{pmatrix} \tag{10.68}$$

which can be solved for $\mathscr{R}$:

$$\mathscr{R} = \frac{u_0 \lambda_1^q - u_s \lambda_2^q}{u_0 \lambda_1^q + u_s \lambda_2^q}. \tag{10.69}$$

Now from (10.66), one of $|\lambda_1|, |\lambda_2|$ must be > 1, so that $|\mathscr{R}| \to 1$ as $q \to \infty$. In other words, in the region of real λ, around $g = (\text{odd})\pi/2$, the system behaves as a mirror. In fact, it is quite easy to get a very good mirror. Suppose that $u_H/u_L = 2$ (approximately the ratio for the commonly-used pair ZnS–MgF$_2$) at normal incidence; then for $g = \pi/2$, (10.67) gives $\xi = -\frac{5}{4}$ and $\lambda_1 = -\frac{1}{2}$, $\lambda_2 = -2$. Thus, for five periods, say, and $u_s/u_0 = 1.5$,

$$\mathscr{R} = \frac{(-\frac{1}{2})^5 - 1.5(-2)^5}{(-\frac{1}{2})^5 + 1.5(-2)^5} = -0.9987. \tag{10.70}$$

The intensity reflexion coefficient is then $\mathscr{R}^2 = 0.9974$. This method allows us to make highly-reflecting mirrors for selected wavelengths, for which the layers have optical thickness of an odd number (usually one) of quarter-wavelengths. It is used routinely for making laser resonator mirrors since the losses achieved (even under real conditions) are much less than in metal mirrors. The region where λ_1 and λ_2 are real extends for a region around $g = (\text{odd})\pi/2$. Its boundaries are given by putting $\lambda_1 = \lambda_2 = 1$, whence $\xi = 1$ and (10.65) leads to

$$-1 = c^2 - \frac{1}{2}\left(\frac{u_H}{u_L} + \frac{u_L}{u_H}\right)s^2 \tag{10.71}$$

which simplifies to

$$\cos g = \pm\left(\frac{u_H - u_L}{u_H + u_L}\right). \tag{10.72}$$

These two solutions define the points a and b in Fig. 10.18; the region of high reflectivity around $(\text{odd})\pi/2$ has width

$$\Delta g = 2\sin^{-1}\left(\frac{u_H - u_L}{u_H + u_L}\right) = 2\sin^{-1}\mathscr{R}_{HL} \tag{10.73}$$

where $\mathscr{R}_{HL}$ is the Fresnel reflectivity of the interface from §5.4.2. Notice that the width of the reflecting region does not depend on the number of periods, q.

In the region where λ_1 and λ_2 are complex, they take the form $\lambda = \exp(\pm i\phi)$ where $\xi = \cos\phi$. As an example, consider $g = (\text{even})\pi/2$, where M_p is already diagonal. Then $\phi = 0$ and we find $\mathscr{R} = (u_0 - u_s)/(u_0 + u_s)$. This is the reflectivity of the substrate as if the multilayer were not there; the multilayer has no effect when the layer optical thickness are multiples of half a wavelength.

Calculation of $\mathscr{R}$ for values of g other than integer multiples of $\pi/2$ is tiresome algebraically, and is best done numerically, directly from (10.60). An example with $u_H/u_L = 2$, $q = 10$ and $u_s/u_0 = 1.5$ gives the result shown in Fig. 10.18. The general features are high reflectivity around $g = (\text{odd})\pi/2$ and low reflectivity around $(\text{even})\pi/2$, with the transition where $\cos g = \pm\mathcal{R}_{HL}$. There are also other values of g which have low reflectivity and which depend on details of the structure.

Finally, we shall once again stress the analogy with the band theory of electrons in crystals. When the period of the crystal is half the wavelength of the electrons, we are exactly in the middle of the band gap. The band gap is then seen as equivalent to the region of high reflectivity (zero transmission) of the multilayer.

10.4.5 Interference filters

An important application of dielectric multilayers is to the design of interference filters. In the preceding section we showed that highly-reflective and non-absorbing wavelength-selective mirrors can be made by using quarter-wave-thick layers of different dielectric media. This idea can be extended to the design of filters with transmission characteristics satisfying almost any specification. We shall use symbols H, L to represent quarter-wave layers of indices μ_H, μ_L respectively.

The most common filter is a narrow-band interference filter based on the properties of the Fabry–Perot étalon (§9.5.1). Suppose that we

Figure 10.17 Eigenvalues of the matrix for a periodic multilayer.

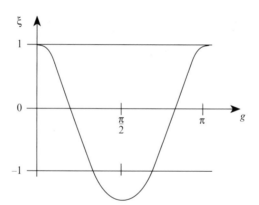

Figure 10.18 Reflectivity of periodic multilayers with 2 and 10 periods.

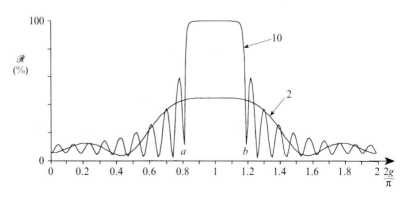

make a pair of reflecting surfaces by using quarter-wave assemblies (HLHL...) and separate them by a spacer corresponding to the first (or higher, m) order of the Fabry–Perot. Ignoring the substrate for simplicity, the amplitude reflexion coefficient at normal incidence for a set of q pairs HL is, from (10.69),

$$\mathscr{R} \simeq \frac{\mu_H^{2q} - \mu_L^{2q}}{\mu_H^{2q} + \mu_L^{2q}}. \tag{10.74}$$

The thickness of the spacer, for order $m = 1$, is given by $t = \lambda/2\mu$, implying a single half-wave layer. We then have a layer system which can be described symbolically as $(HL)^q H^2 (LH)^q$ in which the two consecutive H layers make the half-wave layer. This has a pass band given by (9.56)

$$\frac{\delta\lambda}{\lambda} = \frac{\delta g}{g} = \frac{F^{-\frac{1}{2}}}{2\pi m} = \frac{1 - \mathscr{R}^2}{4\pi\mathscr{R}}$$

$$= \frac{\mu_H^{2q}\mu_L^{2q}}{\pi(\mu_H^{4q} - \mu_L^{4q})}. \tag{10.75}$$

Very accurate coating techniques have been developed to produce such filters (and many others of more intricate design) with many tens of layers. For example, using ZnS ($\mu_H = 2.32$) and MgF$_2$ ($\mu_L = 1.38$), with $q = 5$ (21-layer filter)

$$\frac{\delta\lambda}{\lambda} = 1.7 \times 10^{-3}.$$

For $\lambda = 5000\,\text{Å}$, $\delta\lambda = 9\,\text{Å}$. The filter is non-transmitting in the region of complete reflexion of the quarter-wave periodic assembly, about 1000 Å on each side of the pass band. This is typical of what can be achieved with multilayer filters (Fig. 10.19).

Figure 10.19 (*a*) Transmittance of the interference filter $(HL)^5 H^2 (LH)^5$ on a glass substrate, as a function of $2g/\pi$; (*b*) shows the region $0.995 < 2g/\pi < 1.005$ expanded.

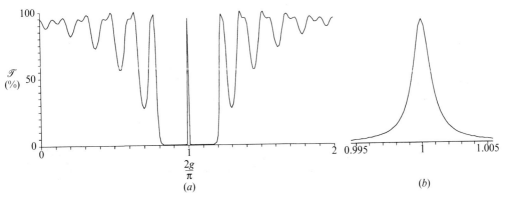

(a) (b)

CHAPTER ELEVEN

Coherence

11.1 Introduction

The coherence of a wave describes the accuracy with which it can be represented by a pure sine wave. So far we have discussed optical effects in terms of waves whose wave-vector $\mathbf{k}$ and frequency ω can be exactly defined; in this chapter we intend to investigate the way in which uncertainties and small fluctuations in $\mathbf{k}$ and ω can affect the observations in optical experiments. Waves that appear to be pure sine waves only if they are observed in a limited space or for a limited period of time are called *partially coherent* waves, and a considerable part of this chapter will be devoted to developing measures of the deviations of such impure waves from their pure counterparts. These measures of the coherence properties of the waves are functions of both time and space, but in the interests of clarity we shall consider them as functions of each variable independently. Fig. 11.1 illustrates, in a very primitive manner, one wave which is partially coherent in time (it appears to be a perfect sine wave only when observed for a limited time) and a second wave which is partially coherent in space (it appears to be a sinusoidal plane-wave only if observed over a limited region of its wavefront).

The understanding of the coherence properties of light has had numerous practical consequences. Amongst these are the technique of Fourier-transform spectroscopy and several methods of making astronomical measurements with high resolution. These are rooted in the understanding of fluctuations in phase and intensity of real light waves, which are discussed in this chapter at a classical level. However, some of the most exciting areas of modern optics have grown out of the application of quantum theory to coherence, and in Chapter 14 we shall see how this leads to new results which are inconsistent with the classical description.

290

11.2 Properties of real light waves

Let us try to define clearly what we know about a real light wave, emitted by a classical monochromatic light source. We know that the light we see at any moment comes from a number of atoms, each making a transition between the same pair of energy levels, but that the emission from any one atom is no way related to that from any other atom. In fact a careful spectroscopic analysis shows us that the light is not really monochromatic in the strict sense of the word; it contains components of various wavelengths within a certain range, called the *linewidth*. If the linewidth is much less than than the average wavelength one uses the term *quasi-monochromatic* for such radiation. The physical reasons for a non-zero linewidth are discussed in more detail in §11.3, but just as an example we recall that at a finite temperature all the atoms in the emitting material (usually a gas) are

Figure 11.1
Schematic partially
coherent waves:
(*a*) perfectly coherent
wave; (*b*) wave with
space-coherence
only; (*c*) wave with
time-coherence only.

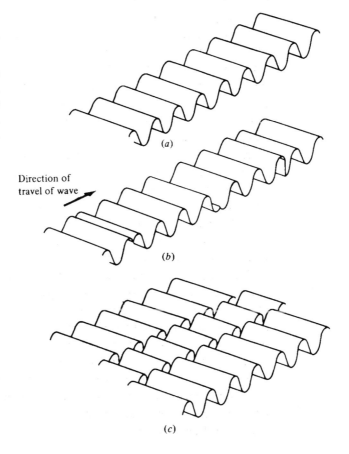

(*a*)

Direction of
travel of wave

(*b*)

(*c*)

moving randomly in various directions, and so the emission from each atom is Doppler-shifted by a different amount.

If we now ask exactly what the light wave looks like, we can answer the question by performing a Fourier synthesis based on the remarks in the previous paragraph. We take a number of sine waves, having frequencies randomly chosen within a specified range – the linewidth of the radiation – and add them together. We have done this in Fig. 11.2, where examples of continuous waves have been generated each from five sine waves with frequencies randomly chosen within a specified interval. What we see is a complicated beat phenomenon; the amplitude of the wave is not a constant, but fluctuates in a rather haphazard fashion. The average length of a beat is related to the range of frequencies involved. If there is no rational relationship between the frequencies themselves, the waveform never repeats itself and is an example of *chaotic light*, §14.2.3.

The wave trains in Fig. 11.2 can also be looked at in a different way. We can consider each beat as an independent wave-group; the complete wave train is then a series of such wave-groups emitted at random intervals. This description turns out to be convenient for some purposes, and we shall show in §11.2.2 that the spectrum for such a process is precisely what we have described above as the spectrum of a quasi-monochromatic light source.

However, this representation does have a danger; the individual wave-groups must not be interpreted as photons, quantum units of

Figure 11.2 Three impure sine waves, showing the amplitude fluctuations resulting from a spread in component frequencies. Each is generated from five components randomly distributed in the range $\pm\frac{1}{2}\epsilon$ about ω_0 where the values of ϵ/ω_0 are as follows: (*a*) 0.04, (*b*) 0.08, (*c*) 0.16.

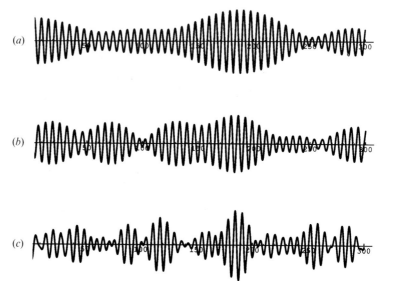

light energy. Apart from the fact that the model has been created by completely classical thinking, and therefore cannot produce a quantized particle, the rate of repetition of the wave-group is determined entirely by the spread of frequencies. Now if the wave-groups were photons, their average rate of occurrence would depend on the *intensity* of the wave, and not the linewidth. So despite the attractiveness of such an idea, it is not consistent with the facts.

11.2.1 The amplitude and phase of quasi-monochromatic light

Let us try to develop the ideas of the previous section a little further with the help of a simple model. It is convenient at this stage to define formally the *average* or *mean* value of a function $g(t)$ during a interval of duration T lasting from $-T/2$ to $T/2$:

$$\langle g \rangle_T = \frac{1}{T} \int_{-T/2}^{T/2} g(t)\, \mathrm{d}t . \tag{11.1}$$

For mathematical convenience we shall work with the complex wave field $f = f^R + \mathrm{i}f^I$; one should remember, however, that observable physical fields are real and are given by f^R. The way in which an associated complex function is generated from a real one was discussed in §4.4.7. The instantaneous intensity of the wave is then defined as $I(t) \equiv |f(t)|^2$; for a pure sine wave, which is described by the complex function $f(t) = a\exp(\mathrm{i}\omega t)$, the intensity is constant.

Now picture a quasi-monochromatic light beam that is represented at a given point in space by the superposition of a large number N of waves with equal amplitude a. Each one has random phase ϕ_n and frequency ω_n randomly chosen within the range $\omega_0 \pm \epsilon/2$ where $\epsilon \ll \omega_0$ (Fig. 11.2). One should note that the actual value of ϕ_n depends on the origin chosen for time. The amplitude and intensity of the combined wave are:

$$f(t) = a\sum_{n=1}^{N} \exp[\mathrm{i}(\omega_n t + \phi_n)] , \tag{11.2}$$

$$I(t) = |f(t)|^2 = a^2|\sum_{n=1}^{N} \exp[\mathrm{i}(\omega_n t + \phi_n)]|^2 , \tag{11.3}$$

which can be written as a double sum

$$I(t) = a^2 \sum_{n}\sum_{m} \exp\{\mathrm{i}[(\omega_n - \omega_m)t + \phi_n - \phi_m]\} . \tag{11.4}$$

What is noticeable about the waves in Fig. 11.2 is that their amplitude and phase fluctuate with a typical beat period $2\pi/\epsilon$, which contains about ω_0/ϵ waves. This effect should be brought out by a comparison of the long-term and short-term averages of (11.4); the average over a interval T_0, long compared with the duration of a beat, should iron out the fluctuations, whereas that over a interval T_1, short compared with the beat, should not. We can evaluate such averages:

$$\langle I(t)\rangle_T = \frac{a^2}{T} \int_{-T/2}^{T/2} \sum_n \sum_m \exp\{i[(\omega_n - \omega_m)t + \phi_n - \phi_m]\}dt . \quad (11.5)$$

When the integration time T is T_0, the term $(\omega_n - \omega_m)t$ can have any value up to ϵT_0 which according to the definition of T_0 can be $\gg 2\pi$. Thus there is a tendency to cancellation of the oscillating parts of the integral, all of them going through many cycles at different rates during the long interval T_0. Only the terms for which $n = m$, all of which have value $e^{i0} = 1$, make a consistent positive contribution to the integral and thus

$$\langle I(t)\rangle_{T_0} = \frac{a^2}{T_0} \int_{-T_0/2}^{T_0/2} \sum_{n=1}^N 1\, dt = a^2 N . \quad (11.6)$$

This equation simply tells us that, because the waves are uncorrelated, the intensity of their sum is equal to the sum of their intensities.

Now let us calculate the short-term average. For this, we write equation (11.5) again, with $T = T_1$. The situation in the integral will be different from that in the long-term average when T_1 is short enough for the maximum phase, $|(\omega_n - \omega_m)T_1|$, to be much less than $\pi/2$. This occurs if $\epsilon T_1 < \pi/2$. Then

$$\begin{aligned}\langle I(t)\rangle_{T_1} &= \frac{a^2}{T_1} \int_{-T_1/2}^{T_1/2} \sum_n \sum_{m=1}^N \exp[i(\phi_n - \phi_m)]dt \\ &= a^2 N + 2a^2 \sum_n \sum_{m<n} \cos(\phi_n - \phi_m) , \quad (11.7)\end{aligned}$$

where in the last line we have separated the terms involving $n = m$ from the rest. We cannot predict the exact value of the second term. We just know that it will generally have a non-zero value, and that this value does not change during the interval T_1. Equation (11.7) therefore tells us that the intensity remains approximately constant during the interval T_1, but has a value differing from the long-term mean $a^2 N$. It is interesting to estimate the size of the fluctuating second term of (11.7), which contains about $\frac{1}{2}N^2$ terms in the $\sum$. Its *square*, when averaged over many intervals, has the expected value

$$\tfrac{1}{2}N^2 \cdot 4a^4 \langle\cos^2(\phi_n - \phi_m)\rangle = a^4 N^2, \quad (11.8)$$

since $\langle \cos^2 \theta \rangle = \frac{1}{2}$. So the fluctuating term has root-mean-square value $a^2 N$. The fluctuations are therefore *comparable with the mean intensity*, also $a^2 N$, even as $N \to \infty$. So we deduce that the short-term mean *fluctuates macroscopically* about the long-term mean $a^2 N$. The critical time $\pi/2\epsilon$, which distinguishes the long term from the short term, is approximately the coherence time of the wave τ_c which will be defined below; it is typically of the order of 10^{-10} s for classical light sources.

We can also show that during the short interval T_1 the wave behaves very much like a pure sine wave. We write the amplitude $f(t)$ (11.2) in the form:

$$f(t) = a \exp(i\omega_0 t) \sum_n \exp[i(\omega_n - \omega_0)t + i\phi_n] . \qquad (11.9)$$

During the period $t < T_1$, $|(\omega_n - \omega_0)t| \ll \pi/2$, and (11.9) can be written

$$f(t) \approx a \exp(i\omega_0 t) \sum_n \exp(i\phi_n) . \qquad (11.10)$$

The sum $\sum \exp(i\phi_n)$ will in general have a non-zero value $\alpha_1 \exp(i\Phi_1)$, where Φ_1 is quite indeterminate. We conclude that during the interval T_1 we can represent the wave by a harmonic wave $a\alpha_1 \exp[i(\omega_0 t + \Phi_1)]$, whose phase Φ_1 and amplitude $a\alpha_1$ are constant but unknown. If we now repeat the measurement for an interval T_1 centred on a later time t_0 we should find the same result with new phase Φ_2 and amplitude $a\alpha_2$ defined by:

$$\alpha_2 \exp(i\Phi_2) = \sum_n \exp[i(\omega_n - \omega_0)t_0 + i\phi_n] , \qquad (11.11)$$

and which in general are quite unpredictably different from Φ_1 and $a\alpha_1$ in view of the ω_ns being randomly chosen. There is no correlation between the phases Φ_1, Φ_2 and amplitudes $a\alpha_1$, $a\alpha_2$ measured in separate intervals of length T_1.

We conclude that measurements made during a short-term T_1 indicate a simple harmonic wave of constant intensity. Observation continued for a period much longer than $\pi/2\epsilon$ will show this short-term intensity to fluctuate and there to be no correlation between the phases measured during intervals separated by much more than T_1.

11.2.2 The spectrum of a random series of wave-groups

We have seen that the beat patterns in Fig. 11.2 can also be described as a random succession of wave-groups. Such a series does indeed have similar spectral characteristics. Consider for example the Gaussian wave-group of §4.4.6:

$$f(t) = A \exp(-i\omega_0 t) \exp(-t^2/2\sigma^2) , \qquad (11.12)$$

whose Fourier transform is

$$F(\omega) = 2\pi A(2\pi\sigma^2)^{\frac{1}{2}} \exp[-(\omega - \omega_0)^2 \sigma^2/2] . \qquad (11.13)$$

A random series of such groups is:

$$f_{\rm r}(t) = \sum_{n=1}^{N} f(t - t_n) , \qquad (11.14)$$

where t_n is the random centre point of the nth wave-group. Now the transform of (11.14) is:

$$F_{\rm r}(\omega) = F(\omega) \sum_n \exp(-i\omega t_n) = F(\omega)\beta \exp[-i\psi(\omega)] . \qquad (11.15)$$

From Parseval's theorem (§4.7.2), it is easy to show that $\beta^2 = N$.

The factor $\beta \exp[-i\psi(\omega)]$ has an indeterminate phase, $\psi(\omega)$, different for each value of ω. So we conclude that the spectrum is like that of the single wave-group (11.12), but has random phase. The *spectral intensity* $J(\omega) \equiv |F_{\rm r}(\omega)|^2$, however, is well defined, having the value

$$J(\omega) = |F_{\rm r}(\omega)|^2 = \beta^2 |F(\omega)|^2 = N|F(\omega)|^2. \qquad (11.16)$$

The series of wave-groups reproduces exactly the spectrum of Fig. 11.2 and is therefore a good physical representation of the wave. Incidentally, one should compare this result to that obtained in §8.3.8 for the Fraunhofer diffraction pattern of a random array of identical apertures.

11.2.3 White light

One limiting case is of particular interest; if we make the wave-group shorter, so that it eventually becomes a δ-function pulse, the function $F(\omega)$ becomes a constant, independent of ω. Thus a series of δ-pulses emitted at random times transforms to a set of all frequencies occurring with random phases. This is white light; it contains all frequencies in equal amounts, and they are completely uncorrelated in phase.

11.3 Physical origin of linewidths

So far we have introduced the width of a spectral line, or the finiteness of a wave train, simply as a parameter to be reckoned with; now we shall enquire briefly into the physical causes of line-broadening.

11.3.1 Natural linewidth

A spectral line has its origin in a quantum transition in which an atom or molecule changes its state from level A to level B, with energies E_A and E_B respectively; a wave of frequency $\omega_0 = (E_A - E_B)/\hbar$ is emitted at the same time (§14.3). However, no energy level except the ground state is an exact stationary state because of fluctuations of the environmental electromagnetic field (§14.2). As a result an atom in level A will decay to a lower level after an average time T_A. According to the uncertainty principle, the value of E_A is therefore uncertain to the extent $\delta E \approx h/T_A$, where h is Planck's constant. The corresponding frequency width of the emitted wave is $\delta\omega = 2\pi(T_A^{-1} + T_B^{-1})$. This is called the *natural linewidth*; it is generally smaller than the Doppler and collision linewidths discussed in the following sections, but can be achieved experimentally under conditions where the environmental effects are neutralized (Haroche and Kleppner, 1989; Foot, 1991).

11.3.2 Doppler broadening

Let us consider radiation from an isolated atom in a gas at temperature T. If the atom, mass m, has velocity v_x along the line of sight while the transition is taking place, the spectral line will appear shifted by the Doppler effect. The distribution of velocities along a particular axis (x) in a perfect gas is Gaussian

$$f(v_x)\mathrm{d}v_x = C \exp\left(\frac{-mv_x^2}{2k_\mathrm{B}T}\right)\mathrm{d}v_x, \tag{11.17}$$

and the Doppler shift in the observed frequency is

$$\omega - \omega_0 = \omega_0 v_x/c, \tag{11.18}$$

so that

$$F(\omega) = C \exp\left[\frac{-m(\omega - \omega_0)^2 c^2}{2\omega_0^2 k_\mathrm{B}T}\right]. \tag{11.19}$$

This effect has broadened an ideally sharp spectral line into a line with a Gaussian profile (see §4.4.6), where

$$\sigma = \omega_0(k_\mathrm{B}T/mc^2)^{\frac{1}{2}}. \tag{11.20}$$

It is common to express spectral linewidths in terms of the *half-width* (§4.4.6) which is 2.36σ for a Gaussian. In terms of wavelength, rather than frequency, we find the half-width to be $1.18\lambda_0(k_B T/mc^2)^{\frac{1}{2}}$ (Fig. 11.3).

As an example, we can take the Kr^{84} line for which $\lambda = 5600\,\text{Å}$, $m = 1.4 \times 10^{-22}\,\text{g}$. At $T = 80\,\text{K}$, (11.20) gives a half-width of $1.6 \times 10^{-11} \approx 0.002\,\text{Å}$, which agrees reasonably with the observed value of $0.003\,\text{Å}$.

11.3.3 Collision broadening

Considering an isolated atom does not give us the whole story. There will always be collisions between the various atoms in a real gas. According to the kinetic theory of gases (see, for example, Jeans, 1982), a particular atom will expect to be free for an average time

$$\tau_1 = (4NvA)^{-1} \tag{11.21}$$

between collisions, where N is the number of molecules per unit volume and A their collision cross-section. Now both N and v depend on temperature T and pressure P, and

$$Nv = P(3/mk_B T)^{\frac{1}{2}} . \tag{11.22}$$

So

$$\tau_1 = bT^{\frac{1}{2}}/P , \tag{11.23}$$

where b is a constant.

Now consider what happens if an emitting atom suffers a collision. We may suppose that the shock of the collision will at the very least destroy phase correlation between the emitted waves before and after the collision. The emission from all the atoms in the gas will therefore appear like a series of uncorrelated bursts of radiation each of average duration τ_1. The actual durations can be assumed to have a Poisson distribution of mean value τ_1; therefore the probability of there being a burst of length between τ and $\tau + \delta\tau$ is

$$p(\tau) = \tau_1^{-1}\exp(-\tau/\tau_1) . \tag{11.24}$$

So the emitted wave consists of wave trains with frequency ω_0 and random phase, starting at random moments and having durations statistically distributed according to (11.24).

Following §11.2.2, we deduce that the spectral intensity of such an emitted wave is the mean spectral intensity of the individual wave trains, the spectral phases being random. The spectrum is thus

$$J(\omega) = \int_0^\infty \left[\int_{-\tau/2}^{\tau/2} \exp(i\omega_0 t) \exp(-i\omega t) dt \right]^2 p(\tau) d\tau . \quad (11.25)$$

In (11.25) the inner integral (in square brackets) is the Fourier transform of a harmonic wave lasting for a duration τ; the outer integral is the statistical average. The inner integral is well-known to us as $\tau \operatorname{sinc}[(\omega - \omega_0)\tau/2]$ (§4.4.2). Thus (11.25) becomes

$$
\begin{aligned}
J(\omega) &= \frac{4}{\tau_1(\omega - \omega_0)^2} \int_0^\infty \exp(-\tau/\tau_1) \sin^2[(\omega - \omega_0)\tau/2] d\tau \quad (11.26) \\
&= 2\tau_1^2/[1 + (\omega - \omega_0)^2 \tau_1^2] . \quad (11.27)
\end{aligned}
$$

The function (11.27) is known as a *Lorentzian function* with a half-width $2\tau_1^{-1}$. We have already met it as the Fourier transform of an exponential in our discussion of multiple-reflexion fringes in §9.5. If we draw (11.27) as a function of ω, we see that the function is superficially similar to the Gaussian but has a much slower decay in its wings (Fig. 11.3).

In practice, temperature and pressure cause both Doppler and collision broadening in various degrees, and observed spectral lines are rarely exactly Gaussian or exactly Lorentzian. Moreover, many spectral lines are multiplets with complicated fine structure, but from the point of view of simple optical coherence theory they can often just be considered as having a single empirical width.

Figure 11.3
Comparison between
Lorentzian and
Gaussian functions.

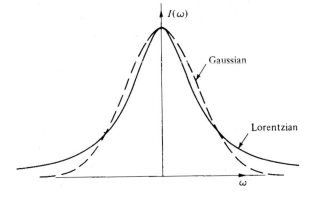

Figure 11.3 Comparison between Lorentzian and Gaussian functions.

11.4 Quantification of the concept of coherence

In the previous sections we have described some of the characteristics of real light waves. In order to understand how they affect optical experiments, it is necessary to develop a quantitative framework to describe their properties statistically. The function γ which will be defined in §11.4.1 is a measure of coherence between two values of a wave field, $f(\mathbf{r}_1, t_1)$ and $f(\mathbf{r}_2, t_2)$. Coherence means roughly that, given $f(\mathbf{r}_1, t_1)$, a recipe exists to estimate the amplitude and phase of $f(\mathbf{r}_2, t_2)$. The better this recipe works on the average the better the coherence, and the closer the function γ is to unity. We shall find it simplest to talk about two limiting cases.

- *Temporal coherence* which measures the coherence between $f(\mathbf{r}, t_1)$ and $f(\mathbf{r}, t_2)$, i.e. between two values of the wave field at the same point $\mathbf{r}$ but different times. Temporal coherence allows us to define a *coherence time* τ_c, the maximum $t_2 - t_1$ for which the recipe works well. As we have seen in §11.2.1, τ_c is intimately connected with the bandwidth of a quasi-monochromatic wave, and we shall show that the degree of temporal coherence is related quantitatively to the spectrum of the wave field.
- *Spatial coherence*, which is a measure of the coherence between $f(\mathbf{r}_1, t)$ and $f(\mathbf{r}_2, t)$, i.e. between two values of the wave field at different points measured at the same time t. In analogy to τ_c one can define a *coherence region* around $\mathbf{r}_1$ for which the recipe is valid. This region need not be circular.

11.4.1 The mutual coherence function

We shall now make these concepts more quantitative by defining a *mutual coherence function* based on the idea of correlation which was introduced in §4.7. We shall write $t_1 = t$ and $t_2 = t + \tau$ and assume that the coherence properties do not change with time and therefore depend only on the difference $\tau = t_2 - t_1$ (this is called the *assumption of stationarity*). The *complex mutual coherence function* Γ is then defined as

$$\Gamma(\mathbf{r}_1, \mathbf{r}_2, \tau) = \langle f(\mathbf{r}_1, t) f^*(\mathbf{r}_2, t + \tau) \rangle , \qquad (11.28)$$

in view of the above assumption Γ does not depend on t. The *complex degree of mutual coherence* is a normalized value defined as

$$\gamma(\mathbf{r}_1, \mathbf{r}_2, \tau) = \Gamma(\mathbf{r}_1, \mathbf{r}_2, \tau)/(I_1 I_2)^{\frac{1}{2}}, \qquad (11.29)$$

where I_1 is the mean intensity at $\mathbf{r}_1$:

$$I_1 \equiv \langle f(\mathbf{r}_1, t) f^*(\mathbf{r}_1, t) \rangle , \qquad (11.30)$$

which is also equal to $\Gamma(\mathbf{r}_1, \mathbf{r}_1, 0)$. A similar definition applies to I_2. Since f is the complex representation of the real function f^R, it can easily be shown that

$$I_1 = 2 \langle [f^R(\mathbf{r}_1, t)]^2 \rangle . \qquad (11.31)$$

Experimentally it is usually easier to work with γ than with Γ since the former can be expressed in terms of relative intensities. To make the physics clear, we shall only study the limiting cases of temporal and spatial coherence defined above.

11.4.2 The optical stethoscope and visibility of interference fringes

The 'optical stethoscope' (Fig 11.4) is a thought-experiment which we have found to be useful in clarifying the concept of coherence. It consists of two single-mode optical fibres (§10.2.4) of equal length, whose ends B_1 and B_2 are suppported close together, maybe a few λ apart. The ends A_1 and A_2 are placed in the quasi-monochromatic wave field, whose coherence properties we want to analyze, which originates in a distant source of small angular size.† A_1 and A_2 can sample this field at any two points we choose, and we assume that the light amplitudes at B_1 and B_2 are the same as those sampled at A_1 and A_2, with equal time delays resulting from the equal lengths of the fibres. B_1 and B_2 radiate as point sources and we observe the interference fringes on a screen a few cm away. If B_1 and B_2 radiate coherently, the interference fringes are clear; if B_1 and B_2 are incoherent, there will be no interference fringes. There can also be an intermediate situation, in which poorly visible fringes can be seen; this occurs when B_1 and B_2 are *partially coherent*. Fig. 11.12 shows some examples, which will be discussed in detail later. If the two points A_1 and A_2 are situated one behind the other, as in (a), in the direction of propagation of the light, they essentially sample the wave at the same place, but at different times separated by $\tau = A_1 A_2 / c$, and the contrast of the fringes measures the temporal coherence. If the points are side-by-side, in the same wavefront but separated by $\mathbf{r}$, as in (b), the contrast measures the spatial coherence.

† The source has to be small to give a reasonably well-defined direction of propagation.

The contrast of interference fringes formed by quasi-monochromatic light can be quantified by defining the *visibility V*:

$$V \equiv \frac{I_{\max} - I_{\min}}{I_{\max} + I_{\min}}, \qquad (11.32)$$

where I is the local intensity at a general point P on the screen. At this point the field is g and the local intensity is $\langle |g^2| \rangle$. Now the field g measured at P is the sum of the fields g_1 radiated by B_1 and g_2 radiated by B_2 at time t:

$$g_1(P,t) = \frac{1}{BP} f(B_1, t - \frac{\overline{B_1 P}}{c}), \qquad (11.33)$$

$$g_2(P,t) = \frac{1}{BP} f(B_2, t - \frac{\overline{B_2 P}}{c}), \qquad (11.34)$$

in which the $1/BP$ term is essentially the same for both fields and will eventually cancel out when the visibility is calculated; it arises because B_1 and B_2 are point-source-like radiators. Hence

$$I(P) = \langle (g_1 + g_2)(g_1^* + g_2^*) \rangle = \langle g_1 g_1^* \rangle + \langle g_2 g_2^* \rangle + \langle g_1 g_2^* \rangle + \langle g_1^* g_2 \rangle. \qquad (11.35)$$

We shall now show that if the intensities $\langle |f(B_1)|^2 \rangle$ and $\langle |f(B_2)|^2 \rangle$ are equal the visibility is a direct measure of the degree of mutual coherence between A_1 and A_2, i.e. $V = |\gamma_{A_1 A_2}(0)|$.† We have

$$
\begin{aligned}
g_1 g_2^* &= \frac{1}{(BP)^2} f(B_1, t - \frac{\overline{B_1 P}}{c}) f^*(B_2, t - \frac{\overline{B_2 P}}{c}) \\
&= \frac{1}{(BP)^2} f(B_1, t - \frac{\overline{B_1 P}}{c}) f^*(B_2, t - \frac{\overline{B_1 P}}{c} - \tau_p), \quad (11.36)
\end{aligned}
$$

where we have defined $\tau_p = (\overline{B_2 P} - \overline{B_1 P})/c$. Since B_1 and B_2 are only a few λ apart, this τ_p is at most a few periods long and therefore much

† The notation $\gamma_{A_1 A_2}(0)$ is short for $\gamma[\mathbf{r}(A_1), \mathbf{r}(A_2), 0]$. In principle, an optical stethoscope could be made for measuring $\gamma_{A_1 A_2}(\tau)$ by using fibres differing in length by $c\tau$.

Figure 11.4 Optical stethoscope. The instrument measures in (*a*) the temporal coherence and in (*b*) the spatial coherence.

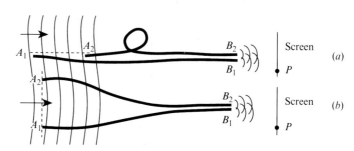

shorter than the coherence time τ_c. We can now use the assumption of stationarity and the fact that during τ_p $f \sim e^{i\omega_0 t}$ to write

$$\langle g_1 g_2^* \rangle = \frac{1}{(BP)^2} \langle f(B_1, t) f^*(B_2, t) \rangle \exp(-i\,\omega_0 \tau_p) . \tag{11.37}$$

Assume for simplicity that the optical stethoscope is constructed so that the intensities incident on the two fibres are equal, so that $I_i = \langle |f(A_i)|^2 \rangle = \langle |f(B_i)|^2 \rangle \equiv I$ and $\langle f(B_1) f^*(B_2) \rangle = \langle f(A_1) f^*(A_2) \rangle$ at all times. We then have from (11.35):

$$I(P) = \frac{1}{(BP)^2}[I_1 + I_2 + \langle f(A_1) f^*(A_2) \rangle \exp(-i\,\omega_0 \tau_p)$$

$$+ \langle f^*(A_1) f(A_2) \rangle \exp(i\,\omega_0 \tau_p)] . \tag{11.38}$$

Since $\langle f(A_1) f^*(A_2) \rangle = \Gamma_{A_1 A_2}(0)$, which can be written as $|\Gamma_{A_1 A_2}(0)|e^{i\Delta}$, we can put (11.38) in the form

$$
\begin{aligned}
(BP)^2 I(P) &= I_1 + I_2 + 2|\Gamma_{A_1 A_2}(0)| \cos(\omega_0 \tau_p + \Delta) \\
&= I_1 + I_2 + 2(I_1 I_2)^{1/2}|\gamma_{A_1 A_2}(0)| \cos(\omega_0 \tau_p + \Delta) \\
&= 2I[1 + |\gamma_{A_1 A_2}(0)| \cos(\omega_0 \tau_p + \Delta)] .
\end{aligned} \tag{11.39}
$$

From the definition (11.32) it now follows that the visibility of the fringes is

$$V = |\gamma_{A_1 A_2}(0)|, \tag{11.40}$$

when the intensities $I_1 = I_2$. The value of Δ in (11.39) can be read from the shift of the fringe pattern from the symmetrical position $(\tau_p = 0)$ on the screen. It is a measure of the actual mean phase difference between the wave fields at A_1 and A_2.

The optical stethoscope thus provides us with direct means of measuring the degree of coherence of a wave field between any two points A_1 and A_2. It can actually be constructed, although not perhaps in the flexible form envisaged above. In later sections, we shall discuss in detail two implementations: the *Fourier transform spectrometer* (§11.6), which measures the temporal coherence using a Michelson interferometer (§9.3.2) and the *Michelson Stellar interferometer* (§11.9.1) which measures the spatial coherence of a light wave. Both were ideas originated by Michelson, and their origins are discussed in his book *Studies in Optics* (Michelson, 1927, 1962).

11.5 Temporal coherence

11.5.1 The degree of temporal coherence

We now return to the situation described in Fig. 11.4(a). Since the source is small, A_2 sees the wave as it was at A_1 a time τ earlier so that $\gamma_{A_1 A_2}(0) = \gamma_{A_1 A_1}(\tau) \equiv \gamma(\tau)$ where $\tau = A_1 A_2 / c$; in this section we can drop the position variable. The visibility of the fringes in the stethoscope is then $V = |\gamma(\tau)|$ where

$$\gamma(\tau) = \frac{\langle f(t) f^*(t + \tau) \rangle}{I} = \frac{\langle f(t) f^*(t + \tau) \rangle}{\langle f(t) f^*(t) \rangle} , \qquad (11.41)$$

which is the *complex degree of temporal coherence.* Now for a pure sine wave we have

$$f(t) = a \exp(i \omega_0 t) , \qquad (11.42)$$

whence

$$\gamma(\tau) = \exp(-i \omega_0 \tau) . \qquad (11.43)$$

It is therefore common to refer to $|\gamma(\tau)|$, whose departure from unity represents the departure of the waveform from a pure sinusoid, as the *degree of temporal coherence.*

For a quasi-monochromatic source $|\gamma(\tau)|$ has a typical form illustrated in Fig. 11.5. By definition, $\gamma(0) = 1$ and as τ increases, $|\gamma(\tau)|$ falls monotonically to zero. For any wave with such a $|\gamma(\tau)|$ we can define the coherence time τ_c as the time at which $|\gamma(\tau)| = 1/e$; for instance the random collection of Gaussian wave groups mentioned in §11.2.2 has $\tau_c = 2\sigma$ (see Problem 11.1). The form shown in Fig. 11.5 is not typical for laser light which will be discussed in §14.6.

It is easy to show from (11.31) – one can check it for (11.42) – that

Figure 11.5 Coherence function for a typical quasi-monochromatic source.

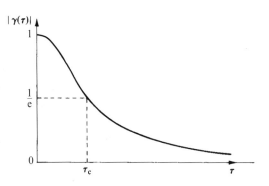

the real part of $\gamma(\tau)$ equals the degree of temporal coherence for the real function $f^R(t)$, i.e.

$$\gamma^R(\tau) = \frac{\langle f^R(t)f^R(t+\tau)\rangle}{\langle [f^R(t)]^2\rangle} = \frac{2\langle f^R(t)f^R(t+\tau)\rangle}{I} . \tag{11.44}$$

11.5.2 Temporal coherence and auto-correlation

The form of $\gamma(t)$ in (11.41) is the same as the auto-correlation function discussed in §4.7.1, when we use (11.1) to express the average values by integrals. When $f(t)$ is real and the average is taken over a long time T, the Wiener–Khinchin theorem (4.85) relates the power spectrum of $f(t)$ to the Fourier transform of $\gamma(\tau)$, which is now a symmetrical real function $\gamma^R(\tau)$. We then have

$$|F(\omega)|^2 = \frac{I}{2}\int_{-\infty}^{\infty} \gamma^R(\tau)e^{-i\omega\tau}d\tau ; \tag{11.45}$$

and

$$|F(-\omega)|^2 = |F(\omega)|^2 . \tag{11.46}$$

The spectral intensity $J(\omega)$ equals $T^{-1}|F(\omega)|^2$, which is independent of the averaging time T. However, as we have already done in (11.16), we shall ignore the factor T^{-1}. Equation (11.45) shows that if $\gamma^R(\tau)$ can be measured, the spectral intensity $J(\omega)$ can be deduced by a Fourier transform; this leads to an important form of spectroscopy, appropriately called *Fourier transform spectroscopy*, or *interferometric spectroscopy*.

11.6 Fourier transform spectroscopy

In 1898 Michelson showed that a two-beam interferometer could be used for spectral analysis. At that time the idea was difficult to implement because of the necessity for a Fourier transform in order to convert the observations into a conventional spectrum, although Michelson made some headway by intuitive methods and even constructed an analogue computer for the purpose of reconstructing spectra. The advent of electronic computers has of course changed the situation. Because of the basic simplicity of construction of a Michelson interferometer, its efficiency in terms of signal-to-noise ratio (§11.6.2) and its ease of application to wavelengths other than the visible (in particular infra-red), Fourier transform spectroscopy has become of considerable practical importance in modern physics and chemistry (see, e.g., Bell, 1972).

As shown in Fig. 11.6 the incident wave is split into two equal parts, each with real amplitude $f^R(t)$, which travel along different paths before recombining. If the two paths have lengths differing by d it is clear that the waves arriving at a given instant at D originated at the source at times separated by $\tau = d/c$. This is essentially the situation described by Fig. 11.4(a). The instrument is adjusted so that the interference fringes form a circular pattern (like Fig. 9.13(b)), at whose centre D the detector is positioned. Ideally, the detector can be as large as the central fringe when d has its maximum value, d_{max}. It measures the intensity of the recombined wave:†

$$I_M(\tau) = \langle [f^R(t) - f^R(t + \tau)]^2 \rangle . \qquad (11.47)$$

We note that $f^R(t)$ is taken as zero outside a certain time interval $-T/2$ to $T/2$ since the measuring time is finite. Using (11.44), (11.47) can be written as

$$I_M(\tau) = I[1 - \gamma^R(\tau)] . \qquad (11.48)$$

Thus the Fourier transform of $[I - I_M(\tau)]$ is equal to that of $I\gamma^R(\tau)$ which, from (11.45), is $2J(\omega)$. The signal I_M is real and symmetrical and so is $J(\omega)$ (11.46), however the part of the spectrum for $\omega < 0$ has no physical significance. Writing the transform out explicitly and replacing τ by d/c we have:

$$J(\omega) = \frac{1}{2c} \int_{-d_{max}}^{d_{max}} [I - I_M(d/c)] \exp[-i\omega d/c] \, \mathrm{d}d . \qquad (11.49)$$

In practice, since the interferogram is symmetrical about $d = 0$, only cosine functions make non-zero contributions to $J(\omega)$. Relying on this

---

† The minus sign in (11.47) results from the unequal phase shift introduced by the beam-splitter into the two optical paths. This effect, which is often ignored in the literature, causes the zero-order fringe to be dark. The final result is independent of the sign used.

Figure 11.6 Michelson interferometer as used for Fourier transform spectroscopy. D is a detector. The path difference is $d = 2(OM_2 - OM_1)$.

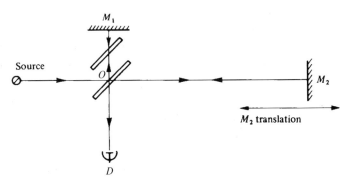

symmetry, the interferogram only needs to be measured for one sign of d (with a short excursion into the other to allow the zero of d to be identified exactly). As a result the integral above can be written in terms of wave-number $k = \omega/c$:

$$J(k) = \frac{1}{c} \int_0^{d_{\max}} [I - I_{\mathrm{M}}(d/c)] \cos(kd) \, \mathrm{d}d \, . \tag{11.50}$$

This equation is the basic algorithm for Fourier transform spectroscopy; the measured data is $I_{\mathrm{M}}(d/c)$ and the derived spectrum is $J(\omega)$ for $\omega > 0$.

In (11.49) we have introduced finite limits to the Fourier integral because there are no data for $d > d_{\max}$ and so the best estimate for this region would be $I_{\mathrm{M}} = I$. This sharp cut-off to the integral results in a spectrum with limited resolution (see Problem 11.5). Furthermore, it introduces 'false detail' into the spectrum, in the same way as will be discussed in §12.3.5, and the technique of apodization (§12.5.1) is often used mathematically to improve the line-shape obtained.

The above treatment can be extended to the case where a material with unknown index of refraction $\mu(\omega)$ is inserted in one arm of the interferometer. One then gets non-symmetrical functions $I_{\mathrm{M}}(\tau)$ and $\gamma^{\mathrm{R}}(\tau)$, and consequently the Fourier transform will have an imaginary part. The value of $\mu(\omega)$ can then be calculated from the ratio between the imaginary and the real parts of the transform. This is called *asymmetric Fourier transform spectroscopy* (Parker, 1990).

11.6.1 Two examples of Fourier spectroscopy

In Figs. 11.7–11.8 we show two examples of Fourier spectroscopy. The first illustrates Michelson's original approach, via the visibility function, and shows its limitations; for this reason we shall work it through in some detail. The second example is typical of a modern commercial Fourier spectrometer.

(*a*) Let us consider a spectral line with fine structure around frequency ω_0. Its spectral intensity can be represented by a δ-function $\delta(\omega - \omega_0)$ convolved with a 'fine-structure function' $s(\omega)$ which, as its name suggests, is limited to a region of frequency $\epsilon \ll \omega_0$. The function $J(\omega)$, which is the spectrum repeated symmetrically about the origin (§11.5.2), is

$$J(\omega) = s(\omega) \otimes \delta(\omega - \omega_0) + s(-\omega) \otimes \delta(\omega + \omega_0) \, . \tag{11.51}$$

Its transform, which from the inverse to (11.45) equals $\pi I \gamma^R(\tau)$, is

$$
\begin{aligned}
\pi I \gamma^R(\tau) &= S(\tau)\exp(-i\omega_0\tau) + S^*(\tau)\exp(i\omega_0\tau) \\
&= 2S^R(\tau)\cos(\omega_0\tau) + 2S^I(\tau)\sin(\omega_0\tau) . \quad (11.52)
\end{aligned}
$$

This gives $I - I_M(\tau)$, (11.48). Now since the fine-structure is restricted to a small frequency region ϵ, the transform S has scale ϵ^{-1}, so that it is appropriate to describe (11.52) as an oscillatory function (fringes) with a slowly varying envelope. Therefore the real and imaginary parts of $S(\tau)$ can be deduced by carefully observing the modulation *and phase* of the fringes. However, the complex function associated to (11.52) is $\pi I \gamma(\tau) = S(\tau)\exp(-i\omega_0\tau)$, giving $V = |\gamma(\tau)| = |S(\tau)|/\pi I$. If we were to measure this visibility only (and reject the phase information) we

Figure 11.7 Parts of the Fourier interferogram $I_M(\tau)$ when the source spectrum is an asymmetrical doublet, to show how the asymmetry is encoded in the fringes. In (a), $g^R(t) = \cos(9t) + \sqrt{2}\cos(11t)$. In (b), $g^R(t) = \sqrt{2}\cos(9t) + \cos(11t)$. The visibility, which is the same in both cases, is shown in (c).

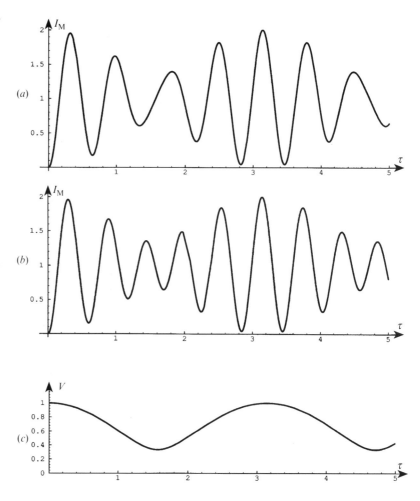

should know $|S(\tau)|$ only and so the fine-structure $f(\omega)$ calculated from it would be wrong.

An example will illustrate this (Fig. 11.7). A source emits a narrow asymmetrical doublet consisting of two waves with intensities a^2 at frequency $(\omega_0 - \epsilon/2)$ and b^2 at $(\omega_0 + \epsilon/2)$:

$$f^R(t) = a \cos[(\omega_0 - \epsilon/2)t + \phi_a] + b \cos[(\omega_0 + \epsilon/2)t + \phi_b] . \quad (11.53)$$

It is easy to show that that its power spectrum is given by (11.51) with

$$s(\omega) = a^2 \delta(\omega - \epsilon/2) + b^2 \delta(\omega + \epsilon/2) . \quad (11.54)$$

From (11.44) and (11.53),

$$\gamma^R(\tau) = \frac{a^2 \cos(\omega_0 - \epsilon/2)\tau + b^2 \cos(\omega_0 + \epsilon/2)\tau}{a^2 + b^2} \quad (11.55)$$

$$= \cos(\epsilon\tau/2)\cos(\omega_0\tau) + \frac{a^2 - b^2}{a^2 + b^2} \sin(\epsilon\tau/2)\sin(\omega_0\tau) . \quad (11.56)$$

Note that the values of ϕ_a and ϕ_b are unimportant. On comparing the coefficients of $\cos(\omega_0\tau)$ and $\sin(\omega_0\tau)$ with those of (11.52), we find the complex $S(\tau)$ to be

$$S(\tau) = \frac{\pi I}{2}[\cos(\epsilon\tau/2) + i\frac{a^2 - b^2}{a^2 + b^2} \sin(\epsilon\tau/2)] , \quad (11.57)$$

whose inverse Fourier transform is indeed the fine structure function (11.54). Equation (11.55) is clearly the real part of the complex function

$$\gamma(\tau) = \frac{a^2 \exp i(\omega_0 - \epsilon/2)\tau + b^2 \exp i(\omega_0 + \epsilon/2)\tau}{a^2 + b^2} , \quad (11.58)$$

Figure 11.8
Interferogram from a broad-band infra-red source, and the calculated spectrum.

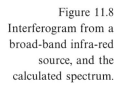

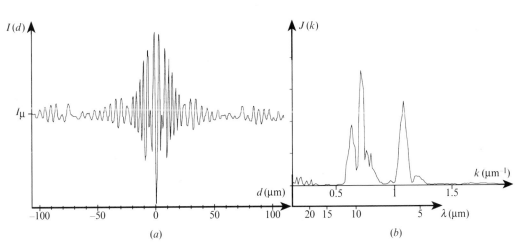

(a)   *(b)*

and therefore

$$V = |\gamma(\tau)| = \frac{|S(\tau)|}{\pi I} = \frac{[a^4 + b^4 + 2a^2b^2 \cos(\epsilon\tau)]^{1/2}}{a^2 + b^2} . \qquad (11.59)$$

But V is invariant on interchanging a and b, and from (11.59) we can only deduce ϵ (from its modulation period) and the ratio $|(a^2 - b^2)/(a^2 + b^2)|$ (from its depth of modulation). This was Michelson's method of analysis, but the question of whether $a^2 > b^2$ or $a^2 < b^2$ can be answered only by studying the *phases* of the fringes themselves. Of course, this is implicit in the Fourier transformation (11.50).

(*b*) The second example (Fig. 11.8) is from an automated Fourier Transform Spectrometer. It measures $\gamma^R(\tau)$ by recording $I_M(d)$ as d scans the region from $-d_1$ to d_{max}. If the instrument is adjusted perfectly, $I_M(d) = I_M(-d)$, but the inclusion of a small negative region allows the zero of d to be determined accurately. The example shows a small central part of an interferogram and the spectrum deduced from it, using (11.50).

11.6.2 Resolution and sensitivity

The resolution limit of Fourier spectroscopy can be seen from the first example above. The modulation of the fringes will only be discernible if d_{max} is sufficient to expose at least half a period of the modulation. Thus from (11.57)

$$\epsilon\tau_{max} = \epsilon d_{max}/c > 2\pi , \qquad (11.60)$$

which is satisfied by $\epsilon_{min} = 2\pi c/d_{max}$. This is most intuitive when expressed in terms of the smallest resolvable wavenumber δk:

$$\delta k = \epsilon_{min}/c = 2\pi/d_{max} ; \qquad (11.61)$$

at a given wavenumber k_0 the resolving power is then

$$k_0/\delta k = d_{max}/\lambda_0 . \qquad (11.62)$$

It is interesting to compare this with a diffraction grating (§9.2.2). The resolving power of the interferometer is the same as the very best that can be achieved with a diffraction grating of the same physical length ($d_{max}/2$)! So once again it is the physical size of the instrument, measured in wavelengths, which determines the resolving power – independently of details. But in a Fourier spectrometer, this resolution can actually be achieved (compare §9.2.2)!

When background radiation at the detector can not be neglected, the Fourier spectrometer has an advantage over conventional spectrometers that use diffraction gratings or prisms. If the output of the latter is measured with a single detector, the spectrum must be scanned in some way, and so for much of time very little light reaches the detector from the source, although background radiation is always received. On the other hand, with the Fourier spectrometer an average of half the input light reaches the detector at any instant (the other half leaves in the direction of the input where, in principle, a second detector can be placed). The Fourier spectrometer then has a distinct advantage (called the *Fellgett advantage*) in the signal to noise attainable. This is the reason for the success of this type of spectrometer in the infra-red region.

A further advantage, the *Jacquinot advantage*, over grating or prism instruments arises in the throughput of radiation, which may be an order of magnitude higher in the Fourier spectrometer. This applies to all spectral regions, including the visible.

11.7 Spatial coherence

The concept of temporal coherence was introduced as an attempt to give a quantitative answer to the following question. At a certain instant of time we measure the phase of a propagating light wave at a certain point. If the wave were a perfect sinusoidal plane-wave, $A \exp(-i \omega_0 t)$, we should then know the phase at any time in the future. But in a real situation, for how long after that instant will an estimate made in the above way be reliable? The gradual disappearance of our knowledge of the phase was seen to result from uncertainty of the exact value of ω_0, and could be related quantitatively to the finite width of the spectral line representing the wave.

The second coherence concept, that of *spatial coherence*, is concerned with the phase relationship at a given instant between waves at various points in a plane normal to the direction of propagation. If the wave were a perfect plane wave, this plane would be a wavefront, and definition of the phase at one point P in it would immediately define the phase at every other point; this can be done for each component wavelength if the wave is not monochromatic. In practice, we can ask the question: if we know the value of the phase at P, how far away from P can we go and still make a correct estimate of the phase?

In a similar way that we found temporal incoherence to be related to uncertainty in the frequency ω_0 of the wave (and hence in the magnitude of the wave-vector, $|k|$), we shall see spatial incoherence to be related to uncertainty in the *direction* of the wave-vector k. And uncertainty in the direction of k arises when the source of the light is not a point source, but is extended.

11.7.1 A qualitative investigation of spatial coherence

We saw in §11.4.2 that if we sample the wave train with our optical stethoscope at two points A_1 and A_2 situated one behind the other we see interference fringes only if the distance A_1A_2 is less than $c\tau_c$. Spatial coherence can be approached in the same way, and can be illustrated by the following simple one-dimensional experiment.

Suppose that an incoherent, quasi-monochromatic source, of linear dimensions a, is used to illuminate a mask $\mathscr{P}$ containing a pair of pinholes P_1 and P_2 separated by x (Fig. 11.9). The appearance of a fringe pattern on a screen indicates coherence between the wave amplitudes at the two pinholes. The source is at distance L and the screen at distance H from the pinholes; for simplicity assume $L, H \gg a, x$, and all angles to be small.

Consider the point S_1 at one end of the source. This point on its own illuminates the pinholes coherently and therefore produces a fringe pattern on the screen. The zero order of the fringe pattern appears at Z_1, corresponding to zero difference between the optical paths $\overline{S_1P_1Z_1}$ and $\overline{S_1P_2Z_1}$. Z_1 lies on the line joining S_1 to the point O half-way between the two pinholes. The period of the interference fringes is given by $H\lambda/x$. Now consider S_2 at the other end of the source. This gives a fringe pattern with the same period, with its zero order at the point Z_2, on the line S_2O. The two sets of fringes overlap, and since S_1 and S_2 are mutually incoherent, tend to cancel one-another out. When Z_1Z_2 is equal to half the fringe spacing, the fringe patterns from S_1 and S_2 will be spatially in antiphase, and so no fringes will be visible on the screen. We can say that the spatial

Figure 11.9 Spatial coherence.

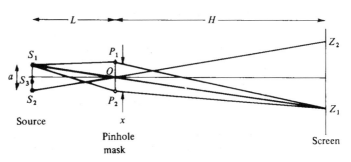

coherence between the two pinholes has disappeared when

$$\tfrac{1}{2}H\lambda/x = Z_1Z_2 = aH/L \; ; \tag{11.63}$$

$$x = L\lambda/2a \; . \tag{11.64}$$

The result can be stated as follows. Because of the size of the source, a, or more usefully its *angular* size, $\alpha = a/L$, coherence between neighbouring points on the mask only occurs if the distance between the points is less than

$$x_c = \lambda/2\alpha \; . \tag{11.65}$$

This maximum distance x_c is called the *coherence distance* in the plane of the pinholes. Notice in particular the reciprocal relationship between x_c and α.

We have neglected, in this discussion, the effect of all points such as S_3 in-between S_1 and S_2, and thereby introduced an error of about 2. This will be corrected in §11.7.3 by a more complete analysis.

When the argument is extended to two dimensions, a source of limited angular dimensions defines a two-dimensional region within which both pinholes must be situated in order to be coherently illuminated. This region is called the *coherence area* or *region*.

The relationship between the coherence area, or strictly the coherence function, and the source dimensions will be shown in §11.7.3 to be that between Fourier transforms, at least when the source has a small angular diameter α. This relationship can be very useful in practice, and is the basis of the technique of 'aperture synthesis' which will be discussed briefly in §11.9.3.

11.7.2 The degree of spatial coherence

We return now to the idea of the optical stethoscope probing a quasi-monochromatic wave field and assume that A_1 and A_2 are approximately on the same wave-front.† To be more exact we assume that τ_c is much longer than τ which is the difference between the times of arrival of the wavefronts at A_1 and A_2. Equation (11.39) is again valid, but now $|\gamma_{A_1A_2}(0)|$ depends only on the lateral distance between A_1 and A_2 since the only effect of a change in their longitudinal distance will be to multiply $\gamma_{A_1A_2}(0)$ by $e^{i\omega_0\tau}$. We can then call $|\gamma_{A_1A_2}(0)|$ the *degree of spatial coherence*. Often $\gamma_{A_1A_2}(0)$, the *complex degree of*

---

† The importance of A_1 and A_2 being w̵ ̵hin a coherence length of the same wavefront will be emphasized in the discussion of aperture synthesis in §11.9.3.

spatial coherence, depends only on the vector **r** connecting A_1 and A_2; it can then be written $\gamma(\mathbf{r})$.

An instrument which implements the above scheme almost exactly is the Michelson stellar interferometer (§11.9.1) where $A_{1,2}$ are the entrance mirrors and $B_{1,2}$ the second pair of mirrors.

11.7.3 The van Cittert–Zernike theorem

This theorem is the spatial equivalent of the Wiener–Khinchin theorem (§4.7.1) and was proved independently by van Cittert and by Zernike. It relates $\gamma(\mathbf{r})$ by a Fourier transform to the intensity distribution $I(\theta_x, \theta_y)$ in the source. To derive the theorem in one dimension, we consider a distant quasi-monochromatic incoherent source of angular extent α (outside which its intensity is zero – Fig. 11.10) illuminating the observation plane. All angles in the figure will be assumed to be small. The amplitude at point S on the source is $g(\theta)$, and the amplitude received at $P(x = 0)$ is then

$$f(0) = \frac{1}{L} \int_{\text{source}} g(\theta) \exp(ik_0 SP) d\theta \ . \tag{11.66}$$

At Q we have

$$f(x) = \frac{1}{L} \int g(\theta) \exp[ik_0(SP + x \sin \theta)] d\theta \ . \tag{11.67}$$

The product $c(x) = f(0)f^*(x)$ is then given by

$$c(x) = \frac{1}{L^2} \int g(\theta) \exp(ik_0 SP) d\theta \int g^*(\theta) \exp[-ik_0(SP + x \sin \theta)] d\theta \tag{11.68}$$

$$= \frac{1}{L^2} \int\int g(\theta) g^*(\theta') \exp[ik_0(SP - S'P)] \exp(-ik_0 x \sin \theta') \, d\theta \, d\theta', \tag{11.69}$$

where we have replaced θ in the second integral in (11.68) by θ' so as to express the product of two integrals as a double integral (cf. 11.4).

Now $\gamma(x)$ can be written from its definition (11.29) in terms of the time-average of $c(x)$, i.e.

$$\gamma(x) = \langle c(x) \rangle / \langle c(0) \rangle \ , \tag{11.70}$$

since the intensities at P and Q, $\langle |f(0)|^2 \rangle$ and $\langle |f(x)|^2 \rangle$ respectively, can both be considered to be equal to $\langle c(0) \rangle$. In calculating $\langle c(x) \rangle$ from (11.69) the only term which has to be averaged over time is $g(\theta) g^*(\theta')$ since all other terms are geometrical. When the source is incoherent, $g(\theta)$ and $g(\theta')$ are uncorrelated. It follows that this average is zero unless $\theta = \theta'$, when it equals $\langle |g(\theta)|^2 \rangle \equiv I(\theta)$, the source intensity at the point S. Introducing this into the double integral we have

$$\langle c(x) \rangle = \frac{1}{L^2} \int_{\text{source}} I(\theta) \exp(-ik_0 x \sin \theta) d\theta . \qquad (11.71)$$

This integral is reminiscent of Fraunhofer diffraction (§8.2). The value of $\gamma(x)$ is

$$\gamma(x) = \langle c(x) \rangle / \langle c(0) \rangle$$
$$= \frac{\int I(\theta) \exp(-ik_0 x \sin \theta) d\theta}{\int I(\theta) d\theta} . \qquad (11.72)$$

Since $I(\theta) = 0$ outside the source, the limits of integration can be written as $\pm\infty$. Then, for small θ,

$$\gamma(x) = \frac{\int_{-\infty}^{\infty} I(\theta) \exp(-ik_0 x \theta) d\theta}{\int_{-\infty}^{\infty} I(\theta) d\theta} ; \qquad (11.73)$$

$\gamma(x)$ is therefore the Fourier transform of I expressed as a function of $(k_0\theta)$, normalized to unity at $x = 0$. This relationship is called the *van Cittert–Zernike theorem*.

The theorem can easily be extended to a two-dimensional source in the same way as any Fourier transform. It then reads as follows. For two points in the observation plane (x, y) connected by $\mathbf{r}$, $\gamma(\mathbf{r})$ is the Fourier transform of $I(k_0\theta_x, k_0\theta_y)$. We shall apply it to a circular star as an example. The source has unit intensity within a circle of small angular diameter α and zero outside it. The correlation function is therefore the Fourier transform which is (§8.2.8)

$$\gamma(r) = \frac{2J_1(k_0 \alpha r / 2)}{k_0 \alpha r / 2} , \qquad (11.74)$$

it has its first zero when $r = 1.22\lambda/\alpha$.

For example, if a star has angular diameter 0.07 arc-sec, or about 3.4×10^{-7} radians, coherence in green light extends throughout a

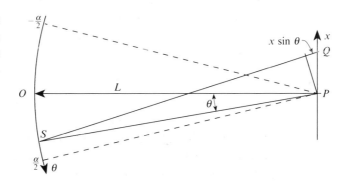

circle of radius about 1.8 m around a given point (Fig. 11.11). After the first zero (11.74) predicts there to be further regions of correlation, both negative and positive, but these result from the sharp cut-off assumed at the edges of the star and are observed only in laboratory experiments.

11.7.4 Partial coherence from a wide-angle source

A particular case in which θ is not small is that of microscope illumination §12.3.6 where the specimen receives light from all directions within a strongly-focused cone whose semi-angle α can approach $\pi/2$. Continuing from (11.72), and using $u = k_0 \sin \theta$, $du = k_0 \cos \theta \, d\theta$, we write for a one-dimensional source

$$\gamma(x) = \frac{\int_{-k_m}^{k_m} [I(\theta)/\cos \theta] \exp(-iux) du}{\int_{-k_m}^{k_m} [I(\theta)/\cos \theta] du}, \tag{11.75}$$

where $k_m = k_0 \sin(\alpha/2)$. For a Lambertian source (a black body, §14.1.2 for example) $I(\theta) \sim \cos \theta$, whence (11.75) gives $\gamma(x) = \mathrm{sinc}(k_m x)$. For the limiting case of an infinite source, $\alpha = \pi$ and $\gamma(x) = \mathrm{sinc}(k_0 x)$. The coherence distance x_c is the first x for which $\gamma(x)$ becomes zero, i.e. $x_c = \lambda/2$. The same calculation for a source infinitely extended in x *and* y follows from the equivalent result for a circular source, $\gamma(r) = 2J_1(k_0 r)/k_0 r$, and gives $r_c = 0.61\lambda$.

11.7.5 A laboratory demonstration of spatial coherence

The van Cittert–Zernike theorem can be illustrated by the experiment of §11.7.1 in which the pair of pinholes P_1 and P_2 with a distance x between them is illuminated by an incoherent source (Fig. 11.9). The visibility (§11.4.2) of the interference fringes then measures the coherence between the fields at the two pinholes, which are equivalent to the two sampling points of the optical stethoscope. The source

Figure 11.11 Correlation function between P and Q as a function of their separation r, when illumination is by a circular source of angular radius α.

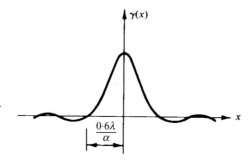

is a slit S of width a, aligned perpendicularly to the line P_1P_2. The function describing the coherence between P_1 and P_2 is therefore:

$$\gamma(x) = \text{sinc}[k_0 x \sin(a/2L)] \approx \text{sinc}\left[\frac{\pi a x}{\lambda L}\right], \tag{11.76}$$

this function being the Fourier transform in (11.72). As a is increased from zero, the coherence between pinholes P_1 and P_2 begins at unity, becomes zero when $a = \lambda L/x$ and has the usual series of weaker maxima and minima as a is increased beyond this value. The interference patterns observed for several values of a are illustrated in Fig. 11.12, in which the visibility clearly follows the same pattern. Notice that the effect of negative values of γ is to shift the pattern by half a fringe ($\phi = \pi$).

The above experiment is the basis of several fundamentally important interferometers that are used to determine the angular dimensions of inaccessible sources such as stars. These will be described in §11.9.

11.8 Fluctuations in light beams, and classical photon statistics

In §11.2.1 we showed that, however intense a light beam might be, its intensity still appears to fluctuate when investigated with a fast enough detector. The argument was completely classical, and the invention of the laser prompted a re-examination of the analysis. In particular, it was hoped to discover basic differences in the statistics of light emitted by lasers and conventional sources. In this section we shall present a simplified account of the classical theory, which has some important applications; the quantum theory of fluctuations will be discussed briefly in Chapter 14. A remarkable feature of the classical results are in fact their similarity to the quantum ones; it is only within the past decade that significant differences between the two have been discovered experimentally.

Before studying the light beam itself, we shall ask what exactly one measures in an experiment to detect fluctuations. The answer, of course, is the electric current from a photodetector. Any treatment of

Figure 11.12 Young's fringes with different degrees of spatial coherence: (a) $\gamma = 0.97$; (b) $\gamma = 0.50$; (c) $\gamma = -0.07$. Note particularly the minimum at the middle of (c), indicating the negative value of the coherence function.

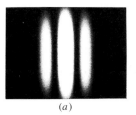

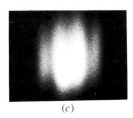

(a)   (b)   (c)

the subject must take into account the fact that we actually observe discrete electrons emitted by, say, a photocathode (other methods of detection, such as semiconductor devices, can be described similarly). In this experiment there are two uncorrelated sources of fluctuation. The first arises because we are observing discrete electron emissions whose *average* rate is proportional to the instantaneous intensity; the second because the instantaneous intensity itself is fluctuating about its long-term mean value. Recall from §11.2.1 that the term 'instaneous intensity' implies an average during a period $T_1 < \tau_c$.

The *mean* number of electrons emitted during a given interval $\delta t < T_1$ is $\bar{n} \equiv \langle n \rangle = \langle I(t) \rangle_{T_1} \eta \, \delta t / \hbar \omega$. Here $\langle I(t) \rangle$ is the mean intensity during δt and η is the *quantum efficiency*, which is the probability of an electron being emitted if a photon of energy $\hbar \omega$ falls on the cathode – typically a few percent. However, the *exact* number of electrons emitted is statistical, being given by a Poisson distribution with the above mean. The probability of n electrons being emitted in δt is then

$$p(n) = \bar{n}^n \exp(-\bar{n}) / n! \, . \qquad (11.77)$$

The variance or mean square fluctuation for the Poisson distribution is well known to be equal to its mean value:

$$\langle (\Delta n)^2 \rangle_{T_1} \equiv \langle n^2 \rangle - \bar{n}^2 = \bar{n} \, . \qquad (11.78)$$

This is one source of fluctuation in the current. Since $\bar{n}$ depends on the mean intensity during the interval, which is itself a fluctuating variable, we can also define $\bar{\bar{n}} = \langle I(t) \rangle_{T_0} \eta \, \delta t / \hbar \omega$, the average of $\bar{n}$ over a very long time T_0, and then the expectation value of $\langle (\Delta n)^2 \rangle_{T_1}$ will equal $\bar{\bar{n}}$.

The second source of fluctuation is that of $\langle I(t) \rangle$ itself, which we have already treated in §11.2.1. There we saw in (11.8) that the mean square difference

$$(\langle I(t) \rangle_{T_1} - \langle I(t) \rangle_{T_0})^2 = a^4 N^2 = [\langle I(t) \rangle_{T_0}]^2 \, . \qquad (11.79)$$

In terms of electrons emitted in time δt this can be written

$$\langle (\Delta \bar{n})^2 \rangle_{T_0} \equiv \langle (\bar{n} - \bar{\bar{n}})^2 \rangle_{T_0} = \bar{\bar{n}}^2 \, . \qquad (11.80)$$

Since $\Delta \bar{n}$ and Δn are not correlated we find that the total variance in photo-electron counts during T_1 is the sum of the individual variances:

$$\langle (n - \bar{\bar{n}})^2 \rangle_{T_0} = \langle (\Delta \bar{n})^2 \rangle_{T_0} + \langle \langle (\Delta n)^2 \rangle_{T_1} \rangle_{T_0} = \bar{\bar{n}}^2 + \bar{\bar{n}} \, . \qquad (11.81)$$

What is remarkable about this equation is that it is identical to the variance in the number of photons in a given state when they are considered as massless Bose–Einstein particles (see, e.g., Landau and

Lifshitz, 1980), which is surely a quantum description. Some inklings of an explanation will be given in Chapter 14.

The above argument, which is completely classical, shows that when a photo-detector is illuminated by a quasi-monochromatic wave, the emission of electrons is not a purely random process, governed by Poisson statistics. Photo-emission events are therefore correlated in some way, this is called *photon bunching*. For a high light intensity $\bar{n} \ll \bar{\bar{n}}^2$, so that n fluctuates in the range 0 to $2\bar{\bar{n}}$.

The concept of photon-bunching, given by the $\bar{\bar{n}}^2$ term, can be understood via the optical stethoscope. We now place two detectors at the exits B_1 and B_2 and correlate their output currents i_1 and i_2 electronically instead of observing the interference pattern. As we have pointed out, these currents are proportional to the mean intensities, $\langle I_1(t) \rangle_{T_1}$ and $\langle I_2(t) \rangle_{T_1}$. Using the model of §11.2.1 we can calculate the correlation between these intensities. Recognizing that the elementary wave j will arrive at A_1 and A_2 at different times and therefore with phase difference ϕ_j related to its direction of propagation we can write from (11.3)

$$
\begin{aligned}
&\langle I_1(t) I_2(t) \rangle \\
&= a^4 \langle \sum_{j=1}^{N} \exp(i\omega_j t) \sum_{j=1}^{N} \exp(-i\omega_j t) \sum_{j=1}^{N} \exp[i(\omega_j t + \phi_j)] \\
&\qquad \sum_{j=1}^{N} \exp[-i(\omega_j t + \phi_j)] \rangle \\
&= a^4 \langle \sum_{j,k,l,m=1}^{N} \exp\{i[(\omega_j - \omega_k + \omega_l - \omega_m)t + (\phi_l - \phi_m)]\} \rangle .
\end{aligned}
\tag{11.82}
$$

When the average is taken over a long time T_0, the only non-zero terms arise from $j = k$ and $l = m$, whence the average $\langle I_1(t) I_2(t) \rangle_{T_0} = a^4 N^2$. However, if $|\phi_l - \phi_m|_{\max} \ll \pi/2$, there is an additional contribution $a^4 N^2$ from terms with $j = m$ and $k = l$. The value of ϕ_j depends on the positions of A_1 and A_2. One can see immediately that if the A_1 and A_2 are one behind the other, so that the phase difference is caused by a time-delay τ, then $\phi_j = \omega_j \tau$ and the second contribution to the average of (11.82) comes in only if $\tau \ll \pi/[2|\omega_l - \omega_m|_{\max}] = \pi/\epsilon \approx \tau_c$, the coherence time. Then let $I_1 \equiv I(t)$, $I_2 \equiv I(t + \tau)$ and we can define the *intensity coherence function* or *second-order coherence function*

$$
\gamma_2(\tau) = \frac{\langle I(t) I(t + \tau) \rangle_{T_0}}{\langle I(t) \rangle_{T_0}^2} .
\tag{11.83}
$$

Note that this definition of γ differs in the denominator from that used, for example, in (11.44). The function $\gamma_2(\tau)$ has the typical form shown in Fig. 11.13, with $\gamma_2(0) = 2$ and $\gamma_2 \to 1$ as $t \to \infty$. Photon-bunching appears as an *excess correlation* for $\tau < \tau_c$. In terms of photo-electron counts, this means that if one observes an electron emission event, there is an higher probability of another one within τ_c than would be expected if the events were completely random (see Fig. 14.20). Experiments of this sort were first done by Morgan and Mandel (1966) and we shall return to the subject in Chapter 14 as a quantum phenomenon.

A similar calculation can be made for spatial coherence. In this case, ϕ_j represents the difference in phase between the waves from a given point on the source on their arrival at A_1 and A_2. For quasi-monochromatic light, this phase difference is $k_0(SP - SQ)$ (from Fig. 11.10) so that the enhanced correlation now arises when A_1 and A_2 are within a coherence region.

One can demonstrate, using this model, the general relationship (problem 11.9) that

$$\gamma_2(\mathbf{r}, \tau) = 1 + |\gamma(\mathbf{r}, \tau)|^2 . \tag{11.84}$$

It follows that coherence, whether temporal or spatial, can be measured by studying the correlation coefficient $\gamma_2(\tau)$ or $\gamma_2(\mathbf{r})$, respectively. Brown and Twiss developed these ideas in 1955–7 and confirmed them with experiments in which the spatial coherence of light from a mercury source was investigated using electronic correlation of the signals from two photocells (Fig. 11.14) (Brown and Twiss, 1956).

It is clear from (11.84) that $\gamma_2(\tau) \geq 1$ for any classical wave. We shall show in Chapter 14 that there also exist quantum forms of light for which $\gamma_2(0) < 1$, and which therefore have no classical equivalents.

Figure 11.13
Temporal intensity
coherence function
$\gamma_2(\tau)$, showing excess
correlation of
fluctuations when
$\tau < \tau_c$.

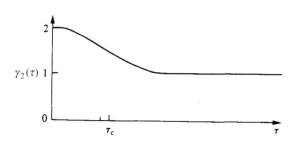

11.9 The application of coherence theory to astronomy

11.9.1 The Michelson stellar interferometer

Michelson used the idea of spatial coherence to design a stellar interferometer which enabled dimensional measurements of stars too small to be resolved by an astronomical telescope.† This instrument can be regarded as a practical realization of the optical stethoscope for the case discussed in §11.7.1. The time difference τ_p mentioned in §11.4.2 is small here because of the small angular size of the image. We saw in the experiment described in §11.7.5 that the visibility of the fringes, measured as a function of the pinhole separation x, was related quantitatively to the size of the source; the van Cittert–Zernike theorem (§11.7.3) shows that $\gamma(x)$, related to the fringe visibility, can be Fourier transformed to yield the stellar intensity distribution in the direction of P_1P_2. But really the fringe visibility measures only $|\gamma(x)|$ and the phase is not known, and so the Fourier transform cannot be performed completely. However, it is often sufficient to assume that a star has a centre of symmetry. This makes the problem soluble; this point is discussed more fully in another connexion in §12.8.

In principle, the Michelson stellar interferometer is constructed by putting a screen over the objective of a telescope and making two holes in it in such a way that their separation is variable. The point spread function of the telescope with the screen (the image produced by a point object – §12.3.1) is shown, simulated, in Fig. 11.16(*a*): this is the diffraction pattern of two holes (producing a circular ring pattern crossed by fringes representing the *two* in the problem).

When a source of finite diameter replaces the point source the continued existence of the fringes depends upon the coherence between the illumination of the two circular holes. The circular pattern is

† Because of the presence of atmospheric turbulence, which is discussed more fully in §12.8, no astronomical telescope can achieve even approximately the Rayleigh limit of resolution (§12.3.1). Since the turbulence only causes the fringes to move around, but does not affect their visibility, the Michelson stellar interferometer was the first instrument to allow this limit to be reached, and even exceeded.

Figure 11.14
Measurement of spatial coherence by correlation of intensity fluctuations.

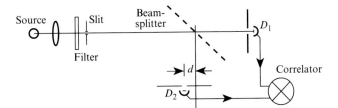

produced by each hole separately, but the straight fringes are joint property of the two. If the separation of the holes is increased, the fringes will become less and less clear, disappearing completely when the separation is $1.22\lambda/\alpha$; this property can be used to measure α, the angular dimension in direction P_1P_2.

If the angular diameter α of the star is very small, the separation of the holes may need to be very large before the fringes disappear. This causes two problems. First, the fringes become extremely closely-spaced when the holes are well apart, and second, a telescope of very large aperture is needed, although the greater part of the expensive mirror is never used. The first difficulty was overcome in an ingenious manner by a mirror system illustrated in Fig. 11.15. The coherence measured is clearly that between the light at mirrors A_1 and A_2, which

Figure 11.15 The Michelson stellar interferometer.

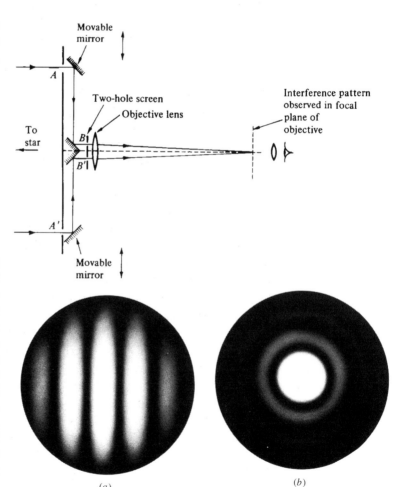

Figure 11.16 (a) Simulation of fringes in the Michelson stellar interferometer compared with the image (b) seen using the full aperture. The area of the small apertures is only 2% of the full aperture, so the exposure of photograph (a) is much longer than that of (b).

(a)   (b)

were mounted on racks to alter their separation. The interference pattern observed is that arising from the two apertures B_1 and B_2, and the scale of the fringes can therefore be made conveniently large by putting the holes B_1 and B_2 close together. Moreover, the dimensions of the pattern do not then change as A_1 and A_2 are separated. As an example, measuring a star of diameter 0.03 arc-sec would require a separation of 4 m in green light.

Michelson made a series of stellar interferometers, the most ambitious of which was built around the 2.5 m Mount Wilson telescope (Michelson, 1927, 1962). It should be clear from the foregoing discussion that the diameter of the telescope objective itself is quite irrelevant to the results of the experiment, and he used the 2.5 m telescope only because its construction was sufficiently rigid to bear the weight of the extra mechanical structures needed. Unfortunately, the inefficient use of light by this instrument and the need to detect the fringes dancing with the atmospheric scintillation limited its use as a visual instrument to stars of exceptional brilliance; the original experiments on it were carried out on the star Betelgeuse in the Orion constellation, and some twenty other stars within its measurement capability were measured by Michelson and his colleagues.

Recently, the optical stellar interferometer has seen a come-back as an astronomical instrument. Modern versions (see Beckers and Merkle, 1992) are constructed with two or more telescopes, each having an aperture of the order of 40 cm diameter, separated by distances reaching more than 100 m. An optical bench allows any pair of images to interfere directly. The differences in optical path from star to interference plane are equalized by means of a computer-controlled mirror system, sometimes called an 'optical trombone', which well describes its construction, and fringes can be observed even with white light. The path difference is modulated periodically at several hundred hertz, so that the interference pattern can be recorded within a characteristic fluctuation time of the atmosphere (see §12.8). The large apertures allow fringe visibilities to be measured with stars much weaker than Michelson could investigate. As well as determinations of stellar angular dimensions, parallax measurements (which allow the absolute distances of stars to be determined) and critical tests of the general theory of relativity (§2.9) are envisaged.

In the intervening years, the idea of the Michelson stellar interferometer inspired the development of aperture synthesis, one of the most fruitful techniques in radio astronomy, which will be discussed further in §11.9.3.

11.9.2 Brown and Twiss's intensity interferometer

Brown and Twiss built a stellar interferometer based on intensity correlations; see §11.8 (Brown, 1974). It consisted of two searchlight mirrors with variable separation **r** and a photo-cell at the focus of each. The two were focused on the star in question and the correlation between fluctuations in the photo-currents measured as a function of **r**. From the data, $\gamma_2(\mathbf{r})$ in (11.84), the dimensions of the coherence region, and thus the stellar dimensions, could be deduced. Since the correlation was carried out electronically, and the typical fluctuation frequency is τ_c^{-1}, it was necessary to use an optical filter to reduce this frequency to a value which was within the capabilities of current electronics, about 100 MHz; a lot of light was thus thrown away since this band-width corresponds to about 10^{-2}Å. On the other hand, the frequency of 100 MHz is far above typical atmospheric scintillation frequencies, so that the intensity interferometer is immune to that problem. Using a maximum baseline of about 200 m, giving a resolution limit at 0.5 μm of 5×10^{-4} arc-sec, the technique has been used to measure some 200 stars, which essentially exhaust the stellar objects of sufficient brilliance for its application.

11.9.3 Aperture synthesis

In astronomy carried out at radio frequencies it is possible to record the actual value (amplitude and phase) of the electric field of the waves received from stellar objects, and not just its intensity. One does this by mixing the current from an antenna with that from a local oscillator of frequency close to that of interest, thereby creating a low-frequency beat signal which can be recorded. The band-width of the beat signal is limited by the frequency response of the recording electronics. This opens up new possibilities, amongst them the capability of building up apertures sequentially in time, the signals from the various parts being added together coherently *post factum*. The technique is called *aperture synthesis* and is a direct application of the van Cittert–Zernike theorem (§11.7.3). It is important to realise that the mixing technique makes the radiation received essentially quasi-monochromatic, since the band-width of the low-frequency signal is much smaller than the frequency of the local oscillator. For example, if the wavelength chosen is 10 cm, and the bandwidth 100 MHz, the ratio ϵ/ω_0 is 0.03 (cf Fig. 11.2). More detailed information on the techniques is given by Perley, Schwab and Bridle (1986) and by Rohlfs (1986). Recently, successful attempts have been made to carry out similar synthesis in the infra-red region, using stablized lasers as local oscillators (Beckers and Merckle, 1992).

Consider as a basis the Michelson stellar interferometer, in which we observe the interference pattern between the signals received at two apertures. Here we use two sensitive point-like† radio receivers at $\mathbf{r}_0$ and $\mathbf{r}_0 + \mathbf{r}$ separated by vector $\mathbf{r}$, the length of which can be changed by the observer. We record the values of the two signals, $E^R(\mathbf{r}_0, t)$ and $E^R(\mathbf{r}_0 + \mathbf{r}, t)$, for a period of time, calculate the associated complex fields and then the complex coherence function $\gamma(\mathbf{r})$. For a quasi-monochromatic field, derivation of the associated complex field is straightforward; the imaginary part is just the real field one quarter of a period (i.e. $\pi/2\omega_0$) earlier (Problem 11.11).

The geometrical considerations are as follows. As the earth rotates, during one day the vector $\mathbf{r}$ traces out a cone in inertial space (Fig. 11.17(b)). When we introduced the complex degree of spatial coherence $\gamma_{A_1 A_2}(\tau = 0)$ in §11.7.2 and showed that its Fourier transform is the source image, we required for simplicity that A_1 and A_2 be within a coherence length of the same wavefront. This is clearly not the case in an aperture-synthesis system, because both antennae are constrained to be on the Earth's surface. However, if the source lies in the direction of unit vector $\hat{\mathbf{s}}$, we can relate the signal at A_2 to that at the equivalent point in the wavefront through A_1 by introducing a time-delay $\mathbf{r} \cdot \hat{\mathbf{s}}/c$ (Fig. 11.17(b)). Thus the spatial complex degree of

---

† They are usually parabolic dishes focusing the radiation from a particular direction onto the antennae, so as to achieve maximum sensitivity in the direction of interest. The idea is like the blazed grating, §9.2.5.

Figure 11.17 Aperture synthesis in radio astronomy; (a) shows the relative positions of the antennae A_1 and A_2 on the Earth, and (b) shows how $\mathbf{r}$ traces out a diurnal cone.

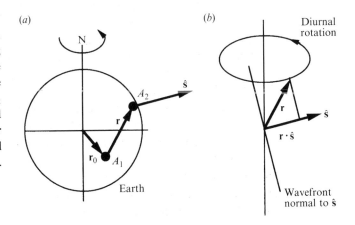

coherence is

$$\gamma(\mathbf{r}) = \frac{\langle E(\mathbf{r}_0, t) E^*(\mathbf{r}_0 + \mathbf{r},\, t - \mathbf{r} \cdot \hat{\mathbf{s}}/c) \rangle}{[\langle |E(\mathbf{r}_0, t)|^2 \rangle \langle |E(\mathbf{r}_0 + \mathbf{r}, t)|^2 \rangle]^{\frac{1}{2}}}. \tag{11.85}$$

Since $\gamma(\mathbf{r})$ is the transform of the source intensity, which is a real function and therefore $\gamma(\mathbf{r}) = \gamma^*(-\mathbf{r})$, collection of data for a given $|\mathbf{r}|$ is complete in 12 hours, although for part of the day a given source might be obscured by the Earth. In successive half-days the spacing between the antennae can be progressively changed up to some maximum depending on the length of the observatory (which may be kilometres) and the astronomers' patience. One then uses the van Cittert–Zernike theorem to compute, from the values of $\gamma(\mathbf{r})$, the intensity picture of the stable† radio universe at the wavelength chosen, with a resolution limited only by the maximum dimension. For example, using a maximum baseline of 4 km, at 2 cm wavelength an angular resolution of 5×10^{-6} rad, approximately 1 arc-sec, is achieved. This is about the same as the seeing limit of of an optical telescope. Practical aperture synthesis observatories use more than two apertures, but the same principle is involved.

Taking this to its extreme, observatories the world over now co-operate in taking simultaneous measurements of the same sources synchronized to very accurate atomic clocks. The results are then processed together at a central laboratory. This is called *very long baseline interferometry* (Kellerman and Thompson, 1988; Perley *et al.*, 1986). With aperture synthesis, the radio universe can be investigated to a resolution determined by a baseline of intercontinental dimensions. Of course, in this case $\gamma(\mathbf{r})$ is not calculable for all $\mathbf{r}$, but the $\mathbf{r}$ plane is sufficiently well-filled for a good picture to be created. It is like using a telescope with some parts of the mirror obscured; some artifacts result (see §12.7.3) but on the whole the damage is not great, and the resolution is given by the maximum dimension of the aperture.

† Only radio sources which do not change with time can be investigated this way.

CHAPTER TWELVE

Image formation

12.1 Introduction

Most optical systems are used for image formation. Apart from the pinhole camera, all image-forming optical instruments use lenses or mirrors whose properties, in terms of geometrical optics, have already been discussed in Chapter 3. But geometrical optics gives us no idea of any limitations of the capabilities of such instruments and indeed, until the work of Abbe in the middle of the nineteenth century, microscopists thought that the only limit to spatial resolution was their technical capability of grinding and polishing lenses. But (it now seems obvious) the basic scale is the wavelength of light, although recently several imaging methods have been devised which achieve resolution considerably in excess of this limit. The relationship is again like that between classical and quantum mechanics. Classical mechanics predicts no basic limitation to measurement accuracy; it arises in quantum mechanics in the form of the Heisenberg uncertainty principle.

This chapter describes the way in which wave optics are used to describe image formation by a single lens (and by extension, any optical system). The theory is based on Fraunhofer diffraction (Chapter 8) and leads naturally to an understanding of the limits to image quality and some of the ways of extending them.

12.2 The diffraction theory of image formation

In 1867 Abbe proposed a rather intuitive method of describing the image of a periodic object, which brought out clearly the limit to resolution and its relationship to the wavelength. We shall first describe his simple method, and later formalize it in terms of a double Fourier transform. The outcome of this understanding is to suggest various methods of improving images, in particular to create contrast from phase variations which are normally invisible.

12.2.1 Abbe theory: the image of an infinite periodic object

We have seen in §8.3.4 that if parallel light falls normally upon a diffraction grating several orders of diffraction are produced (Fig. 12.1). Let us place the grating in plane $\mathcal{O}$ and form its image using the diffracted light. In the paraxial approximation each order can be considered as a plane-wave, and the set of plane-waves can be refracted by a lens so that they converge individually to a set of points S in the focal plane $\mathcal{F}$ of the lens and then continue so that they all overlap again in the plane $\mathcal{I}$. Here they form a complicated interference pattern; this pattern is the image.

The advantage of taking a diffraction grating as an object is that the process of image formation can easily be seen to consist of two distinct stages. First, we have the stage between $\mathcal{O}$ and $\mathcal{F}$. In the latter plane we have produced the Fraunhofer diffraction pattern of the object. Second, we have the stage between $\mathcal{F}$ and $\mathcal{I}$. The orders $S_2, S_1, \ldots, S_{-2}$ behave like a set of equally-spaced point sources and the image is their diffraction pattern. Thus the process of image formation appears to consist of two diffraction processes, applied sequentially.

The second diffraction process in this example can also be analyzed

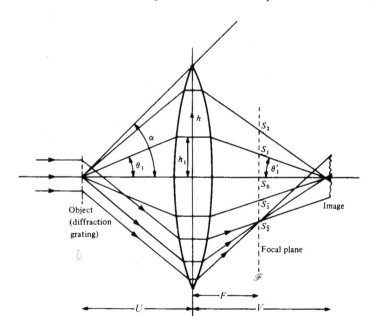

Figure 12.1 Formation of the image of a diffraction grating. Five orders of diffraction are shown, producing five foci S in the plane $\mathcal{F}$. The complete ray bundle is only shown for the -2 order. The angular semi-aperture of the lens is α.

without difficulty. Each pair of orders S_j and S_{-j} produces Young's fringes in the plane $\mathscr{I}$. If the object grating has spacing d, the order S_j appears at angle θ_j given, for small angles, by

$$\theta_j \approx \sin \theta_j = j\lambda/d . \tag{12.1}$$

The small-angle approximation will be seen in §12.2.2 to be unnecessary. By simple geometry one can see from Fig. 12.1 that

$$\theta_j \approx \tan \theta_j = h/U , \tag{12.2}$$

$$\theta'_j \approx \tan \theta'_j = h/V , \tag{12.3}$$

and so

$$\theta'_j \approx U\theta_j/V. \tag{12.4}$$

The waves from the first orders, S_1 and S_{-1} converge on the image at angles $\pm\theta'_1$ and thus form periodic fringes with spacing

$$d' = \lambda/\sin \theta'_1 \approx \lambda V/\theta_1 U = Vd/U . \tag{12.5}$$

Thus a magnified image has been produced; the magnification is $m = V/U$. Fringes from the higher orders produce harmonics of this periodic pattern, with spacings d'/j, and contribute to determining the detailed structure of the image. *The finest detail observable in the image is determined by the highest order of diffraction which is actually transmitted by the lens.*

The zero order contributes a constant amplitude. This zero-order term is of crucial importance. Without it, the interference pattern of the first orders would appear to have half the period of the image, because we observe intensity, and not amplitude; the function $\sin^2 x$ has half the period of $\sin x$. However, the addition of the constant restores the correct periodicity to the intensity, since $(c + \sin x)^2 = c^2 + 2c \sin x + \sin^2 x$, which has the period of $\sin x$.

The use of an infinite grating as an object has, of course, oversimplified the problem. If we had used a finite grating, there would be subsidiary orders (§8.3.3) which also transmit information. The simple result above therefore needs some modification if the object is more complicated, but it expresses the essence of the resolution limit.

12.2.2 The Abbe sine condition

Although the last section might suggest that a faithful image would be built up only if the angles of diffraction were kept small, Abbe realized that larger angles could be employed if the ratio $\sin \theta / \sin \theta'$

rather than θ/θ' were the same for all values of θ. If we had had, exactly:

$$\frac{\sin \theta_j}{\sin \theta'_j} = m \qquad (12.6)$$

we should then have the period of the fringes in the image

$$d'_j = \lambda/\sin \theta'_j = m\lambda/\sin \theta_j = md_j. \qquad (12.7)$$

The harmonics would then have exactly the right periods to fit the fundamental d'_1; and the image would be as perfect as possible. One important lens having this property is described in §3.9, and forms the basis of high-power microscope objectives.† The Abbe sine condition does *not* state that $\sin \theta/\sin \theta'$ is a constant in any particular imaging system, but *requires* that this condition be met if the system is not to produce aberrations when large angles θ and θ' are used. It is not satisfied, for example, by a simple lens.

12.2.3 Image formation as a double process of diffraction

In §12.2.1 we introduced, in a qualitative manner, the idea that image formation can be considered as a double process of diffraction, and in §12.2.2 we saw the Abbe sine condition to be necessary for its realization. In this section we shall formalize the approach mathematically in one object dimension. There is no particular difficulty in the extension to two dimensions.

The analysis is based on the scalar-wave theory of diffraction, and assumes an object uniformly and coherently illuminated by a plane wave. The wave leaving the object is represented by the complex function $f(x)$ (multiplied, of course, by $e^{-i\omega_0 t}$, which is carried unchanged through all the equations, and will be ignored). The object is imaged by a lens, such that the object and image distances are U and V; the object dimensions are small compared with U (Fig. 12.2). The amplitude of the wave reaching point P in the focal plane $\mathscr{F}$ of the lens is, following the treatment and notation of §8.2 in one dimension, the Fourier transform of $f(x)$ with the phase delay appropriate to the path $\overline{OAP}$:

$$\psi(u) = \exp(ik_0\overline{OAP})F(u) = \exp(ik_0\overline{OAP}) \int_{-\infty}^{\infty} f(x)\exp(-iux)\,dx, \qquad (12.8)$$

† The requirement $\sin \theta/\sin \theta' = $ constant can be deduced from purely geometrical reasoning (see Kingslake, 1978) but the above argument is more physically intuitive and emphasizes its importance.

where $k_0 = 2\pi/\lambda$ and u corresponds to the point P:

$$u = k_0 \sin \theta .$$ (12.9)

Now the amplitude $b(x')$ at Q in the image plane can be calculated using Huygens's principle over the plane $\mathscr{F}$.[†] The optical distance from P to Q is

$$\overline{PQ} = PQ \quad = \quad (PI^2 + x'^2 - 2x'PI \sin \theta')^{\frac{1}{2}}$$
$$\approx \quad PI - x' \sin \theta'$$ (12.10)

when $x' \ll PI$. If the Abbe sine condition (12.6) is obeyed,

$$\sin \theta = m \sin \theta',$$ (12.11)

where m is the magnification. We therefore write, from (12.9)

$$PQ = PI - x'u/mk_0 ,$$ (12.12)

whence the amplitude at Q is

$$b(x') \quad = \quad \int_{-\infty}^{\infty} \psi(u) \exp(ik_0 PQ) \, du$$
$$= \quad \int_{-\infty}^{\infty} \exp(ik_0 PI) \, \psi(u) \exp(-ix'u/m) \, du .$$ (12.13)

This is the second Fourier transform in the problem. Inserting (12.8) into (12.13) we write the relationship between the image $b(x')$ and the object $f(x)$:

$$b(x') \quad = \quad \int_{-\infty}^{\infty} \left\{ \exp[ik_0(\overline{OAP} + PI)] \int_{-\infty}^{\infty} f(x) \exp(-iux) dx \right\}$$
$$\times \exp(-iux'/m) du .$$ (12.14)

[†] Since this involves interference of spherical waves a factor $1/r$ ought to be included, but this has no effect on the physics and will not be included.

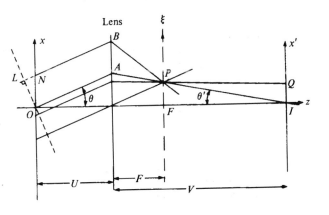

Figure 12.2 Ray diagram for the demonstration of the image–object relationship. The coordinate $\xi \equiv uF/k_0$.

The combined phase factor $\exp[ik_0(\overline{OAP} + PI)]$ appears at first sight to be a function of the point P, and hence of the parameter u. This is indeed true if the planes $\mathcal{O}$ and $\mathcal{I}$ are chosen arbitrarily. But if they are *conjugate* planes then by Fermat's principle (§2.7.2) the optical path from O to I is *independent* of the point P, and the factor can be written as a constant, equal to $\exp(ik_0\overline{OI})$, and can be taken outside the integral. We are left with the integral

$$b(x') = \exp(ik_0\overline{OI}) \int_{-\infty}^{\infty} \left[\int_{-\infty}^{\infty} f(x) \exp(-iux)\mathrm{d}x \right]$$
$$\times \exp(-iux'/m)\mathrm{d}u. \qquad (12.15)$$

The integrals are the same as those involved in the Fourier inversion theorem. From §4.5 we then have:

$$b(x') = \exp(ik_0\overline{OI})f(-x'/m). \qquad (12.16)$$

This equation represents the well-known fact that the image is an inverted copy of the object, magnified by the factor m. The above result, first proved by Zernike, can be stated simply: an optical image can be represented as the Fourier transform of the Fourier transform of the object. It applies exactly only if the lens is well corrected; i.e. it obeys the Abbe sine rule and the optical path $\overline{OPI}$ is completely independent of the point P.†

12.2.4 Illustrations of the diffraction theory of image formation

In the previous section we have shown theoretically that, when the object is illuminated coherently, the imaging process can be considered as a double Fourier transform. We shall now describe some experiments originally carried out by Porter in 1906 which confirm this result. They are done in an imaging system, shown in Fig. 12.3, which allows the comparison of the intermediate transform and the final image (see Appendix 2). A partially-transparent mask is illuminated with a parallel coherent beam, and is imaged by a converging lens. We observe the illumination in the focal plane to be the Fraunhofer diffraction pattern of the object, and the image to be the Fourier transform of that diffraction pattern. The first stage, that of the formation of the Fraunhofer diffraction pattern, has been adequately illustrated in Chapter 8. To confirm that the second stage is also a

† In the terms of §3.7, the requirements are equivalent to the absence of spherical aberration (no errors result from non-zero θ for an axial point) and first-order coma (the same for points a short distance x from the axis).

Fourier transform, we can modify the transform in the focal plane by additional masks or obstacles and observe the resultant changes in the final image. Such processes are called *spatial filtering* by analogy with the corresponding process in the time-domain in electrical circuits. Spatial filtering has some very important applications which will be discussed in detail in later sections.

Let us consider first an object consisting of a piece of gauze. It is two-dimensional, and is basically periodic, although there are deviations from exact periodicity as well as defects such as blocked holes. We image it in the system of Fig. 12.3. The diffraction pattern (plane $\mathscr{F}$) is shown in Fig. 12.4(a). It contains well-defined spots, corresponding to the periodic component of the gauze, and an additional light distribution surrounding each of the orders which expresses the non-periodic components. The complete image of the gauze is shown in (b).

We now insert various masks into the plane $\mathscr{F}$, and thereby cut out parts of the diffraction pattern. For example, if the mask transmits only orders on the horizontal axis (c) the image becomes a set of vertical lines (d); this is the object which would have given (c) as its diffraction pattern. Similarly, a mask which transmits only the orders $(0, \pm 1)$, $(\pm 1, 0)$, (e), gives us a different gauze (f). But the irregularities are the same, because they contribute to the diffraction pattern at all points. The zero order alone, together with the region half way out to the next orders, (g), gives us an image in which no gauze is visible, but only the irregularities – particularly the blocked holes. Finally, a small region of the diffraction pattern (i) remote from the centre emphasizes a different aspect of the deviations from exact periodicity, (j).

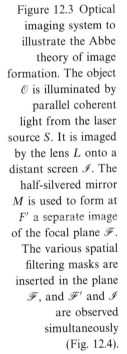

Figure 12.3 Optical imaging system to illustrate the Abbe theory of image formation. The object $\mathscr{O}$ is illuminated by parallel coherent light from the laser source S. It is imaged by the lens L onto a distant screen $\mathscr{I}$. The half-silvered mirror M is used to form at F' a separate image of the focal plane $\mathscr{F}$. The various spatial filtering masks are inserted in the plane $\mathscr{F}$, and $\mathscr{F}'$ and $\mathscr{I}$ are observed simultaneously (Fig. 12.4).

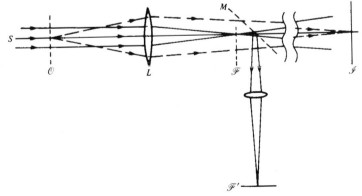

Figure 12.4
Illustrating the Abbe
theory of image
formation. On the
left are the selected
portions of the
diffraction pattern of
a piece of gauze, and
on the right the
corresponding
images.

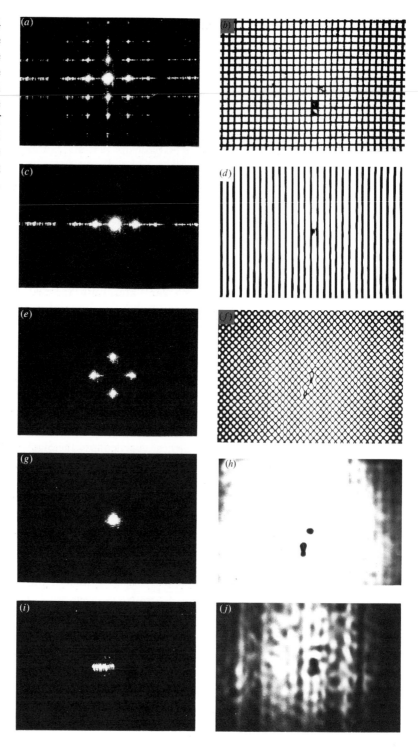

12.2.5 The phase problem

A question that is always asked at this point refers to the possibility of separating the two stages of the image-forming process. Suppose we were to photograph the diffraction pattern in the focal plane and in a separate experiment illuminate the photograph with coherent light and observe its diffraction pattern. Should we not have produced the diffraction pattern of the diffraction pattern and have reconstructed the image? The flaw in the argument concerns the phases of the diffraction pattern. The illumination $\psi(u)$ is a complex quantity containing both amplitudes and phases. Photography records only the intensity $|\psi(u)|^2$ and the phase is lost. A second diffraction process as suggested above would be carried out in ignorance of the phases, and therefore would be unlikely to give the right answer. In fact, the second process would assume all the phases to be zero, and would indeed give the correct image if this were so.

The problem that the phases in a recorded diffraction pattern are generally unknown is called the *phase problem*. Its solution is of central importance to the interpretation of X-ray diffraction patterns, where one wants to create an image of a crystal whose diffraction pattern intensity has been recorded. Several approaches are possible. One is to derive the phases intelligently from information in the diffraction pattern, using some knowledge of the object. This approach was discussed in §8.6, under the heading of 'phase retrieval', and has had remarkable success in recent years.

There is, however, a very important class of diffraction patterns which can be retransformed directly from the diffraction pattern, because all the phases are indeed zero. It follows that all the diffracted waves add constructively at the origin of real space, and so the object must have its maximum value there. In crystallographic terms this means that there is a strongly scattering atom at the origin. Now many molecules are in fact built around a single heavy atom, which can conveniently be considered as the origin. If such a material crystallizes with one such molecule in each unit cell, we have the right conditions for all the diffraction phases to be zero.† In such a case, retransforming the X-ray diffraction pattern as a real function should give a tolerable picture of the original crystal structure.

An experiment of this sort was carried out by W. L. Bragg in

---

† In fact not all the phases need to be zero, but only those of the strong diffraction orders. But this is usually enough for a reasonable initial reconstruction, which can subsequently be improved.

1939, which he called an *X-ray microscope*. Reconstruction of the object was done by making a representation of a section of the X-ray diffraction pattern of the crystal in the form of holes in an opaque plate, Fig. 12.5(a), the area of each hole being proportional to the amplitude of the corresponding X-ray diffraction spot. The Fraunhofer diffraction pattern of the plate was then an image of the structure (b).

This idea is the basis of the *heavy atom* method of solving crystal structure problems, which is applicable to any molecule that can be crystallized with a single or a few heavy atoms in the same positions in each unit cell. These atoms dominate the diffraction pattern and essentially fix the phases of all the diffracted waves. With this information about the phases, the structure of the unit cell can be found; this was the method used to elucidate the structures of proteins, haemoglobin and myoglobin by Perutz and Kendrew, and is still the only one available for the most complicated molecules (Woolfson, 1970; Blundell and Johnson, 1976).

Figure 12.5 (a) Set of holes representing orders of X-ray diffraction from diopside, $(CaMg(SiO_3)_2)$; (b) diffraction pattern of (a) representing a projection of the atoms in the crystal structure; (c) diagram of structure to compare with (b). (Reconstruction of Bragg's experiment by Harburn, 1972).

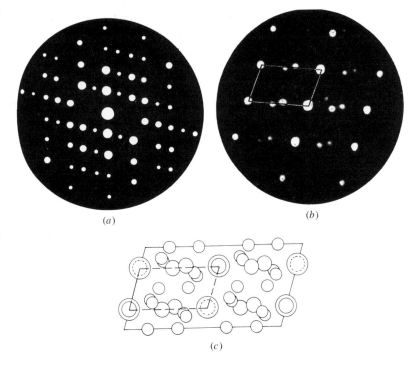

(a)

(b)

(c)

12.3 Resolution limit of optical instruments

The light which forms the image in an optical system is limited angularly by the aperture stop (§3.4.2). In this section we shall use the Abbe theory of image formation in an attempt to understand how the size of the aperture stop and the coherence of the illumination affect the characteristics of the image, and in particular how they limit the resolution attainable. It will appear that the limits of perfect coherence and perfect incoherence of the illumination can be treated fairly clearly; the intermediate case of partially-coherent illumination is complicated and the results can only be indicated in rather general terms.

12.3.1 Rayleigh's criterion for an incoherent object

The simplest and best-known resolution criterion is that due to Rayleigh and applies to the case of a self-luminous or incoherently illuminated object; it is usually applied to an astronomical telescope, because stars certainly fulfil the requirements of self-luminosity and incoherence; but it applies equally well to a microscope observing, for example, a fluorescent object.

If we consider a single point on the object, we have seen in §7.2.7 that we observe in the image plane the Fraunhofer diffraction pattern of the aperture stop, on a scale determined by the image distance. This diffraction pattern is called the *point spread function.* An extended object can be considered as a collection of such points, and each one produces a similar point spread function in the image plane; because the sources are incoherent we add *intensities* of the various patterns to get the final image. The image is therefore the convolution of the object intensity and the point spread function.

The Rayleigh resolution criterion arises when we consider two neighbouring points on the object, separated by a small angle. If the aperture has diameter D, its diffraction pattern, expressed as a function of the angle θ, has intensity (§8.2.7)

$$I(\theta) = [2J_1(\tfrac{1}{2}k_0 D \sin \theta)/(\tfrac{1}{2}k_0 D \sin \theta)]^2 . \qquad (12.17)$$

Rayleigh considered that the two points on the object are distinguishable if the central maximum of one lies outside the first minimum of the other. Now the function (12.17) has its first zero at that of $J_1(x)$, at $x = 3.83$. Then

$$\tfrac{1}{2}k_0 D \sin \theta_1 = \pi D \sin \theta_1 / \lambda = 3.83 . \qquad (12.18)$$

The angle θ_1 is the minimum angular separation of resolvable incoherent sources; since $\theta_1 \ll 1$ the resolution limit is thus

$$\theta_{min} = \theta_1 = 3.83\lambda/\pi D = 1.22\lambda/D \quad \text{(Rayleigh).} \tag{12.19}$$

Notice that only the *angular* separation of the sources enters the result.†

This is the best-known resolution criterion, but fails under some circumstances which we shall meet later. An alternative, which corresponds well with what the human eye can resolve because of its superb sensitivity to intensity *differences*, is the *Sparrow criterion*. This considers two point images to be resolved if their joint intensity function has a minimum on the line joining their centres. If the two points have equal intensities, the Sparrow criterion then indicates θ_{min} when

$$\left(\frac{\mathrm{d}^2 I}{\mathrm{d}\theta^2}\right)_{\theta = \theta_{min}/2} = 0. \tag{12.20}$$

Without entering into details of the differentiation of Bessel functions, this gives

$$\theta_{min} = 0.95\lambda/D \quad \text{(Sparrow).} \tag{12.21}$$

12.3.2 A coherently-illuminated object

Next we consider the resolution problem when the sources are coherent. If the object consists of two points emitting with the same phase, we must add the *amplitudes* of their point spread functions:

$$A(\theta) = J_1(\tfrac{1}{2}k_0 D \sin\theta)/(\tfrac{1}{2}k_0 D \sin\theta). \tag{12.22}$$

The Rayleigh criterion gives the same result as (12.19) because the zeros of the point spread functions have not changed; but the points are not resolved. On the other hand, the Sparrow criterion gives $\theta_{min} = 1.46\lambda/D$. The reason that the Sparrow criterion gives here a larger θ_{min} is illustrated by Figs. 12.6 and 12.7. We show first the intensity as a function of position on a line through the images of two incoherent sources at $\theta = 0$ and $\theta = \theta_{min}$ for the two cases (12.19) and (12.21). The function shown is $I(\theta) + I(\theta - \theta_{min})$. The Rayleigh resolution is clearly more than adequate. We then look

† When two equally-intense points are separated by this angle, the intensity measured along the line joining them has a minimum half way between them of value $8/\pi^2$ times the maximum at each point. The Rayleigh criterion is often interpreted under other conditions as that separation which gives a minimum with this value between the two maxima. We shall not use this interpretation here, preferring that of Sparrow as an alternative.

at the equivalent situation when the sources are coherent; we add amplitudes *before* squaring to find the intensity, $[A(\theta) + A(\theta - \theta_{\min})]^2$, which is illustrated in Fig. 12.7 for the two criteria. Clearly the Rayleigh separation is inadequate.

This argument would indicate that incoherent illumination results in better resolution. But it is not always true; we assumed a particular phase relation between the sources in order to demonstrate the result. If the two sources had phases differing by π, we should have written the joint intensity as $[A(\theta)$ *minus* $A(\theta - \theta_{\min})]^2$, which always has minimum intensity at the mid-point, *however close the sources*. But of course, as the sources become closer, the image gets weaker because of destructive interference. Just the same, this has an important practical application in what are called *phase-shift masks*, used in photo-lithography of micro-electronic devices to enhance the resolution of closely-separated units on a complex mask. Alternate units are covered with a transparent film which introduces the necessary π phase change to ensure that a dark line appears between their images (Levenson, 1993). However, *in general* it can indeed be said that incoherent illumination results in the better resolution. Fig. 12.8, in which images of a pair of pinholes have been formed under various coherence conditions, demonstrates the above argument. One should notice in particular in this figure that the separation of the two coherent antiphase images in (*e*) is quite different from their true separation; it

Figure 12.6 Addition of the images of two pinholes incoherently illuminated. The thin lines show the intensity curves and the thick line their sum: (*a*) Rayleigh and (*b*) Sparrow separations.

Figure 12.7 Addition of the images of two pinholes coherently illuminated. The thin lines show the amplitude curves and the thick line the square of their sum: (*a*) Rayleigh and (*b*) Sparrow separations.

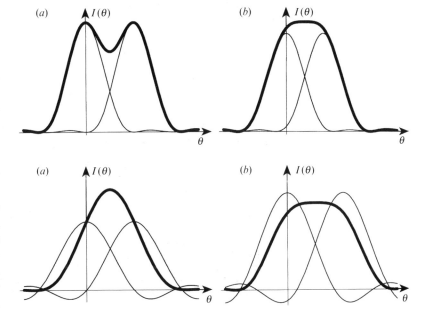

is actually determined by the aperture diameter! See Problem 12.5. A detailed discussion of the effect of coherence on imaging is given by Goodman (1985).

12.3.3 Application of the Abbe theory to resolution

Despite the conclusions of the previous section, most microscopes work with coherent or partially coherent illumination, because of the small dimensions of the object and the practical difficulties (§12.3.6) of producing truly spatially-incoherent light. The Abbe theory discussed in §12.2.1 applies to coherent illumination and is therefore fairly appropriate to a discussion of resolution by a microscope.

We therefore return to the model of a periodic object. The resolution that can be obtained with a given lens or imaging system is, as mentioned in §12.2.1, limited by the highest order of diffraction which the finite aperture of the lens will admit. If the object has period d, the first order appears at angle θ given by

$$\sin \theta_1 = \lambda/d. \tag{12.23}$$

In order to image an object with such a period, the angular semi-aperture α of the lens must be greater than θ_1. Thus the smallest period which can be imaged is given by

$$d_{\min} = \lambda/\sin \alpha. \tag{12.24}$$

It is usual to recall from §3.8 the possible immersion of the object in a medium of refractive index μ, where the wavelength is λ/μ, and to write $d_{\min}$ in terms of the *numerical aperture* $\mathrm{NA} \equiv \mu \sin \alpha$;

$$d_{\min} = \lambda/\mu \sin \alpha = \lambda/\mathrm{NA} \tag{12.25}$$

Figure 12.8 Comparison between coherent and incoherent imaging of a pair of pinholes. In (a)–(c) the aperture is chosen so that the pinhole separation corresponds to the Rayleigh limit, with illumination (a) incoherent, (b) coherent (same phase), (c) coherent (antiphase). In (d) and (e) the same pair of pinholes is imaged through an aperture for which they are at the Sparrow limit: (d) incoherent and (e) coherent (antiphase) illumination. The lower row shows weaker exposures of corresponding images in the upper row.

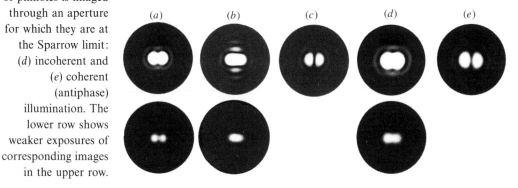

is the coherent imaging resolution in this case.†

Now we have assumed in the above discussion that the illumination is parallel to the axis, and acceptance of the zero and two first orders is necessary to form an image with the correct period. In fact, the correct period will be imaged if the zero order and one first order alone pass through the lens. So we can improve the resolution by illuminating the object with light travelling at angle α to the axis, so that the zero order just passes through; then the condition for the first order on one side to pass through as well is that

$$d_{\min} = \lambda/2\mu \sin\alpha = \lambda/2\,\mathrm{NA}, \qquad (12.26)$$

where we have used the result for Fraunhofer diffraction in oblique illumination from §8.2.2. This result represents the best that can be achieved with a given lens, and is usually quoted as the ultimate resolution limit. In order to implement it one usually illuminates the object isotropically with a cone of light having semi-angle at least α, as shown in Fig. 12.9, in order to get high resolution in all directions.

---

† Microscope objectives are usually labelled with two numbers: one is the magnification at a standard image distance of 200 or 250 mm, and the second is the NA. The thickness of cover glass for which aberrations are corrected is also given on high magnification objectives.

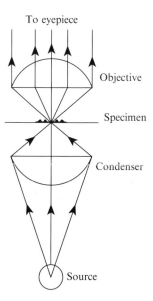

Figure 12.9 Conical illumination of a specimen to get the highest microscopic resolution.

To eyepiece

Objective

Specimen

Condenser

Source

12.3.4 Illustration of coherent resolution

Figure 12.10
Resolution according
to the Abbe theory.
(*a*) Diffraction
pattern of set of
holes shown in (*b*).
The circles indicate
the apertures used to
limit the transform.
(*b*)–(*g*) Images of
object shown in (*b*),
with different
numerical apertures.
The apertures used
are shown as circles
superimposed on the
diffraction pattern
shown in (*a*). (*h*) The
image formed by the
part of the
diffraction pattern
between the second
and third circle from
the centre. This
illustrates
apodization (§12.5.1).
The image is sharper
than the object, but
contains false detail.

This theory may be illustrated by the apparatus described in Fig. 12.3. The laser source S was replaced by a conventional mercury arc, in order to show the effects of partially-coherent illumination, which is very important in practical microscopy. We can now investigate the changes that occur in an image if the optical transform is limited in some way. For example, suppose that we have a general object as shown in Fig. 12.10(*b*); its transform is shown in (*a*). We then place a series of successively smaller holes over the transform, and we can then see how the image is affected. The succession of diagrams is explained in the caption.

The resolution limit imposed by a finite aperture can also be considered as an application of the convolution theorem (§4.6). In a coherently illuminated system, restriction of the optical transform by a finite aperture results in a point spread function in the image plane whose *amplitude* (not intensity as in §12.3.1) must must be convoluted

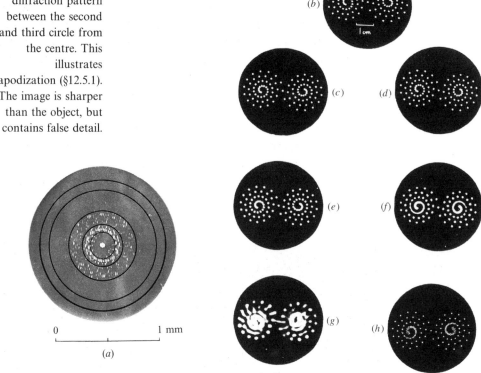

Figure 12.10
Resolution according to the Abbe theory. (*a*) Diffraction pattern of set of holes shown in (*b*). The circles indicate the apertures used to limit the transform. (*b*)–(*g*) Images of object shown in (*b*), with different numerical apertures. The apertures used are shown as circles superimposed on the diffraction pattern shown in (*a*). (*h*) The image formed by the part of the diffraction pattern between the second and third circle from the centre. This illustrates apodization (§12.5.1). The image is sharper than the object, but contains false detail.

with that of the object when the image is formed. The result is, once again, a blurring of the image; but because amplitudes are involved, neighbouring parts of the image can interfere. The result is more complicated than in the incoherent case, and *false detail* (§12.3.5) can be produced.

12.3.5 False detail

As we saw in the previous section, coherent illumination can result in the production of false detail. In many cases, this may be finer than the limit of resolution (examine Fig. 12.8 in detail, for example). The use of an optical instrument near to its limit of resolution is always liable to produce effects of this sort; when the Abbe theory was first announced, many microscopists adduced such effects as evidence that the theory was unacceptable (§1.6.3). Even today, when the theory is fully accepted, it is sometimes forgotten in dealing with images produced by, for example, the electron microscope.

The formation of false detail can be conveniently illustrated in the framework of Fig. 12.4. Suppose that the focal plane stop of the instrument limits the transform to the centre five orders only (Fig. 12.11(*a*)). The image is then illustrated by (*b*). Notice the formation of bright spots on the crosses of the gauze wires. One can easily see quantitatively the origin of these spots by retransforming the zero and first orders only of a square wave, which is a reasonable model for one dimension of the gauze.

12.3.6 The importance of the condenser

As far as geometrical optics is concerned, the condenser in a microscope merely serves to illuminate the specimen strongly. According to the wave theory, however, the coherence of the incident light is important, and the condenser therefore has as much importance as

Figure 12.11 False detail produced by imaging with a restricted region of the diffraction pattern, Fig. 12.4(*a*).

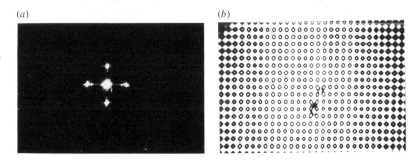

(*a*)   (*b*)

any other part of the optical system.† The reason for this can best be expressed in terms of coherence. Ideally, as we shall show below, the object should be illuminated in completely incoherent light, which we could obtain by a general external illumination from a large source such as the sky. But this would be very weak, and we increase the intensity by using a lens to focus a source of light onto the object. An image, however, cannot be perfect, and each point on the source gives an image of finite size on the object. In other words, neighbouring points on the object are illuminated by partially-coherent light. The poorer the quality of the condenser, the more false detail is to be expected.

In practice, two forms of illumination are widely used. The first is called *critical illumination* and is obtained by forming an image of the source in Fig. 12.12(*a*) directly on the object by means of a condenser. This arrangement, however, has the defect that irregularities of the source can affect the image formed. An arrangement that does not have this defect is called *Köhler illumination* and is shown in Fig. 12.12(*b*). An extended source is used, and although any one point on the source gives parallel coherent illumination at a certain angle, the total illumination from all points on the source is indeed almost incoherent (§11.7.4). This is because the individual coherent plane-waves have random phases and various directions of propagation and therefore add up with different relative phases at each point in the field. The position of the object is such that the condenser

† In a reflecting microscope, the light enters through the objective, which therefore doubles as condenser.

Figure 12.12 Types of incoherent illumination: (*a*) critical; (*b*) Köhler.

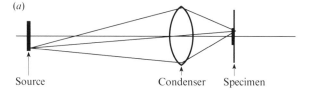

(*a*)

Source   Condenser   Specimen

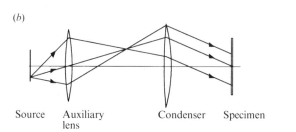

(*b*)

Source   Auxiliary   Condenser   Specimen
           lens

approximately images the auxiliary lens onto it. One would expect this lens to be reasonably uniformly illuminated if it is not too close to the source, even if the latter is patchy.

For either of the above condensing systems the illumination system results in a field with spatial coherence distance r_c. In particular, Köhler illumination corresponds to the situation described in §11.7.4, where r_c was shown to be $0.61\lambda/\mathrm{NA_c}$, where $\mathrm{NA_c}$ is the numerical aperture of the condenser. In the case of critical illumination, each uncorrelated point on the source produces an image in the object plane which is about the size of the Airy disc (§8.2.7) and this results in approximately the same value of r_c. Aberrations in the condenser always increase r_c above this value.

If $\mathrm{NA_c}$ is larger than that of the objective, and its optical quality is good, r_c is smaller than the resolution limit, so that neighbouring resolvable points are substantially uncorrelated. As a result, the resolution limit is given by the Rayleigh or Sparrow criteria and false detail is avoided. Reducing $\mathrm{NA_c}$ often improves the contrast of an image, but increases false detail.

12.4 Applications of the Abbe theory: spatial filtering

Optical instruments can be used without more than a cursory knowledge of how they work, but a physicist should know more than this. He can then fully appreciate their limitations, can find the conditions under which they can be best used and, most important, may find ways of extending their use to problems that cannot be solved by conventional means. The procedures which will be described in this section are known under the general name of *spatial filtering* techniques. They can be expressed in terms of operations carried out by inserting masks affecting the phase and amplitude of the light in the back focal plane $\mathscr{F}_2$ of the lens, which is the plane in which the Fourier transform of a coherently illuminated object would be observed. Because they essentially modify the Fourier transform or spatial frequency spectrum of the image, the name 'spatial filtering' arose by analogy with electronic filtering which is used to modify the temporal frequency spectrum of the signal. When incoherent illumination is used, the Fourier transform in the plane $\mathscr{F}_2$ can not, of course, be recognized, but the principles to be discussed below still apply. The methods used to achieve them are usually only approximate.

12.4.1 Dark-ground imaging

Suppose that we wish to observe a very small non-luminous object. If we use the ordinary method of illumination, in which the incident light bathes the specimen and enters the objective, it is likely that the amount of light scattered will be so small that it will be negligible compared with that contained in the undeviated beam and the object will not be seen. We can avoid this difficulty by arranging that the incident light is directed obliquely at the specimen so that it does not enter the objective; this method has been used widely for observation of Brownian motion. The method is adequate if we merely want to know the position of a scattering object, for example; it is equivalent to forming an image by using only a small, off-centre, part of the transform, and this will not give much information about the nature of the object. For a reasonable image of the object we must use as much as possible of the transform, and this is achieved in practice as shown in Fig. 12.13 by cutting out directly-transmitted (*a*) or reflected light (*b*) as completely as possible without affecting the rest. This procedure is also useful for visualizing a phase object which has little absorption (§8.2.5).

The principle of the method can be illustrated quite simply in the apparatus of Fig. 12.3 by placing a small black spot over the central peak of the transform. We have chosen as object a pattern of holes punched in a thin transparent film, shown in Fig. 12.14(*a*); since the film is not optically uniform the transform is rather diffuse. A small

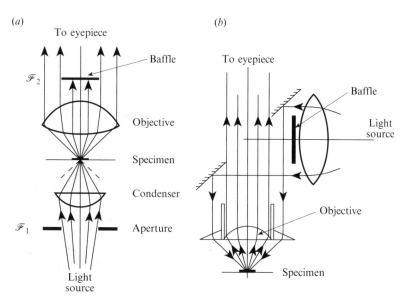

Figure 12.13 Examples of practical systems for dark-ground imaging: (*a*) in transmission, where light directly transmitted from the condenser is blocked by a baffle after the objective; (*b*) in reflexion, where the specimen is illuminated by a hollow cone of light outside the objective.

ink-spot on a piece of glass is then placed over its centre, as shown in (*b*), and the final image (*c*) can be compared with (*a*), which was obtained when the ink-spot was absent. Although the edges of the holes were visible in the unfiltered image, the contrast is improved considerably by the dark-ground filter.

12.4.2 Phase-contrast microscopy

Phase-contrast microscopy is another method of creating contrast from a phase object. Since large phase variations are usually visible because they result in refraction effects, phase-contrast microscopy is mainly useful for small phase variations. It can be explained as follows, in a way which compares it with the dark-ground method.

Suppose that we represent the light amplitude transmitted by an object by a vector in the complex plane. In a phase object (§8.2.5) the vectors representing the complex amplitude at various points on the object are all equal in length, but have different phase angles. In Fig. 12.15(*a*), OA_1, OA_2, OA_3 are typical vectors. In a perfect imaging system, the corresponding image points have the same complex amplitudes (apart from a constant factor, which can be ignored), and therefore their intensities are equal; no contrast is observed. Let us picture each vector OA as the sum of a constant OP and the remainder PA, where the vector sum of all the PAs is zero. Because the vectors OA have different directions, the PAs all have different lengths. Now the vector OP by definition has constant amplitude and phase at all points, and therefore corresponds to the Fourier transform of a δ-function at the origin of the Fourier plane; this is what we have called the zero order of diffraction. In the dark-ground method, we obstruct the zero order, and therefore subtract the vector OP from each of the vectors OA. The remaining vectors PA have different lengths and therefore intensity contrast is achieved as shown in Fig. 12.15(*b*). The *Zernike phase-contrast* method involves changing the phase of the vector OP by $\pi/2$, and therefore replacing it by the vector PP'. The

Figure 12.14 (*a*) Image of a pattern of holes in cellophane sheet; (*b*) diffraction pattern of (*a*) with a small opaque spot over its centre; (*c*) image formed from (*b*).

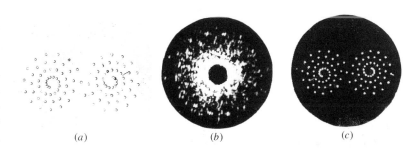

(*a*) (*b*) (*c*)

new image-point vectors $P'A$, once again have different lengths, as in (*c*). This method has the advantage that all the light transmitted by the object is used in forming the image. It is clear that the exact value of the phase change is not very important, so that white light can be used.

The phase-contrast method can be described analytically, when the phase variations $\phi(x) \ll 1$, by expanding the complex transmission function:

$$f(x) = A \exp[i\phi(x)] \approx A + iA\phi(x), \quad |f(x)|^2 = A^2 . \quad (12.27)$$

Changing the phase of the zero order (the term independent of x) by $\pi/2$ gives the function

$$f_1(x) = iA + iA\phi(x) \approx iA \exp[\phi(x)] ; \quad (12.28)$$

$$|f_1(x)|^2 = A^2 \exp[2\phi(x)] \approx A^2[1 + 2\phi(x)], \quad (12.29)$$

which has a real variation of intensity, linearly dependent on ϕ. Fig. 12.16 shows a simulation using the apparatus of Fig. 12.3.

In practice, the application to an incoherently illuminated object is not so simple, since there is no precise transform whose zero order can be identified. A compromise is necessary, and is effected as follows (Fig. 12.17). The illuminating beam is limited by an annular opening in the focal plane below the condenser, and a real image of this opening is formed in the back focal plane of the objective, $\mathscr{F}_2$. The phase plate, inserted in $\mathscr{F}_2$, is a thin transparent evaporated ring of optical thickness $\lambda/4$ whose dimensions match those of the image of the annulus. All undeviated light from the specimen must therefore pass through this plate. The final image is formed by interference between the undeviated light passing through the phase plate and the deviated light which passes by the side of it. The ideal conditions are only approximately satisfied, for some of the deviated light will also pass through the phase plate, giving rise to characteristic halos around phase steps. The phase plate is constructed by the vacuum deposition of a dielectric material such as cryolite, Na_3AlF_6, onto a glass support.

Figure 12.15 Vector diagrams illustrating (*a*) normal imaging of a phase object, (*b*) dark ground, and (*c*) phase-contrast.

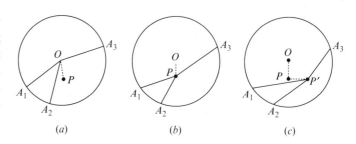

(*a*)   (*b*)   (*c*)

When the object produces only small differences of phase, the zero order of its diffraction pattern is outstandingly strong and changing its phase produces too great a difference in the image. Therefore the phase plate is commonly made to transmit only 10–20% of the light. It looks like a small dark ring on a clear background, and is clearly producing a compromise between dark-ground and phase-contrast images.

12.4.3 Schlieren method

An alternative method of creating contrast from a phase object is to cut off the central peak by a knife-edge, thereby cutting off half the

Figure 12.16 Phase-contrast imaging. (*a*) Fraunhofer diffraction pattern of a phase mask similar to Fig. 12.14(*a*). A spatial filter, consisting of a tranparent plate with a small hole whose size is indicated by the white spot, is placed over the zero order, resulting in image (*b*). Notice that non-uniformities in the thickness of the mask material are highlighted as well as the holes.

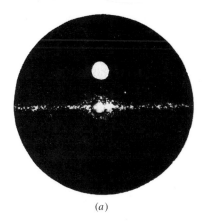
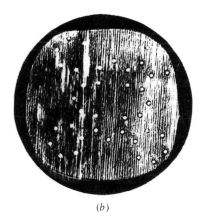

(*a*) (*b*)

Figure 12.17 Optics of phase-contrast microscope. The phase plate is coincident with the image of the annular ring formed in the condenser-objective lens system. Köhler illumination is used.

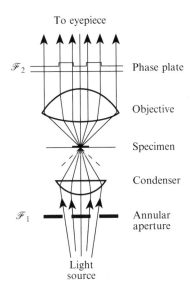

To eyepiece

$\mathscr{F}_2$ — Phase plate

Objective

Specimen

Condenser

$\mathscr{F}_1$ — Annular aperture

Light source

transform as well. In practice the object to be studied is placed in a coherent parallel beam which is brought to a focus by a lens accurately corrected for spherical aberration. A knife-edge is then translated in the focal plane of the lens until it just overlaps the focus. A clear image of the object can then be seen (Fig. 12.19). We have illustrated this method by the same object that we used for dark-ground illumination in Fig. 12.14(*a*); the image now has some defects that will be discussed later.

The schlieren method has two important applications. First, it can be used as a critical test of lens quality, for if a lens suffers from aberrations there will not be a sharp focus and it will not be possible to put the knife-edge in a position to cut off only half the transform. Thus, if we observe the lens itself as the knife-edge traverses the focal plane, the intensity of illumination across the surface will appear to change (Fig. 12.20); we can then deduce what modifications to make to the lens. This is called the *Foucault knife-edge test*; it can also be used to locate the focus of a lens extremely accurately.

A second use of the schlieren method is important in fluid dynamics. A wind tunnel in which the density of air is constant (and hence the refractive index is constant too) is an object with neither phase nor amplitude variations. Waves or other disturbances in the tunnel will modify the density and refractive index in a non-uniform way, and thus produce a phase object. By using the schlieren technique, the phase variations can be studied visually as changes in intensity in

Figure 12.18 Filters in the Fourier plane representing (from left to right) dark-ground, phase-contrast, schlieren and diffraction contrast.

Figure 12.19 (*a*) Cutting out half the diffraction pattern in Fig. 12.14(*b*) gives the image (*b*).

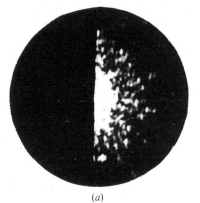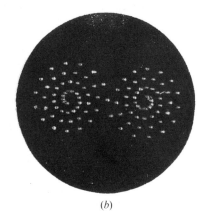

(*a*)   (*b*)

the final image (Fig.12.21). An important difference between phase-contrast and schlieren systems is that the latter operate in one axis only. Schlieren systems are often adjusted to attenuate but not remove the zero-order light; this practice increases the sensitivity substantially.

12.4.4 Diffraction contrast

Spatial-filtering techniques are widely used in the electron microscope as well as the optical microscope. In the former, the numerical aperture is very small, because of the impossibility of fully-correcting electron lens aberrations. When one looks at crystalline matter with an electron

Figure 12.20
Appearance of a lens suffering from spherical aberration when subject to the Foucault knife-edge test: (*a*) when the knife-edge is in the marginal focal plane; (*b*) when the knife-edge is in the paraxial focal plane. Note for example that in (*b*) only light from the lower part of the lens reaches the eye, therefore this part looks bright.

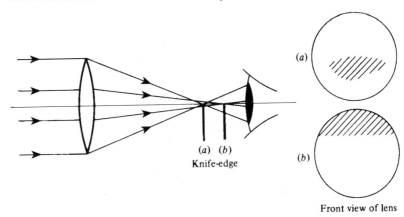

Front view of lens

Figure 12.21
Schlieren pattern of bullet-shaped object at Mach number 3.62. (From Binder, 1973.)

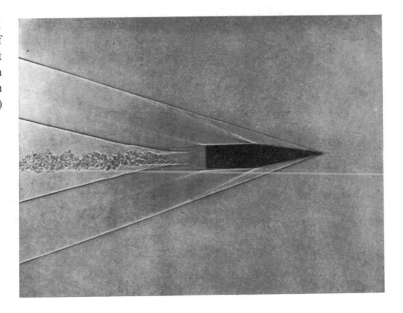

microscope it is therefore possible to use only a very limited region of the Fourier transform which may contain only one order of diffraction and a limited region around it. As we saw in Fig. 12.4(*h*) and (*j*), this is sufficient to make visible structure on a scale larger than the unit cell. The technique called *diffraction contrast* or *dark field* imaging uses a spatial filter consisting of an aperture that selects the region of a single order of diffraction only (usually not the zero order). If, for example, we have a polycrystalline sample, and image it through an off-centre aperture, only those crystallites which have diffraction spots lying within the aperture appear bright in the image; the rest are dark. This can be seen as an optical simulation in Fig. 12.22. In order to see the periodicity (atomic structure) of the sample one needs to use an aperture which selects several orders of diffraction. An image formed this way is called a *lattice image*; there is great danger of getting misleading structures in such images by an unwise choice of orders.

12.4.5 An analytical example illustrating dark-ground, schlieren and phase-contrast systems

In this section we shall calculate the intensity distributions in the image of a simple one-dimensional phase object, when the filters discussed in §§12.4.1–12.4.3 are used. They are summarized diagramatically in Fig. 12.18. Although the examples treat the microscopic methods

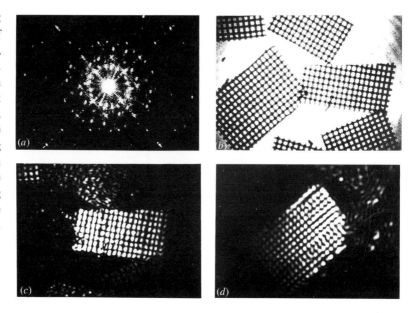

Figure 12.22 Simulation of diffraction contrast, using coherent optics. (*a*) Diffraction pattern of a mask representing a 'polycrystal'; (*b*) image formed using the whole diffraction pattern; (*c*) and (*d*) images formed using selected areas of the diffraction pattern.

quantitatively, it should be stressed that their use is mainly qualitative – to visualize phase objects but rarely to analyse them quantitatively.

The object can be called a *phase slit*; it is a transparent field containing a narrow strip of different optical phase from its surroundings. In one dimension, x, normal to the length of the strip, we describe such an object by:

$$f(x) = \exp[i\phi(x)] \qquad (12.30)$$

where $\phi(x) = \beta$ when $|x| \le a$ and is zero otherwise. This function can be written as the sum of a uniform field and the difference in the region of the strip:

$$f(x) = 1 + (e^{i\beta} - 1)g(x) \qquad (12.31)$$

where $g(x) = \text{rect}(x/2a)$ would represent a normal transmitting slit of width $2a$. The transform of the function written this way is:

$$F(u) = \delta(u) + 2a(e^{i\beta} - 1)\text{sinc}(au). \qquad (12.32)$$

First, let us consider the effect of dark-ground illumination (§12.4.1). In this technique we eliminate the zero-order component; this is the $\delta(u)$ and a narrow region of negligible width at the centre of the sinc function. After such filtering, the transform is to a good approximation,

$$F_1(u) = 0 + 2a(e^{i\beta} - 1)\text{sinc}(au) \qquad (12.33)$$

and the resultant image, the transform of $F_1(u)$, is

$$f_1(x) = (e^{i\beta} - 1)g(x). \qquad (12.34)$$

Recalling the definition of $g(x)$ in (12.31) we see that the slit appears bright on a dark background. In fact, its intensity is dependent on the phase β:

$$I_1(x) = |f_1(x)|^2 = 2(1 - \cos\beta)g(x). \qquad (12.35)$$

It is reassuring to notice that the image would disappear were $\beta = 0, 2\pi, \dots$ because then there is no physical difference between the slit and its surround.

Using the same example we can illustrate the schlieren method, §12.4.3. In this case the filter cuts out the $\delta(u)$ and all the transform for $u < 0$, leaving us with

$$F_2(u) = 2a(e^{i\beta} - 1)\text{sinc}(au)\, D(u), \qquad (12.36)$$

where $D(u)$ is the step function: $D(u) = 1$ when $u > 0$, otherwise 0. Using the convolution theorem, the transform of (12.36) is

$$f_2(x) = (e^{i\beta} - 1)g(x) \otimes d(x),\qquad(12.37)$$

$$\text{where }\ d(x) = \int_{-\infty}^{\infty} D(u)e^{-iux}\mathrm{d}u = \frac{1}{ix}\qquad(12.38)$$

is the transform of the step function.† Evaluating the convolution (12.37) directly for the slit function $g(x)$ gives us

$$
\begin{aligned}
g(x) \otimes d(x) &= -i\int_{-\infty}^{\infty}\frac{g(x-x')}{x'}\mathrm{d}x' \\
&= -i\int_{x-a}^{x+a}\frac{\mathrm{d}x'}{x'} = -i\ln\left|\frac{x+a}{x-a}\right|.
\end{aligned}\qquad(12.39)
$$

The image intensity is then

$$I_2(x) = |f_2(x)|^2 = 2(1 - \cos\beta)\left[\ln|(x+a)/(x-a)|\right]^2,\qquad(12.40)$$

which is illustrated by Fig. 12.23. In this example, the schlieren method clearly emphasises the edges of the slit, which are discontinuities in object phase. In general it can be shown to highlight phase gradients in the direction normal to the knife edge; this effect can be seen in the example in Fig. 12.19.

Finally, this model can be used to illustrate the phase-contrast method (§12.4.2). The transform (12.32)

$$F(u) = \delta(u) + 2a(e^{i\beta} - 1)\mathrm{sinc}(au)\qquad(12.41)$$

† This transform will be discussed in more detail in §13.4.2.

Figure 12.23
Schlieren image of
the phase slit.

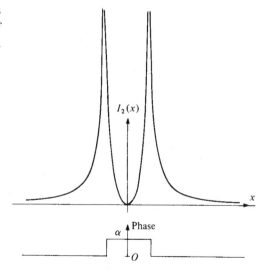

is filtered by the phase plate which changes the phase of the $k = 0$ component by $\pi/2$ (i.e. multiplication by i):

$$F_3(u) = i\,\delta(u) + 2a(e^{i\beta} - 1)\text{sinc}(au). \tag{12.42}$$

The image amplitude is the transform of this:

$$f_3(x) = i + (e^{i\beta} - 1)g(x), \tag{12.43}$$

which has value i in the region $|x| > a$ and value

$$i - 1 + e^{i\beta} = (\cos\beta - 1) + i(\sin\beta + 1) \tag{12.44}$$

within the slit. The intensity contrast is maximum when $\beta = 3\pi/4$.

12.4.6 The interference microscope

When we want a quantitative complex analysis of a phase object, we use an *interference microscope*. This form of microscope is constructed around a two-beam interferometer, and if incoherent illumination is to be used, it is clear that the interference fringes must be localized in the object. Many types of interferometer can be used, and we shall give just one example; others are given by Krug, Rienitz and Schulz (1968) and in textbooks on interferometry. Interference microscopy is not a spatial-filtering technique, but we include it in this section because of its complementary relationship to the techniques described in §§12.4.1–12.4.4.

The interference microscope that we shall describe here uses a a *shearing interferometer*, a type of interferometer not specifically described in Chapter 9, which produces an interference pattern between an image field and the same field displaced linearly by a known distance.† The image therefore appears double, but the translation may be so small that this does not detract from its sharpness. If, in addition, there is a difference of π between the phases of the two images, destructive interference occurs and the combined field is dark in the absence of phase variations. Regions of the image which have differences of phase within the displacement vector then appear bright on the dark background, and the technique is appropriately called *differential interference contrast*. It finds many applications ranging from microelectronic wafer inspection to biology.

In the form of the differential interference microscope due to Nomarski, the small displacement is achieved using the differing optical

---

† The Jamin interferometer (§9.3.1) is an example of an interferometer which can easily be made to perform this function.

properties of a crystal for two orthogonal polarizations. We shall describe it as a transmission microscope, although it is commonly used in a reflecting form. Critical illumination is assumed (§12.3.6). The illumination at two points on the object corresponds, at the entrance to the condenser, to two plane-waves travelling in different directions. These are obtained from a single linearly-polarized plane wave by passing it through a thin crystal device (*Wollaston prism*) with construction similar to the Babinet Compensator (§6.8.2). This is made from a uniaxial crystal in the form of two opposed thin wedges having orthogonal optic axes *OA*, Fig. 12.24(*a*). The initial plane wave is polarized at 45° to these axes, so that the ordinary and extraordinary waves have equal amplitudes. When the angle α of the wedge is small, the angular deviation of the plane wave is $(\mu - 1)\alpha$, where the appropriate μ must be used. The effect of the double wedge is therefore to produce an angular separation between the two orthogonally-polarized waves of $2\alpha(\mu_e - \mu_o)$, which corresponds to a distance $2F\alpha(\mu_e - \mu_o)$ in the object plane where F is the focal length of the condenser and the objective. After transmission through the object and the objective, the two waves are recombined by a second similar Wollaston prism and, since they are mutually coherent (they originated from a single plane-wave component of the illumination), interference between them can be produced by an analyzer crossed with the polarizer, Fig. 12.24(*b*). If the sample introduces no phase difference between the two components, the recombined wave is a plane-wave with polarization orthogonal to the analyzer, and the field is dark. Any phase difference introduced will make the light elliptically polarized and some light will pass the analyzer. The orthogonality of polarizer and analyzer has therefore introduced the required π phase shift, independently of the wavelength.

What does the image represent? If the transmission function of the object is $f(x, y)$, which is assumed to be independent of polarization,† the interference image intensity is

$$I(x, y) = |f(x, y) - f(x + \delta x, y)|^2 \otimes p(x, y) \qquad (12.45)$$

where δx is the translation vector between the two images and $p(x, y)$ is the point spread function (§12.3.1) of the microscope objective, the illumination being assumed to be incoherent. We can expand (12.45) for small δx and get

$$f(x + \delta x, y) \approx f(x, y) + \delta x \frac{\partial f}{\partial x}, \qquad (12.46)$$

† Otherwise, a polarizing microscope would be more suitable for its investigation.

$$I(x, y) = \delta x^2 \left|\frac{\partial f}{\partial x}\right|^2 \otimes p(x, y). \tag{12.47}$$

Now we write, for $f = |f| \exp[i\phi(x, y)]$,

$$\frac{\partial f}{\partial x} = if \frac{\partial \phi}{\partial x} + \exp(i\phi)\frac{\partial |f|}{\partial x}. \tag{12.48}$$

A phase object is assumed to be dominated by phase variations, so we neglect the second term and get

$$I(x, y) = |f|^2 \delta x^2 \left|\frac{\partial \phi}{\partial x}\right|^2 \otimes p(x, y). \tag{12.49}$$

The image therefore highlights phase gradients in the direction of the displacement vector. The phase slit of §12.4.5 would therefore appear as two bright lines along its edges unless δx is parallel to it. If the slit is very narrow, the lines may merge; one is tempted to add their intensities to give a single line in this case, but one should be wary of doing so because close to the resolution limit the illumination may not be spatially incoherent. As we saw in §12.4.3, the schlieren system also emphasizes phase gradients. However, this microscopic technique uses incoherent illumination, and is therefore more suitable for high-resolution work.

Figure 12.24 Nomarski's differential interference contrast microscope. (*a*) The Wollaston prism with the optic axes *OA*, (*b*) schematic ray diagram of the complete microscope. *W* indicates a Wollaston prism, *P* the polarizer and *A* the analyzer.

(*a*)

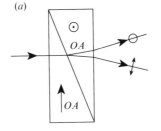

(*b*)

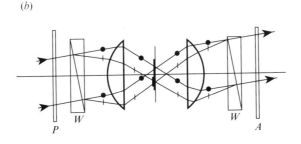

12.5 Improving the resolution

None of the techniques described in the preceding sections improves the spatial resolution beyond the $\lambda/2\text{NA}$ limit; in fact some of them, by restricting in some way the region of the Fourier plane used to form the image, actually spoil the resolution (schlieren, for example). The question therefore arises: is the limit of $\lambda/2\text{NA}$ fundamental?

This limit indeed has the aura of a fundamental limitation. It was used by Heisenberg to illustrate the quantum-mechanical uncertainty principle in his famous 'γ-ray microscope' thought-experiment as follows. Suppose we wish to determine the position of a point particle in the field of a microscope as accurately as possible. In order to do this, first choose a microscope with a high NA, and use waves with the shortest wavelength available (γ-rays). Then, to make the determination, we must scatter at least one photon off the object, and that photon must enter the lens of the microscope. But there is no way of knowing at which angle the photon entered the lens; all we know is that, after scattering off the point object, the photon had some direction within the cone of semi-angle α which determines the NA of the lens (Fig. 12.25). If the photon has wavenumber k_0, its x-component after scattering must therefore lie in the range $-k_0 \sin\alpha \leq k_x \leq k_0 \sin\alpha$. Thus the uncertainty $\delta k_x = 2k_0 \sin\alpha$. Now from the theory of resolution of the microscope, (12.26) gives an uncertainty in position of the image (the point spread function) $\delta x = \lambda/2\text{NA} = \lambda/2\sin\alpha$. Thus

$$\delta x \, \delta k_x = 2\pi \; ; \tag{12.50}$$

which can be written in the form

$$\delta x \, \delta p_x = h \,. \tag{12.51}$$

This is the usual form of the uncertainty principle (§14.2).

Now we can look for ways of 'circumventing' the uncertainty principle, which might lead us to ways of improving the resolution in microscopy. One profitable idea is to use many photons. Every observed photon must enter the lens, so that δk_x is unchanged. But a statistical assembly of n such photons will have a total uncertainty of about $2n^{\frac{1}{2}}k_0 \sin\alpha$, so that we might expect $\delta x \approx \lambda/(2n^{\frac{1}{2}}\text{NA})$. If n

Figure 12.25 The γ-ray microscope thought-experiment.

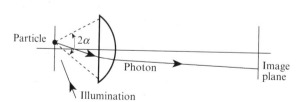

is very large, this could represent a substantial improvement in the resolution.

There are several ways in which optical resolution in excess of the Abbe limit can be achieved, and they do indeed involve large numbers of photons. One of them is to use evanescent waves. As we saw in Chapter 7, an electromagnetic wave must satisfy the (scalar) wave equation (7.5)

$$\nabla^2 \psi + k_0^2 \psi = 0. \tag{12.52}$$

So far we tacitly assumed that the solution is wavelike in all directions, but if it can be made to be evanescent in one of them, say z, so that ψ can be written in the form $\exp[i(xk_x + yk_y) \pm 2\pi z/a]$, then substituting in (12.52)

$$k_x^2 + k_y^2 - \frac{4\pi^2}{a^2} = k_0^2 = \frac{4\pi^2}{\lambda^2}. \tag{12.53}$$

If $a \ll \lambda$, k_x or k_y will be almost as large as $2\pi/a$, considerably exceeding k_0. As a result, the resolution δx or $\delta y \approx a$ will be much less than the wavelength. The process of 'making the wave evanescent' implies application of boundary conditions which require evanescent waves for their satisfaction.

We shall briefly discuss three ideas which use these concepts, two of which have led to practical working instruments. However, this field still invites new inventions.

12.5.1 Apodization

Apodization is a spatial filtering technique in which the Fourier transform is tailored, by inserting a mask in the plane of the lens, to 'improve' the point spread function (Jacquinot and Roizen-Dossier, 1964). One application is to suppressing the diffraction rings of the Airy function, done with the help of a mask which is approximately Gaussian in form.† This spoils the resolution a little (Problem 12.9). Another application, which is of interest to us here, improves the resolution a little. Suppose we obstruct the centre of the lens, leaving only a narrow annular aperture around its periphery. The point spread function is then the Fourier transform of this ring aperture which is, in the terms of §12.3.1 and Appendix 1,

$$I(\theta) = J_0^2(\tfrac{1}{2}k_0 D \sin \theta), \tag{12.54}$$

---

† Apodization is also used to improve the line-shape in Fourier transform spectroscopy (§11.6) in essentially the same way.

which has its first zero at $\frac{1}{2}k_0 D \sin\theta = 2.40$. Applying the Rayleigh and Sparrow criteria (§12.3.1) we find $\theta_{\min} = 0.76\lambda/D$ and $0.69\lambda/D$ respectively, which we can compare with $1.22\lambda/D$ and $0.95\lambda/D$ for the full aperture – an improvement of some 40% in resolution. Since the diffraction rings due to the J_0 function are rather strong (Fig. 12.26), this method has few direct applications, but it can be considered as a two-dimensional analogue of the Michelson stellar interferometer (§11.9.1). It has the same disadvantage of being very wasteful of light. In terms of the uncertainty principle argument, we have maximized δk_x by selecting only photons which entered around the edge of the lens and thus minimized δx. An example is shown in Fig. 12.10(h).

12.5.2 Super-resolution

An illuminating idea which was proposed by Toraldo di Francia (1952) shows that theoretically the Abbe limit can be exceeded without limit, although at the expense of photon efficiency.

Suppose that a lens has radius $R_0 = N\lambda$. Then, apodizing it with an thin annular ring aperture around its periphery, a point spread function $J_0(uR_0)$ is achieved, where the argument in (12.54) is $u = k_0 \sin\theta$. The Rayleigh resolution limit is $\delta u = 2.4/R_0$. Now add a second concentric ring aperture with radius $R_1 < R_0$ and phase π, with its transmission factor chosen so that its point spread function cancels that of the first ring (because of the phase π) at some u_0 *less than* $2.4/R_0$. Since there is now a dark ring at u_0, the Rayleigh resolution is improved. But the point spread functions *almost* cancel at $u = 0$ also, so the central peak is very weak; most of the light is diffracted to the outer regions of the field, to places where the point spread functions add. One can now repeat the process with a second pair of annular apertures at R_2 and R_3, both $< R_1$, which also give together a zero at the same u_0. The amplitude transmissions of these rings would be chosen so that the second pair cancels the first one at another chosen radius $u_1 > u_0$, which serves to push the light out to greater radii. In principle, the process can be continued until there are about $N/2$ rings. When one does this, one indeed finds that there is a very weak central peak surrounded by a virtually dark field, and the *choice* of u_0 determines how narrow the central peak is.

Where has the light energy gone? Mathematically, the large intensity has been moved to the region $u > k_0$, which is unobservable; it corresponds to evanescent waves which never reach the image. The light is therefore reflected back to the object, in analogy with critical reflexion. Because of its poor light efficiency, this idea is of no practical use, except possibly for extremely small objects lying completely within the outer bright ring, but illustrates the fact that with enough photons available, one can circumvent the Abbe criterion. An example of a point spread function achieved this way is shown in Fig. 12.26(*c*).

12.5.3 Confocal scanning microscopy

A practical technique which achieves somewhat higher resolution than the limit $\lambda/2NA$ and is based on a conventional microscope is *scanning confocal microscopy* (Wilson and Sheppard, 1984). It is illustrated schematically in Fig. 12.27. This shows a transmitted light imaging system (it can also be implemented in reflected light) in which the object is illuminated by the image of a point source. The light transmitted by the object is concentrated by a second lens onto a pinhole H, after which lies a detector which measures the power received. The object is then translated through the system mechanically and its image displayed electronically from the detector output.

The approach to understanding the resolution is to calculate the point spread function, as it appears on the display. We therefore picture a point object having a transmission function $\delta(x)\,\delta(y)$. The illumination system produces an amplitude point spread function $s_1(x)$ in the object plane, so that the amplitude at distance x from the axis is:

$$A(x) = A_0 s_1(x). \tag{12.55}$$

As the object point scans through this, it behaves as a point source itself which is situated at x and has amplitude $A_0 s_1(x)$. This is imaged by the objective lens, with magnification m, onto the plane of H so that it appears with its centre at $x' = -mx$. The objective has amplitude point spread function in the pinhole plane $s_2(x'/m)$. The amplitude in this plane is therefore $A_0 s_1(x) s_2[(x' - mx)/m]$, and that at the pinhole, $x' = 0$, is $A_0 s_1(x) s_2(-x)$. If the two lenses are identical, the point spread function for amplitude is thus $s^2(x)$ and that for intensity $s^4(x)$.

Figure 12.26
Comparison of the point spread functions for (*a*) a circular aperture, (*b*) an annular aperture and (*c*) a set of five concentric annular apertures designed to give a super-resolved image.

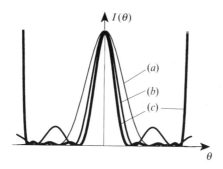

Putting in the form for a diffraction limited lens of numerical aperture = NA, $s(x) = 2J_1(k_0 x \, \text{NA})/(k_0 x \, \text{NA})$, we find a narrower point image than the best the equivalent microscope can do; this is not shown by the Rayleigh criterion (see footnote to §12.3.1 on p. 338) because the zeros are unchanged, but the Sparrow criterion gives

$$d_{\min} = \lambda/3.1 \, \text{NA} \qquad (12.56)$$

since the second derivative of $[J_1(x)/x]^4$ is zero at $x = 1.08$. Fig. 12.27 shows the very high resolution attainable; notice that an object with *periodic* detail has been used for this demonstration. Clearly, further improvement could be obtained by using annular apodization (Problem 12.10).

Now the pinhole H collects only a fraction of the light transmitted by the object, so the improvement of resolution occurs at the price of a reduction in efficiency. For a higher price, more resolution could theoretically be obtained by returning the light through the object again before it reaches the pinhole. We should point out that the full resolution gain will only be seen if the imaging is truly incoherent, when the sample fluoresces as a result of the incident light. Otherwise, imaging is partially coherent (illumination by a point source). It is for this reason that the treatment above has been in terms of amplitudes, so that it can readily be adapted to an object with general amplitude transmission function $f(x)$ instead of the δ-function. But as we saw in §12.3.2, we cannot quote a general result for coherent resolution since it depends on the phase behaviour of the object.

The confocal microscope has other qualities besides the improvement in resolution, which are possibly more important. If the object is moved axially out of the plane passing through the point image, it is illuminated by a patch of light and not a point, and the second imaging stage is also out of focus. Thus very little light reaches the detector, and the system has very limited depth of focus. This feature can be employed to build up images of three-dimensional objects, which leads to many practical applications.

12.5.4 Near-field microscopy

A very different imaging system which achieves a resolution an order of magnitude better than the wavelength is the *near-field microscope*. Such high resolution was first achieved with light waves by Lewis *et al.* (1984) and Pohl, Denk and Lanz (1984), and the most recent developments are described by Betzig *et al.* (1991, 1992) and Moyer and Paesler (1995). In this instrument, illustrated in Fig. 12.28, a point source with lateral dimensions $a \ll \lambda$ is scanned across the object and the light scattered or transmitted by it is collected by a lens and focused onto a detector. The image is built up sequentially as the object is scanned. Because the light source has dimensions $\ll \lambda$, the wave in the region $z \ll \lambda$ is evanescent, and therefore the resolution indicated by (12.53) can be achieved. The source has, of course, to be brought within distance a of the object, which is a distinct disadvantage, but there is always a price to be paid for resolution; the situation is similar to the scanning tunnel electron microscope (Binnig and Rohrer, 1985) in this respect. The original instrument used laser light focused through a tiny pinhole; recent versions employ extremely thin hollow fibres or small electro-luminescent crystals. Use of the evanescent wave results in very poor luminous efficiency, so we return to the demand for many photons!

12.6 Holography

Since all the information concerning the image of an object is contained in its diffraction pattern, it is tempting to ask how to record this information on a photographic plate and then use it to reconstruct the image. The difficulty involved is the recording of the relative phases of

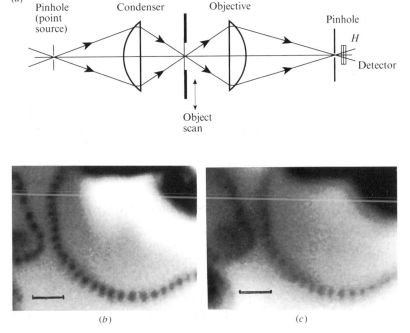

Figure 12.27
(*a*) Optical layout of a confocal scanning microscope.
(*b*) Confocal and
(*c*) conventional fluorescence images of a spore of Dawsonia superba. The scale lines show 1 μm. Photograph courtesy of V. Sarafis and C. Thoni, made at Leica Lasertechnik, Heidelberg.

the different parts of the optical transform (§12.2.5) since photography can record only intensities. The germ of a solution was suggested in 1948 by D. Gabor and had some limited success at that time; but the invention of the laser subsequently enabled the operation to be carried through completely successfully.

Figure 12.28
(a) Schematic optics of the near-field microscope, showing the extruded fibre tip as inset; and (b) images recorded with probe-to-object distances $a < .005$, $a = 0.005,\ 0.010$, $0.025,\ 0.10$, and $0.40\ \mu$m. Photograph courtesy of E. Betzig.

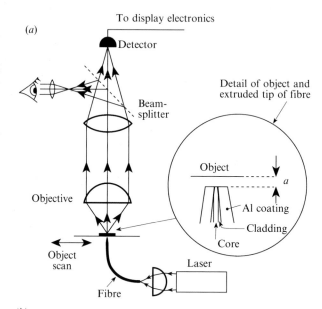

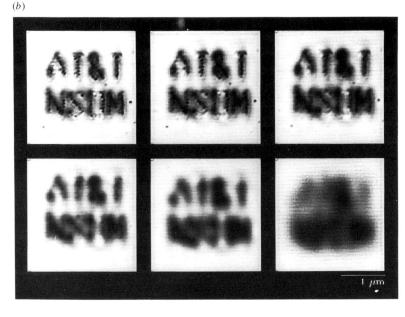

12.6.1 Gabor's method

Gabor's reason for trying to overcome this difficulty was to solve the problem of aberrations in electron microscope imaging. The resolution obtainable in the electron microscope is not limited by the wavelength (~ 0.1 Å) but by the aberrations of the electron lenses; these cannot be corrected and therefore a very small numerical aperture would normally be used. Gabor thought that a better image might be reconstructed if the aberrated image could be recorded and re-imaged with light using a lens system designed to correct for the aberrations of the electron lenses. To demonstrate this idea, he simulated his solution optically. The phase problem was solved by using an object that consisted of a small amount of opaque detail on a large transparent background; the background would produce a strong zero order and the variations in phase of the diffraction pattern would be recorded as variations in intensity, as in the X-ray microscope (§12.2.5). The intensity would be greatest where the phase of the diffraction pattern was the same as that of the background and least where there was a phase difference of π. The idea has only recently been applied to electron-microscope images (Tonomura, 1986) but was developed successfully for optical imaging in the 1960s when lasers became available.

12.6.2 Application of the laser

The idea of the hologram was implemented successfully, initially by Leith and Upatnieks in 1960, by using the intense beam of coherent light that can be produced by a laser.† The experimental set-up is quite simple (Fig. 12.29). A spatially-coherent laser beam is divided, either in wavefront or amplitude, so that one part falls directly on a photographic plate, and the other falls on the object to be recorded, which scatters light onto the same plate. The two waves, called the *reference wave* and the *object wave* respectively, interfere and the interference pattern is recorded by the plate. It is necessary to reduce relative movements of the various components to amplitudes considerably less than one wavelength to avoid blurring the interference fringes. Reconstruction of the image is carried out by illuminating the developed plate with a light wave that is identical, or at least very similar, to the original reference wave. Two images are usually observed. We shall first give a qualitative interpretation of the recording

---

† In fact, their first demonstration was with ordinary sources of light, but the laser made life much easier.

and reconstruction processes, and afterwards discuss them in a more quantitative manner.

The process can be described in general terms by considering the hologram as analogous to a diffraction grating (§§8.3.4, 9.2). Suppose that we photograph the hologram of a point scatterer – Fig. 12.30(a). The point generates a spherical object wave, and this interferes with the plane reference wave. The result is a set of curved fringes (b) which look like an off-centre part of a zone plate (§7.3.4). The hologram is photographed and the plate developed. To reconstruct the image we illuminate the hologram with a plane-wave identical to the original reference wave (c). We can consider each part of the hologram individually as a diffraction grating with a certain local line spacing. Illumination by the plane reference wave gives rise to a zero order and two first orders of diffraction, at angles θ_1, θ_{-1} which depend on the local spacing of the fringes. It is not difficult to see that the -1 orders intersect and form a real image of the point scatterer, and the $+1$ orders form a virtual image at a position identical to the original point. The images are localized in three dimensions because they are formed by the intersection of waves coming from different directions.

Two other important points are brought out by this model. First, the reconstructed point is more accurately defined in position if a large area of the plate is used, causing the reconstruction orders to meet at a considerable angle; the resolution is therefore a function of the size of the hologram. Secondly, the fringes are sinusoidal, since only two waves interfere. If the plate records this function faithfully, only zero and first orders will be produced on reconstruction, and only the above two images are produced. This approach is also useful in understanding the effects of altering the angle of incidence, the wavelength or the degree of convergence of the reference wave used for the reconstruction (Problem 12.11).

Figure 12.29
Example of a
holographic
recording set-up.

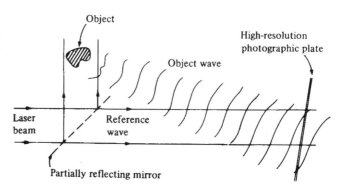

In a more quantitative fashion, we can now see how both the
amplitude and the phase of the scattered light are recorded in the
hologram and how the reconstruction works. Suppose that at a general
point (x, y) in the plate the scattered light has amplitude $a(x, y)$ and
phase $\phi(x, y)$. Furthermore, we shall assume that the reference wave
is not necessarily a plane-wave, but has uniform amplitude A and
phase $\phi_0(x, y)$ at the general point. Then the total wave amplitude at
(x, y) is

$$\psi(x, y) = A \exp[i\phi_0(x, y)] + a \exp[i\phi(x, y)] , \tag{12.57}$$

and the corresponding intensity

$$I(x, y) = |\psi(x, y)|^2 = A^2 + a^2 + 2Aa \cos[\phi(x, y) - \phi_0(x, y)] . \tag{12.58}$$

Figure 12.30
Formation and
reconstruction of the
hologram of a point
object: (a) spherical
wave from the object
interferes with plane
reference wave;
(b) fringes recorded
on the photographic
plate; (c) the plate
behaves as a
diffraction grating
with non-uniform
line spacing.

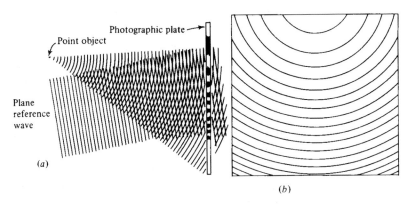

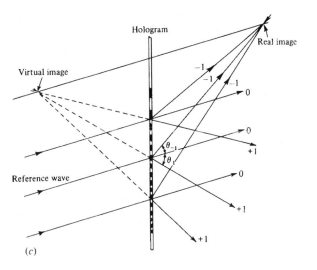

One usually arranges in holography for a to be much smaller than A, in which case the term a^2 can be neglected and

$$I(x, y) \approx A^2 + 2Aa \, \cos[\phi(x, y) - \phi_0(x, y)] \, . \qquad (12.59)$$

The photograph of this is the hologram. It consists of a set of interference fringes with sinusoidal profile and phase $\phi - \phi_0$. The visibility of the fringes is $2a/A$. Since A is a constant and ϕ_0 is known, both $a(x, y)$ and $\phi(x, y)$ are thus recorded in the hologram. The need for coherent light to record the hologram should now be clear, since the phase *difference* $\phi - \phi_0$ is recorded in the interference pattern.

To deduce the form of the reconstruction, we assume that the interference pattern (12.59) is photographed on a plate whose amplitude transmission $\mathscr{T}(x, y)$ after development is linearly† related to the exposure intensity $I(x, y)$:

$$\mathscr{T}(x, y) = 1 - \alpha I(x, y). \qquad (12.60)$$

The hologram is illuminated by a wave identical to the original reference wave $A \exp[i\phi_0(x, y)]$ and so the transmitted amplitude is:

$$A\mathscr{T}(x, y) \exp[i\phi_0(x, y)] = [1 - \alpha I(x, y)]A \exp[i\phi_0(x, y)] \qquad (12.61)$$

$$= A(1 - \alpha A^2) \exp[i\phi_0(x, y)] \qquad (a)$$
$$-\alpha A^2 a(x, y) \exp[i\phi(x, y)] \qquad (b)$$
$$-\alpha A^2 a(x, y) \exp\{i[\phi_0(x, y) - \phi(x, y)]\} \, . \qquad (c)$$

The three terms in the above equation are interpreted as follows.

(a) An attenuated continuation of the reference wave (the zero order).
(b) The first order is the virtual image. Apart from the constant multiplier αA^2, the reconstructed wave is *exactly* the same as the object wave and so the light appears to come from a virtual object perfectly reconstructed. Because the complete complex wave $a(x, y)$ has been reconstructed, the reconstruction looks exactly like the object from every direction, and so appears three dimensional.
(c) The -1 order is the conjugate image. This wave is the *complex conjugate* of the object wave if ϕ_0 is a constant, and then gives a real mirror image of the object. Otherwise it is distorted.

This is not the place for practical details on the production of holograms, for which the reader is referred to texts such as Collier,

† It is always possible to find a *limited* range of intensities for which this is true. This is another reason for making $a^2 \ll A^2$, so that the range of I is not too large.

Burckhardt & Lin (1971) and Hariharan, (1989); we shall only mention a few points that arise directly from the above discussion.

The intensity ratio between the object beam and the reference beam, a^2/A^2, has been required to be small; in general a ratio of 1:5 is sufficient, although for some purposes even 1:2 can be tolerated. Perfect reconstruction requires the photographic plate to record the light intensity linearly. However, the condition can be relaxed quite considerably for many purposes, since the main effect of non-linearity in the plate is to create second- and higher-order reconstructions which are usually separated in space from the main images, although under some conditions may interfere with them. Another obvious requirement is for high spatial resolution of the photographic plate. If the reference beam and the object beam are separated by angle θ, the period of the fringes in the hologram is approximately $\lambda/\sin\theta$. For, say, $\theta = 30°$ this period is only about $1\,\mu\text{m}$ with the common helium–neon laser. To record fringes on this scale, the plate must be capable of resolving less than $0.5\,\mu\text{m}$, a very stringent requirement which needs special high-resolution photographic plates to fulfil it. These plates are usually very insensitive and, as we shall see, very inefficient.

12.6.3 Phase and volume holograms

The reader will no doubt remember from the discussion of diffraction gratings in §9.2 how poor is the efficiency of an amplitude grating. This is essentially what we have created in a hologram, and the argument in §9.2.4 can be repeated for a sinusoidal grating to show that the diffraction efficiency $\eta \approx a^2/12A^2$, which is very small. The answer, as with the diffraction grating, lies in the use of *phase holograms*. There are several practical methods of replacing the amplitude transmission $\mathcal{T}(x, y)$ by a proportional refractive index field $\mu(x, y)$. These include bleaching a developed amplitude hologram (chemically replacing absorbing Ag metal by a transparent complex such as AgCl whose presence locally modifies the refractive index of the emulsion) and the use of gels or polymers in which the degree of cross-linking is modified by exposure to light. Other devices, such as thermoplastic media and electro-optic crystals can also be employed to record phase holograms; all these methods improve the diffraction efficiency considerably (look at your credit card for an example).

Once we have techniques to make non-absorbing holograms, it is also possible to create *volume holograms*. These are usually polymers or crystals which record the incident light intensity in three dimensions as a local modulation to the refractive index (see §13.5, for example) which is then 'fixed' (i.e. made insensitive to further light exposure). The volume holographic medium replaces the holographic plate in Fig. 12.30 and records the complete spatial fringe pattern, which creates a three-dimensional diffraction grating.

Reconstruction of the image occurs when a plane reconstruction wave is diffracted by this grating. Here we have the same problem as we met in the case of the acousto-optic effect (§8.4.5) except that in this case the grating is stationary and so the 'acoustic' frequency Ω is zero. We saw there that the volume grating diffracted the wave only if it had exactly the right angle to obey Bragg's law of diffraction, as in Fig. 8.27. Suppose, for simplicity, that we form the hologram of a plane-wave such that the angle between it and the reference wave is 2α. Then the fringes in the volume hologram are planar with spacing $\Lambda = \lambda/2 \sin \alpha$. Bragg's law then tells us that diffraction occurs when the reference beam is at angle α, i.e. when it is in the same direction as it was when the hologram was formed. Otherwise *there is no reconstruction at all* and the hologram appears transparent to the wave. Contrast this with the two-dimensional situation, where use of a reconstruction wave differing from the reference wave just gives rise to a distorted image.

The complete absence of reconstruction when the reconstruction wave is not in its original direction allows many holograms (each representing a different image) to be recorded simultaneously in the same medium, each one with a different reference wave, and the possibility of viewing each image separately by choosing the appropriate individual reconstruction wave. This has led to the idea of the *holographic memory* in which a great deal of information (in the form of three-dimensional images) is stored in a crystal, with the possible of quick access via the appropriate reconstruction beam.

12.6.4 Holographic interferometry

Holographic reconstructions have two main advantages over ordinary photographs. They are three dimensional, and they contain phase information (§12.4.2). The possibility of recording phase information has allowed the development of *holographic interferometry* in which an object can be compared interferometrically with a holographic recording of itself at an earlier time. If any changes – of optical density or dimensions, for example – have occurred since the recording was made, the differences will be apparent as interference fringes. It is not our intention to go into practical details here; an example is shown in

Fig. 12.31 in which the growth of a crystal of a transparent material within an optical cell is observed by recording the pattern of changes in optical density. The hologram was recorded before the crystal started to grow, so that the interferogram refers to the crystal alone. The details of the exact shape of the experimental cell are irrelevant to an interpretation of the photograph, since only *changes* in the optical density are observed. Other applications of interferometric holography include vibration analysis and aerodynamic experiments (see the books on holography referenced earlier, and also Jones and Wykes, 1989).

12.6.5 Applications of the Abbe theory involving holographic filters

The spatial-filtering techniques discussed in §12.4 involve very simple geometrical filters such as the dark-ground stop and the phase plate. Holography allows us to record the complex amplitude of a wave field, and the use of holographic filters adds another dimension to spatial filtering.

Suppose that the object $f(x)$ is imaged in the usual spatial-filtering system (Fig. 12.3), in which a lens of focal length L is employed. In the focal plane of the lens we see the complex Fourier transform $F(u)$ on a scale ξ where $u = k_0 \xi / L$ and the final image is its transform $f(-x/m)$, where m, the magnification of the system, will henceforth be assumed to be unity. In general we can insert a filter $H(u)$ in the focal plane; it multiplies the transform $F(u)$ and the resultant image is then the transform of $F(u)H(u)$ which is $f(-x) \otimes h(-x)$. There are several interesting applications of this result, when $H(u)$ is complex.

Figure 12.31
Holographic
interferogram
showing a growing
crystal of solid
helium at a
temperature of 0.5 K.

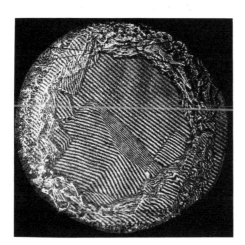

In practice, a complex filter $G(u)$ is recorded on a photographic plate holographically as a real function $H(u)$, either in an experiment (Fig. 12.32) or from a computer-generated drawing. For a plane reference beam at angle β to the optical axis, the filter transmission is proportional to the intensity (12.59):

$$\begin{aligned} H(u) \propto I(u) &= |G(u) + A \exp(iuL \sin \beta)|^2 \\ &\approx A^2 + AG(u) \exp(-iuL \sin \beta) \\ &\quad + AG^*(u) \exp(iuL \sin \beta). \end{aligned} \tag{12.62}$$

The action of this filter on the image of $f(x)$ is to produce an image

$$\begin{aligned} f_1(x) &= A^2 f(-x) + Af(-x) \otimes g(-x) \otimes \delta(x - L \sin \beta) \\ &\quad + Af(-x) \otimes g^*(x) \otimes \delta(x + L \sin \beta). \end{aligned} \tag{12.63}$$

This expression contains three identifiable terms:

(a) $A^2 f(-x)$ is just the image of $f(x)$;
(b) the second term is the convolution of f and g, centred on the point $x = L \sin \beta$;
(c) the third term is the correlation of f and g centred on the point $x = -L \sin \beta$.

The three image functions are physically separated in the image plane because they are centred on different points. One application of such a system is to the identification of complicated patterns such as fingerprints. Suppose that $g(x)$ corresponds to a known object, and $f(x)$ to an unknown object. If the two are identical, the correlation function has a strong central spot, which is weaker or absent if the two are not identical.

Figure 12.32
Production of a
holographic filter.

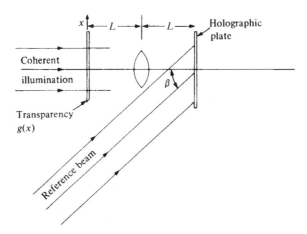

Another application is the deblurring of photographs or other pictures. As was pointed out in §4.6.1, blurring is equivalent to convolution by a 'blur function'. We call this function $b(x)$. Then a blurred image of $f(x)$ is $f(x) \otimes b(x)$. The transform of this blurred image is $F(u)B(u)$. What could be simpler, if $b(x)$ is known, than generating the filter $1/B(u)$ by computer and multiplying $F(u)B(u)$ by it to restore $F(u)$ and the sharp image $f(-x)$ in the image plane? This idea would allow the deblurring of, for example, out-of-focus photographs, pictures of moving objects and clinical X-ray photographs (which are essentially shadows from a finite source). But it is deceptively simple. What happens when $B(u)$ goes through zero? Almost all transforms have zeros, and their presence results in irretrievably lost information in the transform $F(u)B(u)$. Some methods have been developed to deal with such problems, and have been moderately successful, but are outside the discussion here (see, for example, Guenther, 1990).

12.7 Advanced topic: interferometic imaging in astronomy

12.7.1 Radio astronomy

It was an important event for astronomy when Jansky, in 1932, discovered cosmic radiation at 14.6 m wavelength. Based on this discovery and developments in electronics during the Second World War, radio astronomy has rapidly developed since the 1950s and has taken advantage of many optical principles, adapted to wavelengths between several millimetres and many metres. Many radio telescopes have been constructed from a large parabolic reflector with an aerial at its focus. Apart from the fact that the detector samples only one point and therefore has to scan through the image, the concept is similar to a reflecting optical telescope; in particular their angular resolution is calculated in the same way (§12.3.1). The problem is size. In terms of the wavelength, even the largest structures that can be built are small compared with what can be achieved with visible light, and therefore alternative techniques based on interferometry have proved invaluable in the attainment of high spatial resolution. The most important technique today in radio astronomy is that of aperture synthesis, which was described in §11.9.3 because of its connexion with coherence theory, but to fill in the picture we shall briefly discuss below the development of radio interferometer arrays, which have in recent years been extended to sub-millimetre wavelengths.

12.7.2 The two-antenna interferometer

The basic radio interferometer consists of two antennae separated by a horizontal distance L, Fig. 12.33(a). We shall consider each antenna as a point receiver, even if it is actually backed by a paraboloid of dimensions $\ll L$ to increase its sensitivity. In order to calculate the point spread function, we suppose that the pair observes a single point source of radiation at angle θ to the zenith in the vertical plane containing both antennae. Each antenna sees the same signal from the source, which we represent as $f(t)$ at one antenna, that at the other being advanced by $L \sin \theta / c$ in time. Their outputs are therefore proportional to $f(t)$ and $f(t + L \sin \theta / c)$ respectively. The signals are then combined coherently, a time delay of T being introduced into the signal from the second antenna. Then we have the interferometric signal $s(t) = f(t) + f(t + L \sin \theta / c - T)$. Clearly, the maximum signal $2f(t)$ is obtained when the delay $T = L \sin \theta / c \equiv T_0$. If we assume the source to be quasi-monochromatic, with frequency ω, a delay of $T = T_0 + m\pi / \omega$ will give

$$s(t) = f(t) + f(t - m\pi / \omega). \tag{12.64}$$

When the delay T is scanned we observe interference, destructive when m is odd and constructive when m is even – in other words, Young's fringes with period $2\pi / \omega$ – until the delay become longer than the coherence time. As the Earth rotates, if the antennae are situated along a line in the east–west direction, θ changes with time and thus T_0 is scanned. A signal of this sort (obtained in the 1950s) is shown in Fig. 12.33(b). Scanning by changing the delay T artificially is required in the north–south direction.

The angular resolution to which the position of the point source can be measured is given by the angular equivalent of half a fringe, i.e. $L \delta(\sin \theta) / \lambda = \frac{1}{2}$, giving

$$\delta\theta_{\min} = \lambda / (2L \cos \theta). \tag{12.65}$$

This is the same result as we obtained in §9.1.1 for the optical Young's experiment. For example, if $L = 1$ km, and $\lambda = 1$ m, the angular resolution is 10^{-3} rad, which is about the same as the unaided eye. However, it is also equivalent in resolution (but not in sensitivity) to a parabolic dish of diameter 1.2 km, which cannot be constructed. The problem is that it is difficult to interpret the signal obtained when the source is not a point source. We therefore continue the optical analogy by adopting the idea of the diffraction grating.

12.7.3 Diffraction gratings and aerial arrays

The use of a regular array of slits instead of the two in Young's experiment made interpretation of the resultant spectral patterns considerably easier, as we discussed in §§9.1–9.2. The same principle applies to antenna arrays. We imagine a linear array of $N+1$ antennae, spaced by distance L, observing the same source as before. The combined signal is now

$$s(t) = \sum_{n=0}^{N} f(t + nL \sin \theta/c).$$ (12.66)

Once again, if we compensate the terms $nL \sin \theta/c$ by introducing a time-delay $T_n = nT_0$ into the signal from the nth antenna, we get the maximum signal $s(t) = (N+1)f(t)$. Now T_n is varied (either by using the diurnal variation of T_0 or otherwise) in such a way that $T_n = n(T_0 + \tau)$. For the quasi-monochromatic source, $f(t) = a\exp(-i\omega t)$,

$$s(t, \tau) = \sum_{n=0}^{N} f(t - n\tau) = a\exp(-i\omega t) \sum_{n=0}^{N} \exp(in\omega\tau),$$ (12.67)

the same expression as we obtained for the diffraction pattern of a linear array of $N+1$ slits in §8.3.3, with $ud \equiv \omega\tau$. We saw there that the

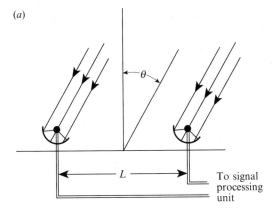

Figure 12.33
(*a*) Geometry of a
simple radio
interferometer.
(*b*) Observed signal
as Cygnus and
Casseopeia pass
through the zenith
plane. (From Smith,
1962.)

(*a*)

To signal
processing
unit

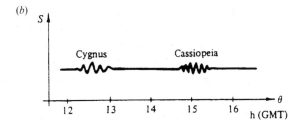

(*b*)

Cygnus   Cassiopeia

12   13   14   15   16
h (GMT)

diffraction pattern consists of strong principal maxima separated by $N-1$ subsidiary weak maxima (see Fig. 8.15). This is the point spread function for the antenna array, considered as an imaging instrument. The width of the principal maximum (the zero order, whose position is independent of wavelength) then follows as $\tau_{min} = 2\pi/N\omega$. The angular resolution is obtained by substituting $\tau = L\,\delta(\sin\theta)/c$, whence $\delta\theta_{min} = \lambda/(NL\cos\theta)$, cf (12.65).

An array of antennae therefore gives the same resolution as a pair separated by the complete length NL; however, because the principal maxima are narrow and well-separated, the structure of more complicated sources can more easily be elucidated. In addition, the power $|f(t)|^2$ in the principal maxima increases as N^2. These advantages parallel the way in which the diffraction grating gives a more intense diffraction pattern than Young's slits, and one which is more easily interpreted when the source spectrum is complex.

The point spread function calculated above is only one dimensional; the array has high resolution along its length, but none at right angles. Two orthogonal linear arrays of antennae would reconstruct orthogonal sections through a two-dimensional radio source, with angular resolution given by the lengths of the arrays, but this is still not equivalent to a complete two-dimensional image.

12.7.4 The Mills cross antenna array

It would seem from the above that $(N+1)^2$ antennae, covering the area $L \times L$ would be necessary to achieve the two-dimensional image with the same resolution. The *Mills cross* allows this to be achieved by using two orthogonal one-dimensional arrays, each with $N+1$ antennae. It is instructive to illustrate this by using the equivalent optical diffraction pattern which, as we have seen, is the point spread function obtained by scanning.

The individual antenna is represented optically by a pinhole. The two orthogonal arrays correspond to the mask shown in Fig. 12.34(a). Its diffraction pattern is shown in Fig. 12.34(b); the streaks arise from the two linear arrays independently and they interfere where they cross. If the two arrays have the same phase, the interference is constructive at the crossing points and at each one there is a bright spot, with intensity proportional to $[2(N+1)]^2$. Now suppose that the phase of one of the arrays is changed by π, which can be done easily electronically by reversing the polarity of the signals from one set of antennae. Then, at the crossing points the amplitudes will subtract, giving zero intensity. At other points of the diffraction pattern there is no change since the signal comes from one array only. By subtracting the images obtained with and without the phase change, only the signals from the crossing points remains, and this is equivalent to a set of $(N+1)^2$ antennae filling the complete area.

When the linear arrays are arranged as shown in Fig. 12.34(a), with the holes at *odd* multiples of $d/2$ along the axes, alternate streaks in the diffraction pattern have phases 0 and π.† As a result there is a phase difference of π between the streaks which cross at the points $(u,v) = (m_1, m_2)2\pi/d$, where one of m_1 and m_2 is odd and the other even, and therefore they interfere destructively. The 'image' of a point star as seen by the Mills cross array is obtained by subtracting the intensity of the diffraction pattern with origin at $(1,0)$, for example, from that with origin at $(0,0)$. One can see from the magnified view in Fig. 12.34(b) that these functions differ only in a small region around their origins.

12.8 Advanced topic: astronomical speckle interferometry

The theoretical resolution limit of a telescope, $\theta_{\min} = 1.22\lambda/D$, (§12.3.1), can not be achieved by any Earth-based instrument because of the

---

† The transform of $f(x) = \sum_n \delta(x - d/2 - nd)$ is
$$F(u) = \exp(iud/2)\sum_m \delta(u - 2\pi m/d) = (-1)^m \sum_m \delta(u - 2\pi m/d).$$

Figure 12.34 The principle of the Mills cross array, illustrated optically: (*a*) a pinhole mask representing the antenna array, (*b*) its Fraunhofer diffraction pattern. The ringed central region, containing the (0,0), (−1,0), (0,−1) and (−1,−1) orders is shown magnified as an inset.

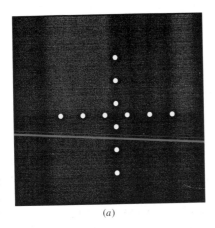

(*a*)

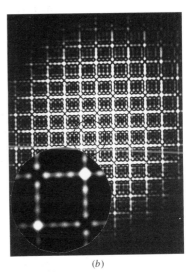

(*b*)

presence of non-uniformities in the atmosphere. Local pressure and temperature variations result in the atmosphere's having a rather poor optical quality and its properties vary widely as a function of the weather, the time and the azimuth angle. Just to get some idea of the parameters involved we can quote some typical deviations from the mean optical thickness of the whole atmosphere. The r.m.s. fluctuation amplitude is between two and three wavelengths of visible light, and it changes randomly in a time of the order of 10 ms. In the spatial dimension, fluctuations are correlated within transverse distances of about 0.1 m and are responsible for the twinkling of small stars. The general smearing effect of atmospheric fluctuations on a stellar image is called by the astronomer the *seeing*, and might be 3 arc-sec on a poor night and 0.5 arc-sec on an exceptionally good, still night. The telescope therefore acts as if it were a collection of small independent telescopes, each of which has diameter (of order 0.1 m) such that θ_{min} is the seeing. This should be compared with the Rayleigh resolution limit for, say, a 2 m telescope which is about 0.05 arc-sec. The resolution that can be achieved with a very large telescope therefore seems to be no better than that from a telescope of diameter 10 cm; only the brightness of the image is greater with the larger telescope.

Two major inventions attempted to overcome the resolution limit set by the atmosphere by using multiple telescopes: the Michelson stellar interferometer (§11.9.1) and the Brown–Twiss interferometer (§11.9.2). Recently two new techniques have been introduced in an attempt to overcome the problem of atmospheric degradation of single-telescope images – speckle interferometry (described below) and adaptive optics (wavefront correction by a flexible mirror: see Pearson, Freeman and Reynolds, 1979, and Tyson, 1991).

Because of their relationship to the theme of this chapter we shall describe the original form of speckle interferometry (Labeyrie, 1976) and two of its extensions. The technique arose from careful observation of 'instantaneous' photographs of stellar images. With the introduction of image-intensifier tubes it has become possible to photograph images through a narrow-band filter at a high magnification using an exposure time less than the 10 ms stability time of the atmospheric fluctuations. This is sufficient to see detail at the Rayleigh resolution limit without it being blurred by atmospheric motion. Such images have an overall size of the order of the seeing, but contain a wealth of fine detail. Three examples of 'instantaneous' photographs are shown in Fig. 12.35. There are obvious differences in their detailed structure and these differences represent real differences in the objects. The method of speckle interferometry separates the atmospheric and object contributions to these images by using a series of exposures during which the atmosphere changes from exposure to exposure, but the star remains invariant.

Suppose, first, that the telescope was used to observe an ideal point star at time t. The image, photographed through the atmosphere, has an intensity distribution $p(\mathbf{r}, t)$, (where $\mathbf{r} \equiv (x, y)$), which is the instantaneous atmospherically-degraded point spread function of the telescope. This is actually illustrated by Fig. 12.35(c), in which one can see that it is like a random collection of sharp spots. If the atmosphere had been homogeneous, the extended star would have given an ideal image of intensity $o(\mathbf{r})$ with resolution limited only by the finite aperture of the telescope. In the presence of the real atmosphere the composite image is the convolution of $o(\mathbf{r})$ with the point spread function $p(\mathbf{r})$:

$$i(\mathbf{r}, t) = o(\mathbf{r}) \otimes p(\mathbf{r}, t). \tag{12.68}$$

In the original technique, this image is photographed at time t_j under conditions such that the photographic film has amplitude transmission proportional to the exposure intensity (as in holography, §12.6). Subsequently the developed photograph is used as a mask in a diffractometer. One records on a second film the intensity of its Fraunhofer diffraction pattern, the Fourier transform of $i(\mathbf{r}, t_j)$. The process is repeated for a series of exposures at times t_j, the transforms being superimposed on one another on the second film. Today, video-recording and digital analysis have replaced the photography and the diffraction, but the result is the same.

Figure 12.35 Speckle images (above) and corresponding spatial power spectra (12.72) (below). From left to right: (a, d) Betelgeuse (resolved disc), (b, e) Capella (resolved binary) and (c, f) an unresolved reference star. (From Labeyrie, 1976.)

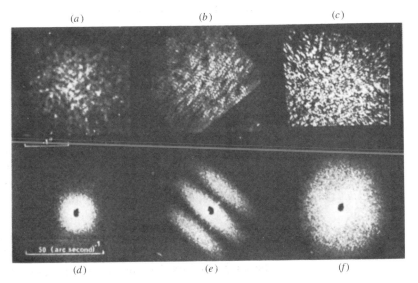

(a)   (b)   (c)

(d)   (e)   (f)

50 (arc second)$^{-1}$

Now the transform of $i(\mathbf{r}, t_j)$, (12.68), is

$$I(\mathbf{u}, t_j) = O(\mathbf{u})P(\mathbf{u}, t_j), \qquad (12.69)$$

where $\mathbf{u}$ is the vector (u, v), and its intensity is

$$|I(\mathbf{u}, t_j)|^2 = |O(\mathbf{u})|^2 |P(\mathbf{u}, t_j)|^2. \qquad (12.70)$$

The summation for a long series of t_js gives:

$$\sum_j |I(\mathbf{u}, t_j)|^2 = |O(\mathbf{u})|^2 \sum_j |P(\mathbf{u}, t_j)|^2. \qquad (12.71)$$

Since $|P(\mathbf{u}, t_j)|^2$ is a random function in which the detail is continuously changing, the summation becomes smoother and smoother as more terms are added (§8.3.8). Finally, we have, when enough terms have been added to make $\sum |P(\mathbf{u}, t_j)|^2$ smooth enough

$$\sum_j |I(\mathbf{u}, t_j)|^2 = |O(\mathbf{u})|^2 \times \text{(a smooth function)}. \qquad (12.72)$$

The smooth function can be determined by observing an unresolvable star. In this way the intensity of the Fourier transform $|O(\mathbf{u})|^2$ can be measured. If this function is retransformed, we get the spatial auto-correlation function of the stellar image, which reveals simple structural features (such as stellar diameters or separation of binary components); but a true stellar image cannot be deduced.

In Fig. 12.35 we show three examples of speckle transforms obtained by this technique. The upper row shows examples of single exposures from the series of some hundred speckle patterns, and the lower row the summed spatial transforms (12.72). The 'smooth function' is shown in (f) which corresponds to an unresolvable point star with angular diameter less than 0.02 arc-sec, the Rayleigh limit of the telescope. The other examples are resolvable stars; transform (e) in particular exhibits Young's fringes which reveal the star to be a binary.

The basic technique of speckle interferometry suffers from the phase problem, but the loss of phase information occurred in this case *after* the speckle images were recorded and it can therefore be retrieved. Several techniques have been devised for this purpose and today are highly developed, so that true diffraction-limited images can really be obtained. We shall briefly discuss two basic methods.

The first method is due to Knox and Thompson (1974). It is generally implemented digitally, although it is not difficult to imagine an method of carrying out parts of it by shearing interferometry (§12.4.6).

In this method, the transform $I(\mathbf{u}, t_j)$ interferes with itself after translation by a small shift vector $\delta\mathbf{u}$. The size of the shift should be small, but not very small, compared with the structure observed in $|O(\mathbf{u})|^2$, (12.72). This gives us the product

$$I(\mathbf{u}, t_j)\, I^*(\mathbf{u} + \delta\mathbf{u}, t_j) = O(\mathbf{u})\, O^*(\mathbf{u} + \delta\mathbf{u}) \cdot P(\mathbf{u}, t_j)\, P^*(\mathbf{u} + \delta\mathbf{u}, t_j). \quad (12.73)$$

Introducing the phases explicitly in the form $O = |O|\exp(i\phi_O)$ etc., and using $\delta\phi = \phi(\mathbf{u}) - \phi(\mathbf{u} + \delta\mathbf{u})$, we have

$$|I(\mathbf{u}, t_j)|\, |I(\mathbf{u} + \delta\mathbf{u}, t_j)|\exp(i\,\delta\phi_I) = |O(\mathbf{u},)|\, |O(\mathbf{u} + \delta\mathbf{u})|\exp(i\,\delta\phi_O)$$
$$\times |P(\mathbf{u}, t_j)|\, |P(\mathbf{u} + \delta\mathbf{u}, t_j)|\exp(i\,\delta\phi_P). \quad (12.74)$$

As before, (12.74) is summed over a large number of frames. The important feature is that, statistically, $\langle\delta\phi_P\rangle = 0$. When $\delta\mathbf{u}$ is small, as defined above, $|O(\mathbf{u} + \delta\mathbf{u},)| \approx |O(\mathbf{u})|$ etc., and so

$$\sum_j I(\mathbf{u}, t_j)\, I^*(\mathbf{u} + \delta\mathbf{u}, t_j) = |O(\mathbf{u})|^2 \exp(i\delta\phi_O) \times \text{(a smooth function)},$$

$$(12.75)$$

Figure 12.36 An example of a diffraction-limited image retrieved by speckle masking: (*a*) the long-exposure image of R136 in the 30 Doradus Nebula; (*b*) a single speckle image; (*c*) high-resolution reconstruction of the source. The scale bars show 1 arc-sec. Courtesy of G. Weigelt.

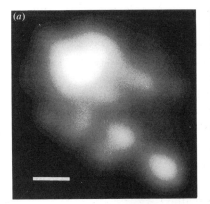
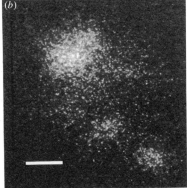
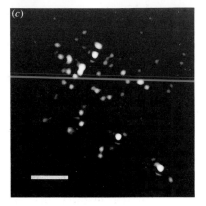

from which, together with (12.72), $\delta\phi_O$ can be determined.

When the above calculation is performed for two orthogonal values of $\delta\mathbf{u} = \Delta_u, \Delta_v$, the phases $\phi_O(\mathbf{u})$ at a grid of points $(m_1\Delta_u, m_2\Delta_v)$ in the (u, v) plane can be calculated by summing the $\delta\phi$s, assuming $\phi_O = 0$ at the origin. As a result the complex $O(\mathbf{u})$ is known, and its Fourier transform gives the complete image with diffraction-limited detail.

Another idea which is in wider use today is called *speckle masking* (Weigelt, 1991). First we note that, if there were another single isolated and unresolvable star in the field of view, one contribution to the spatial auto-correlation would be an *image* of the original star field (convoluted with the unresolvable star, which is essentially a δ-function). Speckle-masking creates such a 'reference star' artificially.

Suppose, as an example, the object $o(\mathbf{r})$ is a binary star whose separation $\mathbf{r}_1$ has been determined by speckle interferometry. Then the product $i(\mathbf{r}) \cdot i(\mathbf{r} + \mathbf{r}_1)$ contains one overlapping point for each speckle and therefore corresponds to $p(\mathbf{r})$. There will be other accidental overlaps in a complex speckle field, which introduce an error that can be corrected statistically. This is treated in the further development of the technique, but we shall ignore it in this discussion. It is now easy to see that, statistically, the correlation between the point spread function and its speckle image is the object function. Using (12.68) we write this as $c_3(\mathbf{r})$:

$$
\begin{aligned}
c_3(\mathbf{r}, t_j) &= i(\mathbf{r}, t_j) \otimes p(-\mathbf{r}, t) \\
&= [o(\mathbf{r}) \otimes p(\mathbf{r}, t_j)] \otimes p(-\mathbf{r}, t_j) \\
&= o(\mathbf{r}) \otimes [p(\mathbf{r}, t_j) \otimes p(-\mathbf{r}, t_j)].
\end{aligned}
\tag{12.76}
$$

When the second term (the auto-correlation of p) is averaged over many frames at times t_j the sharp peak at the origin dominates (see §8.3.8, and Fig. 8.21(c)); this is essentially a δ-function, so that

$$
\sum_j c_3(\mathbf{r}, t_j) = C\, o(\mathbf{r}),
\tag{12.77}
$$

where C is a constant. Thus speckle masking retrieves the image. The tricky point in the technique is the choice of $\mathbf{r}_1$ to get the best approximation to p when we are dealing with an object more complicated than a double star, and often several possibilities are used, the results being averaged. Examples of recent results using this and other techniques can be found in Beckers and Merkle (1992), and are illustrated by Fig. 12.36.

CHAPTER THIRTEEN

The classical theory of dispersion

13.1 Classical dispersion theory

Many aspects of the interaction between radiation and matter can be described quite accurately by a classical theory in which the medium is represented by model atoms consisting of positive and negative parts bound by an attraction which depends linearly on their separation. Although quantum theory is necessary to calculate from first principles the magnitude of the parameters involved, in this chapter we shall show that many optical effects can be interpreted physically in terms of this model by the use of classical mechanics. In §13.5 we shall relax the restriction of linearity. Some of the quantum mechanical foundations will be discussed briefly in Chapter 14, but most are outside the scope of this book (see Yariv, 1989; Loudon, 1983).

The term *dispersion* means the dependence of dielectric response (dielectric constant and refractive index) on frequency of the wave field. This will be the topic of the present section. Afterwards we shall see some of the applications of dielectric response to spatial effects.

13.1.1 The classical atom

Our classical picture of an atom consists of a massive positive nucleus surrounded by a light spherically-symmetrical cloud of electrons with an equal negative charge. We imagine the two as bound together by springs as in Fig. 13.1, so that in equilibrium the centres of mass and charge of the core and electron charge coincide. As a result the static atom has zero dipole moment. When it is disturbed, the electron cloud oscillates about the centre of mass with frequency η determined by the reduced mass m of the atom and the spring constant defined as $m\eta^2$.

This model can be applied to individual atoms and simple molecules; more complicated molecules may have internal dynamics and static dipole moments, but still the model gives considerable physical understanding. In addition it can be used for very small particles. But it only predicts a single resonant frequency, whereas atoms really respond resonantly to a number of discrete frequencies; this fact is usually introduced phenomenologically, as in §13.3.2. However, our main concern is with the interaction between the atom and a wave field having a well-defined frequency ω, and the interaction is strong only if $\omega \approx \eta$; so usually one resonance alone is dominant and the others can be ignored.

We shall show the atom to behave as an oscillating dipole, which therefore loses energy by electromagnetic radiation. In this chapter, we introduce energy loss phenomenologically into the equation of motion of the atom through a damping constant $m\kappa$. This is one of the parameters whose microscopic origin has to be explained by quantum theory.

Having said all of this, we can write down the equation of motion for the displacement x between the centres of mass of the positive nucleus and the electron charge, when F is a force acting equally and oppositely on each of them:

$$m\frac{\mathrm{d}^2 x}{\mathrm{d}t^2} + m\kappa\frac{\mathrm{d}x}{\mathrm{d}t} + m\eta^2 x = F \; . \tag{13.1}$$

If $F = qE$ is the force due to a constant electric field, (13.1) has the solution $x = F/m\eta^2$. Remembering that x is the separation between positive and negative charges, this corresponds to an induced dipole moment

$$p = qx = q^2 E/m\eta^2 \; . \tag{13.2}$$

As a result the electrical polarizability of the atom at zero frequency is

$$\alpha(0) = \frac{p}{\epsilon_0 E} = \frac{q^2}{\epsilon_0 m\eta^2} \; . \tag{13.3}$$

Figure 13.1 The classical atom.

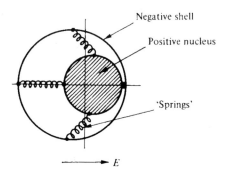

In the same way, we can calculate the effect of an electric field $E = E_0 \exp(-i\omega t)$, using $d/dt \equiv -i\omega$, to be

$$\alpha(\omega) = \frac{q^2}{\epsilon_0 m(\eta^2 - \omega^2 - i\kappa\omega)} . \tag{13.4}$$

Notice that α is complex. This indicates that there is a phase difference between the applied field and the induced dipole moment, which is particularly prominent in the frequency interval of about 2κ around η.

We shall now look at some of the applications of this model. We start with a discussion of the scattering by particles sufficiently well-separated that there is no interference between the waves they scatter (§13.2). Following this, we shall see the application of the model to dense matter (§13.3) where considerations of interference are crucial.

13.2 Rayleigh scattering

When an electromagnetic wave falls on an isolated particle, it is either absorbed or scattered. If the wave frequency ω is well removed from any resonant frequency η, the absorption of the wave is negligible, and only scattering need be considered. *Rayleigh scattering* occurs when the particle size is much smaller than the wavelength, so the wave field it experiences is essentially uniform. The result will be seen to be particularly useful for scattering by isolated atoms or molecules, although it is also applicable to very fine particulate matter and density fluctuations. We write the instantaneous dipole moment (13.2)

$$p(t) = \alpha\epsilon_0 E(t). \tag{13.5}$$

If $E(t) = E_0 \exp(-i\omega t)$, $p(t)$ behaves as an oscillating dipole. This, we know, radiates energy at a rate (see equation (5.34))

$$W = \frac{\omega^4 p_0^2}{12\pi\epsilon_0 c^3} = \frac{\omega^4 E_0^2 \alpha^2}{12\pi c^3} . \tag{13.6}$$

If there are N *independent* scattering particles in a cube of unit volume, the total power scattered is just N times the result (13.6). Now the radiant power incident on a face of the cube is the Poynting vector Π (§5.2.1) which has average magnitude $\frac{1}{2}E_0^2\epsilon_0 c$. Therefore the loss of power per unit distance of propagation is

$$\frac{d\Pi}{dz} = -NW = \frac{-\Pi N\omega^4 \alpha^2}{6\pi c^4} . \tag{13.7}$$

This equation has the solution:

$$\Pi = \Pi_0 \exp\left(\frac{-N\omega^4\alpha^2 z}{6\pi c^4}\right) = \Pi_0 \exp(-z/z_0) , \tag{13.8}$$

where $z_0 = 6\pi c^4/N\omega^4\alpha^2$ is a decay distance, telling us that the intensity of light travelling through the scattering region falls to e^{-1} of its initial value in a distance z_0. Before proceeding with an estimate of z_0 for systems, such as gases, where N and α are known, it is important to recall that the calculation has assumed the scattering from the individual particles to be independent, so that the scattered waves are incoherent and the intensities of the scattered waves are simply added. This assumption is very often untrue, and will be examined in more detail in §13.2.3.

13.2.1 Wavelength dependence of scattered radiation

A most striking part of equation (13.8) is the fourth-power dependence on frequency; blue light is scattered about ten times more intensely than red light. This is the reason for the common observation that the sky is blue (weather permitting) during most of the day, but can appear red when one looks directly towards the Sun at dawn or sunset. The sky is blue because we see sunlight scattered by air molecules at all heights and the spectrum is therefore biased strongly to short wavelengths. The redness occurs at daybreak and sundown because at those times the Sun's light, and that reflected from clouds near the horizon, passes horizontally through the atmosphere, and the very long air passage results in the scattering away of a much greater fraction of the blue light than the red. Rayleigh scattering is also responsible for other everyday effects, such as the colours of diluted milk and cigarette smoke, and the glorious sunsets induced by air pollution.

13.2.2 Polarization of scattered radiation

The dipole moment produced in the atom is parallel to the electric vector of the incident light and will re-radiate with a radiation polar diagram as described in §5.3.1 (Fig. 5.2(b)). The intensity radiated along the axis of the dipole is zero. It therefore follows that scattered radiation along a line perpendicular to the incident light is linearly-polarized normal to the plane containing the incident and scattered light. In other directions the light will appear partially polarized. With the aid of a single polaroid sheet (§6.3.2), these conclusions can easily be tested using ordinary sunlight (Fig. 13.2), although polarization is far from complete because of multiple scattering. This effect is commonly used in photography to eliminate the effects of haze (Problem 13.1).

13.2.3 Incoherent and coherent scattering

Next we should like to use (13.8) to calculate the decay distance for clean air at atmospheric pressure, but first we should check whether the assumption of independent scattering by individual molecules applies. It turns out that the mean distance between air molecules under atmospheric conditions is two orders of magnitude *less* than a wavelength of light, so that almost completely *coherent* scattering would be expected. If the medium has uniform density, we shall see in §13.3 that there is no net scattering at all. It is only the *deviations* from uniform density that give rise to scattering. The subject of scattering by density fluctuations can be treated fully by thermodynamics (see, e.g., Landau and Lifshitz, 1980) but we can get an idea of the results by a simple argument. One would expect incoherent Rayleigh scattering to result from independent 'blocks' of material of dimensions of order λ, each one therefore having volume $V \approx \lambda^3$. Larger blocks are not small compared with the wavelength, and smaller ones will not scatter incoherently. Now, in such a volume there are on average NV molecules. In a perfect gas the molecules do not interact with one another and the exact number of molecules in the volume V will be governed by Poisson statistics (§11.3.3). For such statistics, the r.m.s. fluctuation in the number is $(NV)^{\frac{1}{2}}$, and it is these fluctuations which should be considered as the scattering 'particles'. We should therefore consider Rayleigh scattering by 'particles' containing $(NV)^{\frac{1}{2}}$ molecules, which would have polarizability $\alpha(NV)^{\frac{1}{2}}$ and number density $1/V$. Returning to (13.8) we find

$$z_0 = \frac{6\pi c^4}{V^{-1}\omega^4[\alpha(NV)^{\frac{1}{2}}]^2} = \frac{6\pi c^4}{N\omega^4\alpha^2} \tag{13.9}$$

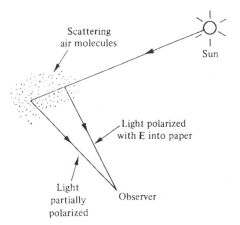

Figure 13.2 Polarization of atmospherically-scattered sunlight.

which is exactly the same result as we obtained for incoherent scattering, (13.8)! Thus the scattering by density fluctuations in a perfect gas is just *the same as if all the molecules were to scatter incoherently*.

To estimate the value of z_0 for scattering by a clean atmosphere we relate the atomic polarizability α to the dielectric constant ϵ of the gas and thus to its refractive index (§5.1.3)

$$\mu = \epsilon^{\frac{1}{2}} = (1 + N\alpha)^{\frac{1}{2}} \approx 1 + N\alpha/2 . \tag{13.10}$$

Thus z_0 can be written, substituting the wavelength $\lambda = 2\pi c/\omega$ into (13.8):

$$z_0 = \frac{3N\lambda^4}{32\pi^3(\mu - 1)^2} . \tag{13.11}$$

Using the values at atmospheric pressure $\mu - 1 = 3 \times 10^{-4}$ and $N = 3 \times 10^{25}$ m^{-3} we find for green light $z_0 \approx 65$ km. At first sight this figure seems surprisingly low, particularly as molecular scattering is often not the only factor which limits visibility through the atmosphere. One frequently finds situations where the meteorological visibility exceeds 100 km and even reaches 200 km. However, one should remember that z_0 corresponds to an attenuation factor of $e^{-1} = 0.37$, and factors of e^{-2} (at $2z_0$) or even $e^{-3} = 0.05$ (at $3z_0$) can be tolerated before a distant view of snow-capped mountains against an azure sky merges into the haze.

Under what conditions might we expect scattering to differ from from the incoherent case? We look for situations where Poisson statistics do not describe the fluctuations satisfactorily. If the medium is relatively incompressible, as is a liquid, the motions of the particles are correlated in such a way that they avoid one another. Density fluctuations are then suppressed and the scattering is less than in the incoherent case, approaching zero in the uniform density limit (§13.3). On the other hand, near the critical point in a fluid, for example, the compressibility diverges and there is a tendency towards local condensation which enhances density fluctuations. We then see excess scattering and the phenomenon of *critical opalescence* (Fig. 13.3).

13.3 Coherent scattering and dispersion

We shall now consider the problem of scattering by a uniformly-dense incompressible medium, where the molecules are much closer than one wavelength and therefore the scattered waves are correlated in phase. In this problem we have to sum the *amplitudes* of the scattered waves. It turns out that a real polarizability α results in no net scattering whatsoever; the material simply refracts the incident wave. But when α is complex, absorption of the incident light occurs.

13.3.1 Refraction as a problem in coherent scattering

Consider scattering by a thin slab of thickness $\delta z \ll \lambda$ in the plane $z = 0$, where z is the axis of propagation of the radiation (Fig. 13.4). In this slab there are N molecules per unit volume, each having polarizability α. Now the oscillating dipoles in the slab will all be excited with the same phase by an incident plane-wave $E = E_0 \exp[i(kz - \omega t)]$ and we can calculate their combined radiation at the point $Q \equiv (0, 0, z)$. A

Figure 13.3 Critical opalescence of CO_2 near its liquid-vapour critical point. (a) $T < T_c$; (b) $T \approx T_c$.

(a)
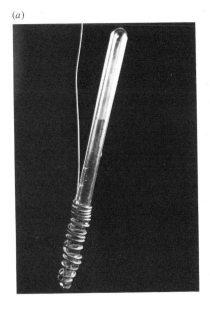

(b)
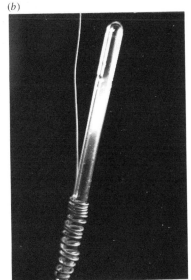

Figure 13.4 Coherent scattering by a dense medium.

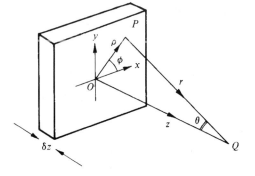

molecule in the slab at point P, $(x, y, 0)$ responds to the incident wave with an oscillating dipole moment of magnitude :

$$p(t) = \alpha \epsilon_0 E_0 \exp(-i\omega t) . \qquad (13.12)$$

From (5.31) its transverse radiation field at Q, $(0, 0, z)$ is

$$e(t) = \frac{\alpha \epsilon_0 \omega^2 E_0 \exp[i(kr - \omega t)] \cos \theta}{4\pi \epsilon_0 c^2 r} , \qquad (13.13)$$

where $r^2 = x^2 + y^2 + z^2 \equiv \rho^2 + z^2$ and θ is the angle between the vector $\mathbf{r}$ and the z-axis. The total field from all the molecules in an elementary volume $dx\, dy\, \delta z$ at this point is (13.13) multiplied by $N dx\, dy\, \delta z$. We can therefore write down the total scattered field δE_Q at Q as the integral of (13.13) over the whole slab:

$$\delta E_Q = \frac{N \alpha \omega^2 E_0 \,\delta z\, \exp(-i\omega t)}{4\pi c^2} \int\!\!\int_{-\infty}^{\infty} \frac{z \exp(ikr)}{r^2} dx\, dy, \qquad (13.14)$$

where $\cos \theta$ has been replaced by z/r. In terms of ρ,

$$\delta E_Q = 2\pi z \frac{N \alpha \omega^2 E_0 \,\delta z\, \exp(-i\omega t)}{4\pi c^2} \int_0^{\infty} \frac{\exp[ik(z^2 + \rho^2)^{\frac{1}{2}}]}{\rho^2 + z^2} \rho \, d\rho .\, (13.15)$$

The integral in (13.15) can be rewritten simply as

$$\int_z^{\infty} \frac{\exp(ikr)}{r} dr, \qquad (13.16)$$

which can easily be evaluated by an amplitude–phase diagram. As explained in §7.4.1, this gives the curve shown in Fig. 13.5 and the integral is seen to be the vector OC which has the value

$$OC = (kz)^{-1} \exp[i(kz + \pi/2)] . \qquad (13.17)$$

Figure 13.5
Amplitude–phase
diagram for the
integral (13.16).

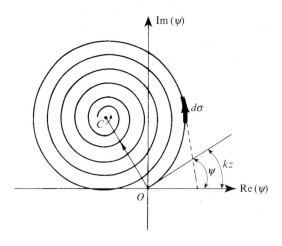

Thus

$$\begin{aligned}
\delta E_Q &= \frac{N\alpha\omega^2 E_0 \delta z}{2c^2 k} \exp[i(kz - \omega t + \pi/2)] \\
&= \tfrac{1}{2} ik N\alpha E_0 \delta z \, \exp[i(kz - \omega t)].
\end{aligned} \tag{13.18}$$

This scattered wave must be added to the unscattered wave which has reached Q; since δz is small, the unscattered wave is negligibly different from the incident wave $E_{Q0} = E_0 \exp[i(kz - \omega t)]$ whence

$$\delta E_Q = \tfrac{1}{2} ik N\alpha \delta z \, E_{Q0}. \tag{13.19}$$

If α is real, the scattered amplitude is in phase quadrature with the direct wave and therefore does not alter its magnitude, but only its phase; in other words, the velocity of the wave is modified, but there is no attenuation (Fig. 13.6). Then

$$\begin{aligned}
E_Q = E_{Q0} + \delta E_Q &= (1 + \tfrac{1}{2} ik N\alpha \delta z) E_{Q0} \\
&\approx \exp(\tfrac{1}{2} ik N\alpha \delta z) E_{Q0}.
\end{aligned} \tag{13.20}$$

If we had inserted a transparent plate with refractive index μ and thickness δz into the beam, we should have increased the optical path by $(\mu - 1)\delta z$ and modified the wave E_{Q0} to

$$E_Q = E_{Q0} \exp[ik\delta z(\mu - 1)]. \tag{13.21}$$

So coherent scattering by the slab has resulted in an effective refractive index

$$\mu = 1 + \tfrac{1}{2} N\alpha. \tag{13.22}$$

This is just the refractive index we have used, for example in (13.10). Thus coherent scattering results in refraction, but not absorption. It might seem that we have discovered nothing new. The importance of this calculation is that it links refraction and scattering, and can be used to derive an effective refractive index for other types of wave, whose scattering behaviour is known – for example, neutrons, §13.3.6.

If the medium is dense, so that μ is not close to unity, we must consider the field which polarizes the molecules as the local field, and not simply the applied field. This makes the treatment more complicated but does not introduce absorption.

Figure 13.6
Amplitude–phase diagram for coherent scattering when there is no absorption.

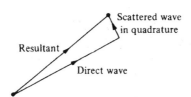

13.3.2 Resonance and anomalous dispersion

At frequencies near the resonance η, (13.14) shows α to be complex, and as a result the statement that the scattered wave is in quadrature with the direct wave is no longer correct. The refractive index is still modified, but absorption may also occur as can be seen from Fig. 13.7. In (13.4) we had polarizability $\alpha(\omega) = q^2/[\epsilon_0 m(\eta^2 - \omega^2 - i\kappa\omega)]$ and thus the refractive index (13.22) is approximately

$$\mu = 1 + \tfrac{1}{2}N\alpha = 1 + \tfrac{1}{2}\Omega^2(\eta^2 - \omega^2 - i\kappa\omega)^{-1} \qquad (13.23)$$

when $N\alpha \ll 1$. Ω is called the *plasma frequency* $(Nq^2/\epsilon_0 m)^{\frac{1}{2}}$, whose significance will be discussed in §13.3.4. Then the real and imaginary parts of μ are

$$\mu_r \;=\; 1 + \frac{\Omega^2(\eta^2 - \omega^2)}{2[(\eta^2 - \omega^2)^2 + \kappa^2\omega^2]} , \qquad (13.24)$$

$$\mu_i \;=\; \frac{\Omega^2\kappa\omega}{2[(\eta^2 - \omega^2)^2 + \kappa^2\omega^2]} . \qquad (13.25)$$

Fig. 13.8 shows the two quantities, $\mu_r(\omega)$ and $\mu_i(\omega)$ schematically. The curves show several important features.

(a) Outside the frequency region $\eta \pm \kappa$, $d\mu_r/d\omega$ is positive and $\mu_i \ll 1$. This is called *normal dispersion* and is typical of all transparent media.

(b) The refractive index becomes large at frequencies just below resonance, and sharply drops to a value less than unity just above the resonance. In the region of sharp change, $d\mu_r/d\omega$ is negative; this is called *anomalous dispersion*.

(c) In the anomalous dispersion region μ_i can not be neglected and there is absorption. We shall show in §13.4 that this is necessary from very general considerations. This is, of course, the absorption corresponding to an emission line in the atomic spectrum.

A real atom has a series of spectral lines at various frequencies, and anomalous dispersion takes place in the region of each one of them.

Figure 13.7 Absorption occurs when the scattered wave is no longer in quadrature with the direct wave.

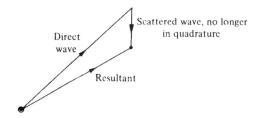

As we presented it here, the model atom has only a single resonance; the multiplicity is taken into account by assuming that it has several resonant states, the jth state having frequency η_j and relative strength N_j. Then, writing the refractive index as a superposition of the effects of all of them,

$$\mu_r = 1 + \frac{q^2}{2\epsilon_0 m} \sum_j \frac{N_j(\eta_j^2 - \omega^2)}{(\eta_j^2 - \omega^2)^2 + \kappa_j^2 \omega^2}. \tag{13.26}$$

The N_js are called *oscillator strengths* and are related to the matrix elements which appear in the quantum-mechanical description (§14.3). Fig. 13.9 shows a typical refractive-index curve.

Figure 13.8
Anomalous
dispersion

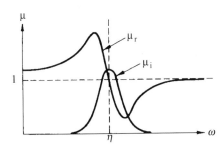

Figure 13.9 Real and imaginary components of the refractive index of sea water at wavelengths between 0.6 and 14 μm, measured using oblique reflectivity in the ∥ and ⊥ polarizations (§5.4.2).

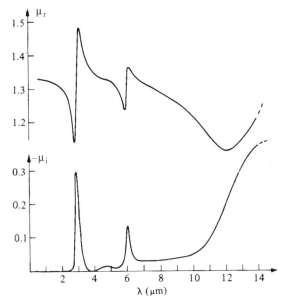

13.3.3 Dispersion remote from an absorption band: X- ray refractive index

In the normal dispersion region, remote from a resonance frequency η, we can neglect the absorption and (13.24) becomes

$$\mu \approx 1 + \frac{\Omega^2}{2(\eta^2 - \omega^2)} . \tag{13.27}$$

In particular, if ω is well above that of the highest resonance in (13.26) we have:

$$\mu \approx 1 - \frac{\Omega^2}{2\omega^2} . \tag{13.28}$$

which shows that the refractive index in the X-ray region is less than unity, but only just so. Substitution of typical values gives $\mu - 1 \approx -10^{-7}$. This allows the use of total *external* reflexion as a method of handling X-rays. Although $v = c/\mu$ is greater than c, the theory of relativity is not contradicted because it is the group velocity, not the phase velocity, at which information and energy are transported (Problem 2.3).

13.3.4 Plasma absorption edge in a free-electron gas

If the electrons in a medium are unbound, for example as a plasma in the ionosphere or as conduction electrons in a simple metal, we can calculate the dispersion by substituting $\eta = 0$. We obtain from (13.23):

$$\epsilon = \mu^2 = 1 + N\alpha = 1 - \frac{\Omega^2}{i\kappa\omega + \omega^2} . \tag{13.29}$$

When the electrons are free, $\kappa \ll \omega$, and

$$\mu \approx (1 - \Omega^2/\omega^2)^{\frac{1}{2}} , \tag{13.30}$$

which shows that for $\omega < \Omega$ the wave is evanescent and the medium is therefore opaque. At frequency Ω there is a transition to a transparent state. This is called the *plasma absorption edge* and is shown in Fig. 13.10. It is particularly sharp in the alkali metals, where it occurs in the near ultra-violet. At the edge, $\mu = 0$ and the wavelength is infinite; the whole plasma oscillates in phase, creating a *collective oscillation*.

13.3.5 Refractive index of a free-electron gas in a magnetic field

A similar calculation to the above can be made in the presence of a constant magnetic field $\mathbf{B}_0$ and shows the origin of the magneto-optic effect discussed in §6.9.3. Returning to the basic mechanical equation (13.1) we can add a term $q\mathbf{B} \times \mathbf{v}$ representing the Lorentz force, but it is now necessary to work in three dimensions because of the vector product. With $\mathbf{B}_0$ and the incident wave-vector in the z-direction, for example:

$$m\frac{d^2x}{dt^2} + m\kappa\frac{dx}{dt} + m\eta^2 x + qB_0\frac{dy}{dt} = F_x = qE_{0x}\exp(-i\omega t) \; ; \quad (13.31)$$

$$m\frac{d^2y}{dt^2} + m\kappa\frac{dy}{dt} + m\eta^2 y - qB_0\frac{dx}{dt} = F_y = qE_{0y}\exp(-i\omega t) \, , \quad (13.32)$$

where (x, y) represents the displacement of the charge. We shall illustrate the effects in the high-frequency region $\omega \gg \kappa, \eta$ only.[†] Clearly $(x, y) = (x_0, y_0)\exp(-i\omega t)$, and so we can replace d/dt by $-i\omega$ whence:

$$-m\omega^2 x_0 - iq\omega B_0 y_0 = qE_{0x} \; ; \quad (13.33)$$

$$-m\omega^2 y_0 + iq\omega B_0 x_0 = qE_{0y} \, . \quad (13.34)$$

These equations are analogous to those for the Foucault pendulum in classical mechanics (see also §9.6). The result is particularly simple for circularly-polarized radiation (§6.2.2) for which $E_{0y} = \pm iE_{0x}$, the upper

---

[†] A wider range of frequencies was treated in Chapter 11 of the second edition of this book, and a fuller discussion is given by Budden (1966) and Altman and Suchy (1991).

Figure 13.10 (a) Broken line – real and imaginary parts of the complex refractive index for a free-electron gas with zero damping. (b) Solid line – measured values for sodium.

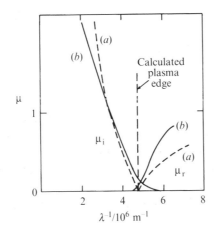

and lower signs representing left- and right-handed senses. Eliminating y_0 we have

$$-\left(\omega^2 m - \frac{q^2 B_0^2}{m}\right) x_0 = E_0 q \left(1 \pm \frac{qB_0}{m\omega}\right). \tag{13.35}$$

Then

$$x_0 = \pm i y_0 = \frac{-E_0 q}{\omega^2 m (1 \mp \omega_c/\omega)} \tag{13.36}$$

where $\omega_c = qB_0/m$ is the *electron cyclotron frequency*. From the charge displacement we calculate the polarization of the medium, $\mathbf{P}_0 = Nq(x_0, y_0)$ and hence the dielectric constant $\epsilon = 1 + \mathbf{P}_0/\epsilon_0 \mathbf{E}_0$:

$$\epsilon = \mu^2 = 1 - \frac{\Omega^2}{\omega^2 (1 \mp \omega_c/\omega)} \tag{13.37}$$

for the two circularly polarized waves, in which the effect of the magnetic field is represented by ω_c. When ω is large the corresponding refractive indices are real and the medium therefore shows a magnetically induced optical activity, which is the Faraday effect.

We can also represent (13.31)–(13.32) by a matrix equation, which will bring us into line with the formalism of Chapter 6. From (13.33) and (13.34), we calculate in the above manner the dielectric tensor ϵ :

$$\epsilon = 1 - \frac{\Omega^2}{\omega^2 - \omega_c^2} \begin{pmatrix} 1 & i\omega_c/\omega & 0 \\ -i\omega_c/\omega & 1 & 0 \\ 0 & 0 & 1 - \omega_c^2/\omega^2 \end{pmatrix} \tag{13.38}$$

where 1 is the unit tensor. This can be compared directly to (6.40) for a uniaxial magneto-optic medium, and its principal values can easily be shown to be (13.37).

13.3.6 Refractive index of a solid for neutron waves

In §13.3 we showed how the refractive index of a medium can be derived by a coherent scattering argument. Scattering cross-sections have been studied for many types of particle, and in this section we shall apply the idea to neutrons whose scattering by solids has been extensively studied as a single particle phenomenon (Squires, 1978). The result is an effective refractive index for neutron waves in the solid.

A plane neutron wave of amplitude $A \exp[i(kz - \omega t)]$ is incident on a single scattering centre, and a spherical wave of the form

$$A \frac{-\bar{b}}{r} \exp[i(kr - \omega t)] \tag{13.39}$$

is scattered, where $\bar{b}$ is called the *scattering length* and has a value typical of nuclear dimensions, 10^{-14} m. To find the equivalent refractive index for the neutron waves, we observe that (13.39) is the same as (13.13) if we substitute $-\bar{b}$ for $\alpha k^2/4\pi$ in the latter. Since neutron waves are scalar, the $\cos\theta$ term is absent and will be ignored. We can now jump to the final result (13.22) and deduce that for N scattering centres per unit volume the refractive index is:

$$\mu_n = 1 - N\frac{2\pi\bar{b}}{k^2} = 1 - \frac{N\bar{b}\lambda^2}{2\pi}. \tag{13.40}$$

A typical wavelength for slow neutrons is 10^{-10} m, and solids have densities N of the order of 10^{30} nuclei per m^3. Thus $\mu_n - 1$ is of order -10^{-5}. It is interesting that μ_n is less than unity. As with X-rays, this allows total external reflexion and the use of grazing angle reflecting optics for focusing and guiding slow neutron beams.

13.4 Dispersion relations

This section will discuss some very general relationships between the real and imaginary parts of response functions such as $\epsilon(\omega)$ that arise because of *causality*, which expresses the self-evident fact that no event can cause observable consequences which precede it in time.

13.4.1 Relationship between the impulse and frequency responses

A convenient way to understand the dynamic response of a system to an external field is to start by investigating the effect of a single impulse of the field.† Provided the response of the system is linear, the effect of a more complicated time-varying field can then be built up by superposing the response to impulses.

Suppose we apply an electric field E to a dielectric for a short time dt. The field impulse is then $E\,dt$, and it causes a polarization $E\,X(t)\,dt$ which is initiated by the field impulse but may die away more slowly. $X(t)$ is called the *impulse response*, and is the temporal polarization response to a unit field impulse applied at $t = 0$. Causality requires that $X(t)$ be zero at negative t. Now let us calculate the polarization

---

† This approach is by no means modern; Newton used it in his analysis of the Moon's motion in the Earth's gravitational field.

caused by a field $E(t)$ by linear superposition of the effects of pulses $E(t')dt'$ at time t':

$$P(t) = \int_{-\infty}^{t} E(t')X(t-t')dt'. \qquad (13.41)$$

Since $X(t)$ is zero for negative t the upper limit of the integral can be replaced by ∞. In the particular case of the oscillatory field $E = E_0 \exp(i\omega t)$, (13.41) becomes, with $t'' \equiv t - t'$,

$$P(t) = E_0 \int_{-\infty}^{\infty} \exp(i\omega t')\,X(t-t')dt' \qquad (13.42)$$

$$= E_0 \exp(i\omega t) \int_{-\infty}^{\infty} \exp(-i\omega t'')\,X(t'')dt'' = E\,\chi(\omega)\,, \qquad (13.43)$$

where $\chi(\omega)$ is the Fourier transform of $X(t)$ and $\chi(\omega)$, being the relationship between P and E, is the polarizability ($\equiv N\alpha(\omega)$). The dielectric constant at ω is then

$$\epsilon_0\epsilon(\omega)E = \epsilon_0 E + P \qquad (13.44)$$

$$\text{whence} \quad \epsilon_0[\epsilon(\omega) - 1] = \chi(\omega)\,. \qquad (13.45)$$

This relationship shows how the frequency response is related to the Fourier transform of the impulse response in the dielectric case.

13.4.2 The Kramers–Kronig relations

By introducing the requirement that the response of any system must be causal, we can now deduce relationships between the real and imaginary frequency response functions, of which (13.24)–(13.25) are examples. We shall define a unit step function $d(t)$ as follows:

$$\begin{aligned} d(t) &= \lim_{s \to 0} \exp(st)\,(\sim 1), \quad \text{when } t < 0 \;; \\ &= 0 \quad \text{when } t \geq 0 \;. \end{aligned} \qquad (13.46)$$

As pointed out in §12.4.5, the step function obtained by putting $s = 0$ does not really have a Fourier transform, but this problem is avoided by letting s be infinitesimal but not zero. Then the transform is

$$D(\omega) = \lim_{s \to 0} (s - i\omega)^{-1} \,. \qquad (13.47)$$

Now since $X(t)$ only starts at $t = 0$, and $d(t)$ finishes at the same time, we can write†

$$X(t)\,d(t) = 0 \,. \qquad (13.48)$$

† We ignore a possible δ- function response as part of $X(t)$ which might not give 0 when multiplied by $d(t)$ at $t > 0$.

Taking the Fourier transform of this equation

$$0 = \chi(\omega) \otimes D(\omega) = \lim_{s \to 0} \int_{-\infty}^{\infty} \frac{\chi(\omega')}{s - i(\omega - \omega')} d\omega'$$

$$= \epsilon_0 \lim_{s \to 0} \int_{-\infty}^{\infty} \frac{\epsilon(\omega') - 1}{s - i(\omega - \omega')} d\omega' . \tag{13.49}$$

As $s \to 0$ there is a singularity at $\omega' = \omega$. We therefore divide the integral into two parts, that from $\omega - s$ to $\omega + s$ and the rest. The first integral can be evaluated straightforwardly since when s is small enough $\epsilon(\omega')$ is constant throughout the range of integration:

$$\lim_{s \to 0} \int_{\omega-s}^{\omega+s} \frac{\epsilon(\omega') - 1}{s - i(\omega - \omega')} d\omega' = [\epsilon(\omega) - 1] \int_{\omega-s}^{\omega+s} \frac{d\omega'}{s - i(\omega - \omega')}$$

$$= \pi[\epsilon(\omega) - 1] ; \tag{13.50}$$

the integral is independent of s. The rest is called the *principal part* of the integral and is denoted by $\mathscr{P} \int$:

$$\lim_{s \to 0} \left(\int_{-\infty}^{\omega-s} + \int_{\omega+s}^{\infty} \right) \frac{\epsilon(\omega') - 1}{s - i(\omega - \omega')} d\omega' \equiv \mathscr{P} \int_{-\infty}^{\infty} \frac{\epsilon(\omega') - 1}{-i(\omega - \omega')} d\omega' . \tag{13.51}$$

Since (13.49) is the sum of (13.50) and (13.51),

$$\epsilon(\omega) = 1 + \frac{1}{\pi} \mathscr{P} \int_{-\infty}^{\infty} \frac{\epsilon(\omega') - 1}{i(\omega - \omega')} d\omega'. \tag{13.52}$$

We can equate real and imaginary parts of (13.52) separately and obtain two integral relationships between $\epsilon_r(\omega)$ and $\epsilon_i(\omega)$:

$$\epsilon_r(\omega) = 1 + \frac{1}{\pi} \mathscr{P} \int_{-\infty}^{\infty} \frac{\epsilon_i(\omega')}{(\omega - \omega')} d\omega'$$

$$\left[= 1 + \frac{2}{\pi} \mathscr{P} \int_{0}^{\infty} \frac{\omega' \epsilon_i(\omega')}{(\omega^2 - \omega'^2)} d\omega' \right] ; \tag{13.53}$$

$$\epsilon_i(\omega) = -\frac{1}{\pi} \mathscr{P} \int_{-\infty}^{\infty} \frac{\epsilon_r(\omega') - 1}{(\omega - \omega')} d\omega'$$

$$\left[= -\frac{2}{\pi} \mathscr{P} \int_{0}^{\infty} \frac{\omega[\epsilon_r(\omega') - 1]}{(\omega^2 - \omega'^2)} d\omega' \right] . \tag{13.54}$$

In the bracketed forms of (13.53) and (13.54) we have used the property $\epsilon(\omega) = \epsilon^*(-\omega)$ of the Fourier transform of a real response to an applied electric field. Equations (13.53) and (13.54) (in either form) are known as the *Kramers–Kronig relations*.

13.5 Advanced topic: non-linear optics

Up to this stage, we have only regarded the polarization of a material by an external field as a linear process. For small enough fields this can be seen as the leading term in the Taylor expansion of $P(E)$:

$$P(E) = P(0) + E \left(\frac{dP}{dE}\right)_0 + \frac{1}{2}E^2 \left(\frac{d^2P}{dE^2}\right)_0 + \frac{1}{6}E^3 \left(\frac{d^3P}{dE^3}\right)_0 + \ldots (13.55)$$

For a material with no static dipole moment $P(0) = 0$; and we write (13.55) as:

$$P(E) \approx \chi E + \chi_2 E^2 + \chi_3 E^3 + \ldots \qquad (13.56)$$

in which χ_n is the nth-order non-linear polarizability or susceptibility. Since the invention of the laser, a wealth of fascinating phenomena has been discovered which use light beams intense enough to require the expansion (13.56) to be carried beyond the linear term (Bloembergen, 1982; Yariv, 1989, 1991). In the following sections we shall briefly describe two of them: *second harmonic generation*, which involves expansion up to the second order, and *four-wave mixing*, which involves the next order.

13.5.1 Second harmonic generation

We now consider the effect of wave field $E = E_0 \cos \omega t$ on the medium.† From (13.56), expanding as far as the second order term,

$$P(E) = \chi E_0 \cos \omega t + \tfrac{1}{4}E_0^2 \chi_2 (\cos 2\omega t + 1) + \ldots. \qquad (13.57)$$

One can see that the harmonic frequency 2ω has been induced, and this will be radiated by the oscillating dipole. This is called *second harmonic generation*. In terms of photons, two photons of frequency ω have combined to form one of frequency 2ω, and so the process belongs to the wider category of *three-wave mixing*. In general higher terms in the expansion will also occur, giving frequencies $n\omega$, but these may not be observable for reasons which will be explained below.

What governs the intensity of the observed harmonic waves? First of all the intensity of the 2ω component in P is proportional to E_0^2 so that the high intensity is necessary to produce observable effects. Secondly, since the value of χ_2 is identically zero in materials whose molecular structure has a centre of symmetry, for the same reasons as discussed in §6.9.1, a medium with sufficiently low crystal symmetry must be chosen.

Given large enough E_0 and non-zero χ_2, second harmonic generation will occur in a small volume of the dielectric. For the effect to increase in proportion to the volume of a sample, we also require the harmonic

---

† It is not appropriate to use the complex exponential representation here because E will be squared and cubed in what follows.

waves generated in different volume elements to add coherently. Now at its point of origin, the 2ω wave is created with a well-defined phase relation to the ω wave which generated it. To maintain this relationship at all other points, which is a requirement for constructive interference, the two waves must propagate in the same direction at the same phase velocity, i.e. $v(\omega) = v(2\omega)$. This is called *phase matching*, and when it is satisfied it becomes quite easy to observe harmonic generation. Most of the crystals used for second harmonic generation are anisotropic, and the anisotropy of $\mathbf{\mu}$ (§6.4.2) can then be used to find directions of propagation in which the refractive indices, and hence velocites, for orthogonal polarizations at the two frequencies are equal, i.e. $\mu_1(\omega) = \mu_2(2\omega)$. This is shown geometrically in Fig. 13.11 for a biaxial crystal. The same mechanism can be used to mix light waves of different frequencies (Problem 13.6), but clearly can not be used to generate, say, a third harmonic simultaneously with a second.

Phase matching can best be appreciated by representing the condition of equal velocities as a requirement for conservation of both energy and linear momentum when the two photons combine in the crystal to create a single one. Clearly, energy is conserved because $2\hbar\omega = \hbar\omega + \hbar\omega$. Conservation of momentum for two photons travelling in the same direction then requires $k(2\omega) = 2k(\omega)$ which implies equal refractive indices at the two frequencies. We could possibly consider the interaction of two waves of the same frequency but different directions, with wave-vectors $\mathbf{k}_1(\omega)$ and $\mathbf{k}_2(\omega)$, and combine the vectors as in Fig. 13.12(a) so that the resultant has the right magnitude $k(2\omega)$. However, this does not work in transparent media having

Figure 13.11
Matching refractive indices at ω and 2ω in a biaxial crystal. The outer branch at ω intersects the inner branch at 2ω along the curve shown, so that phase-velocity matching is achieved in directions such as D.

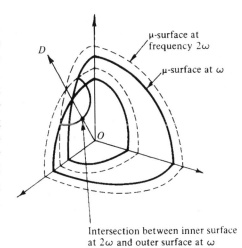

normal dispersion because then $k(2\omega) > 2k(\omega)$.† On the other hand, if we can find a way of adding a fixed vector $\mathbf{k}_0$, it is possible to satisfy the vector equation as shown in Fig. 13.12(b). This can be done by modulating the medium periodically, and provides an alternative method of phase matching (see Bloembergen, 1982).

13.5.2 Four-wave mixing

When a non-linear material is illuminated by a pair of coherent waves travelling in different directions, they result in a phase grating within the material, which in turn can diffract a third wave (not necessarily coherent with the first two) into a fourth one. This process is called *four-wave mixing* and has been demonstrated in several crystals, particularly $BaTiO_3$. The effect can be understood in terms of the third-order non-linear suscepibility χ_3.

The two coherent waves, each with time variation $E = \frac{1}{2}E_0 \cos \omega t$, create an interference pattern or standing wave within the crystal. The pattern consists of regions near the nodes where the local field is essentially zero, and regions near the antinodes where its value is close to $E_0 \cos \omega t$. The response to an additional weak field is different in the two parts. Retaining terms up to third order in (13.56), the polarization is

$$P = \chi E + \chi_2 E^2 + \chi_3 E^3. \tag{13.58}$$

Now we shall add a small extra field e. We then have, up to first order in e:

$$P + \delta P = \chi(E + e) + \chi_2(E^2 + 2Ee) + \chi_3(E^3 + 3E^2 e), \tag{13.59}$$
$$\delta P = e(\chi + 2\chi_2 E + 3\chi_3 E^2). \tag{13.60}$$

Now if e is incoherent with E, the ratio $\delta P/e$ is obtained by averaging the bracketed term in (13.60) over many periods of ω. The average of $E_0 \cos \omega t$ is of course zero; that of $E_0^2 \cos^2 \omega t$ is $E_0^2/2$ and so we have, on average,

$$\delta P/e = (\chi + \frac{3}{2}E_0^2 \chi_3). \tag{13.61}$$

† In any case, the overlap volume of two waves travelling in different directions in the crystal would be very small.

Figure 13.12 Vector diagrams for conservation of wave-vector in second harmonic generation. (a) Hypothetical situation with two input waves in different directions; (b) interaction of two parallel waves in a periodically modulated medium.

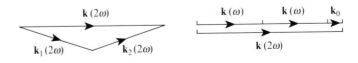

The refractive index seen by the field e is thus

$$\begin{aligned}
\mu_d &= (1 + \delta P/\epsilon_0 e)^{\frac{1}{2}} \\
&= [1 + (\chi + \tfrac{3}{2}\chi_3 E_0^2)/\epsilon_0]^{\frac{1}{2}}
\end{aligned} \tag{13.62}$$

which depends on the field E_0, and is therefore different at the nodes and antinodes of the standing wave pattern created by the first two waves. The wave e therefore sees the standing wave 'written' into the crystal as a phase grating (§8.4.5), which diffracts it in the same way as the volume hologram, according to Bragg's law (§12.6.3). A physical description of the type of mechanism involved in the writing process is given by Pepper, Feinberg and Kukhtarev (1990).

The description of four-wave mixing in terms of momentum conservation is straightforward. For simplicity we shall assume the medium to be isotropic. The main waves – those with amplitude $\tfrac{1}{2}E_0$ – have wave-vectors $\mathbf{k}_1$ and $\mathbf{k}_2$ which have the same lengths. Their interference pattern is represented by the vector $\mathbf{K} = \mathbf{k}_2 - \mathbf{k}_1$. When the third wave $\mathbf{k}_3$ is incident it is diffracted to $\mathbf{k}_4 = \mathbf{k}_3 \pm \mathbf{K}$, where the sign is chosen so that $|\mathbf{k}_3| = |\mathbf{k}_4|$ in order to conserve energy. The equations are represented in Fig. 13.13, where of course $\mathbf{k}_1$, $\mathbf{k}_2$ and $\mathbf{k}_3$ are not necessarily co-planar.

13.5.3 Phase-conjugate mirror

When the wave we have designated by e is *coherent* with E_0, the situation is richer, because of the greater variety of terms that can interfere. A fascinating outcome is the concept of the *phase-conjugate mirror*, which reflects an incident wave as its spatial complex conjugate. This idea has been applied to other branches of physics as far afield as superfluid ^{3}He, and is discussed further by Yariv (1991) and Pepper (1986).

Figure 13.13 Vector diagram for four-wave mixing.

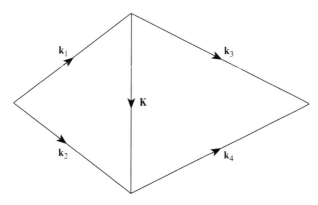

Our approach is once again to calculate the polarization arising from the simultaneous incidence of the three waves – E_1 with amplitude $\frac{1}{2}E_0$ and wave-vector $\mathbf{k}_1$, E_2 with $\frac{1}{2}E_0$ and $\mathbf{k}_2$, E_3 with e and $\mathbf{k}_3$ – and to look for all the terms having the correct $k(\omega)$ to propagate in the medium. We shall treat the case which is specifically different from §13.5.2, that in which E_1 and E_2 are counter-propagating, i.e. $\mathbf{k}_1 = -\mathbf{k}_2$. These waves are called *pump beams*. The fields of the three waves are

$$E_1 = \tfrac{1}{2}E_0 \cos(\mathbf{k}_1 \cdot \mathbf{r} - \omega t)$$
$$E_2 = \tfrac{1}{2}E_0 \cos(\mathbf{k}_2 \cdot \mathbf{r} - \omega t)$$
$$E_3 = e \cos(\mathbf{k}_3 \cdot \mathbf{r} - \omega t + \phi_3) . \tag{13.63}$$

We shall see below that these generate, *inter alia*, a wave with wave-vector $\mathbf{k}_4 = -\mathbf{k}_3$ and amplitude $e\cos(-\mathbf{k}_3 \cdot \mathbf{r} - \omega t - \phi_3)$ whose phase is the spatial complex conjugate of E_3. Clearly, from Fig. 13.14(*a*), momentum is conserved.

The third-order polarization term in (13.58) which interests us is $\chi_3 E^3 = \chi_3(E_1 + E_2 + E_3)^3$. Substituting (13.63) and replacing the cosines by their complex exponential representations to facilitate the multiplications gives an expression with about 70 sinusoidal terms! These include terms which are not relevant to the present discussion:

(*a*) waves with frequency 3ω which are third harmonics,

(*b*) waves in which $\omega/|\mathbf{k}|$ does not equal the wave velocity and therefore do not propagate,

(*c*) waves with arguments like those of E_1, E_2 and E_3 which arise essentially from the modification of the refractive index which we saw in (13.62).

Figure 13.14 Vector diagram for phase-conjugate reflexion and its interpretation in terms of a volume grating.

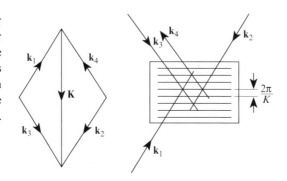

The terms which are relevant are those involving all three of the incident waves and therefore have amplitude $E_0^2 e$,

$$\frac{1}{4}\chi_3 E_0^2 e\{\cos[(\mathbf{k}_1 + \mathbf{k}_2 + \mathbf{k}_3) \cdot \mathbf{r} - 3\omega t + \phi_3]$$
$$+ \cos[(\mathbf{k}_1 - \mathbf{k}_2 + \mathbf{k}_3) \cdot \mathbf{r} - \omega t + \phi_3]$$
$$+ \cos[(\mathbf{k}_1 + \mathbf{k}_2 - \mathbf{k}_3) \cdot \mathbf{r} - \omega t - \phi_3]\}. \quad (13.64)$$

When we put $\mathbf{k}_1 = -\mathbf{k}_2$ we see that the first is of type (a) above, the second has wave-vector $2\mathbf{k}_1 + \mathbf{k}_3$ which does not have modulus $|\mathbf{k}_1|$ appropriate to frequency ω and is therefore of type (b) and does not radiate a propagating wave. We are left with the third term only, and this has wave-vector $-\mathbf{k}_3$ which has the right magnitude for frequency ω. Its phase is $-\phi_3$, and so its complex form $\exp[i(-\mathbf{k}_3 \cdot \mathbf{r} - \omega t - \phi_3)]$ shows it to be the spatial complex conjugate of E_3.

The situation can also be simulated statically by forming a hologram (§12.6) of $\mathbf{k}_3$ using $\mathbf{k}_1$ as reference beam. Then, it follows from the treatment in (12.57)–(12.61) that if the hologram is reconstructed by using the wave $-\mathbf{k}_1$, instead of $\mathbf{k}_1$ the complex conjugate reconstruction $\mathbf{k}_4$ is produced (Fig. 13.14(b)).

A phase-conjugate mirror is compared with an ordinary mirror in Fig. 13.15, both from the geometrical and Huygens points of view.

Figure 13.15 Comparison between phase-conjugate and ordinary mirrors; (a) according to geometrical optics; (b) according to Huygens's principle. In (b) the incident wavefronts are indicated by continuous circles, the reflected wavefronts by broken circles.

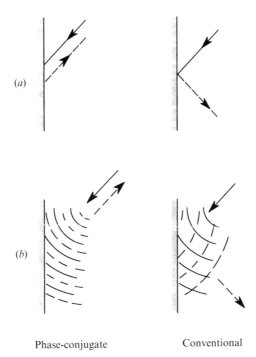

Phase-conjugate   Conventional

It reflects the wavefront back the way it came, like a collection of atomically-sized cat's-eye reflectors which also reverse the sign of the phase. The wavefront will then return to its source exactly the same way as it came, and because of the phase reversal all the waves will get back with exactly the same phase as they started with! If the source has spatial structure, each point on it can be considered independently. Suppose that the wave is distorted by some bad optics on the way to the mirror. The distortions will be incorporated in ϕ_3 at each point, but since the reflected wave has phase $-\phi_3$, the distortions will be corrected on the way back so that the image is perfect. This should be contrasted with an ordinary mirror; in the same experiment, the reflected wave has the same phase ϕ_3 as the incident one and the distortion will be doubled when it passes through the bad optics a second time. An experiment which demonstrates this recorrection of distorted wavefronts is illustrated in Fig. 13.16. In this experiment, the bad optics are a multi-mode fibre, which of course distorts the wave out of all recognition at its far end. But when the wave is returned by a phase-conjugate mirror, the image is restored.

Figure 13.16 (a) Experiment to show compensation of distorting optics by a phase-conjugate mirror. The object is a slide T, and the light transmitted by it enters one end of a multi-mode fibre. At the the far end of the fibre the light (4) is concentrated onto the phase-conjugate mirror crystal from which it is returned (3) to the fibre. The image at S, observed with the aid of the beam-splitter BS, is shown in (b). From Fischer and Sternklar (1985), courtesy of B. Fischer.

(a)

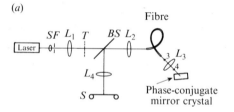

(b)

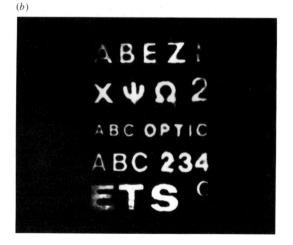

CHAPTER FOURTEEN

Quantum optics and lasers

14.1 Quantization of the electromagnetic field

14.1.1 The 'ultra-violet catastrophe'

At the end of the nineteenth century it began to be clear that classical ideas could not explain all physical phenomena (§1.5). One of the most notable problems arose from a consideration of the electromagnetic wave spectrum in a cavity. Essentially, the point was as follows. Suppose that we have a cubic reflecting cavity, with side L, made out of a highly-conducting metal. (The cubic shape is chosen for simplicity only; it is not critical.) Any electromagnetic wave which satisfies the boundary conditions $\mathbf{E}_\parallel = 0$ on the inner surface of the cavity is one of its normal modes. We can find such normal modes easily; for example, the standing wave

$$E_y = \cos z k_z \; \cos x k_x \, e^{-i\omega t} \tag{14.1}$$

has zero value on the planes $x = \pm L/2$, $z = \pm L/2$ provided that $Lk_x = l\pi$, $Lk_z = n\pi$, where l and n are odd integers. There are also sine solutions leading to even integers. Since the field is in the y-direction, its component parallel to the planes $y = \pm L/2$ is always zero. Now (14.1) is the superposition of four plane-waves; it can be written

$$
\begin{aligned}
E_y &= \tfrac{1}{2} e^{-i\omega t} [\cos(x k_x + z k_z) + \cos(x k_x - z k_z)] \\
&= \tfrac{1}{4} \{ \exp[i(x k_x + z k_z - \omega t)] + \exp[i(-x k_x - z k_z - \omega t)] \\
&\quad + \exp[i(x k_x - z k_z - \omega t)] + \exp[i(-x k_x + z k_z - \omega t)] \}
\end{aligned} \tag{14.2}
$$

which all require

$$\omega^2 = (k_x^2 + k_z^2)c^2 = \pi^2 c^2 L^{-2}(l^2 + n^2) . \tag{14.3}$$

The allowed frequencies of electromagnetic waves in this cavity, when polarizations E_x and E_z are added are

$$\omega^2 = \pi^2 c^2 L^{-2}(l^2 + m^2 + n^2) \qquad (14.4)$$

in which l, m and n are positive† integers, at least two of which must be non-zero. There are two independent polarizations which give the same values of l, m and n. *There is no upper limit to l, m and n.* The number of modes is infinite, and their density (number of modes possible in a given interval of ω) *increases* with ω. Now, according to the classical equipartition theorem of Boltzmann, every normal mode in thermal equilibrium has energy $k_B T$ ($\frac{1}{2}k_B T$ for each degree of freedom, of which an oscillator mode has two; see §14.2) and so the total energy inside the cavity must be infinite, its density increasing without limit at higher frequencies, towards the ultra-violet. This was an absurd conclusion, and was called the *ultra-violet catastrophe*. Rayleigh and Jeans, amongst others, tried hard to find a solution. Experimental data (Fig. 14.1) on the spectrum of a black body (a cavity with a small inspection hole in it) showed a radiation density increasing with frequency at the red end of the spectrum, in accordance with (14.4), which leads to a density which increases like ω^2 as we shall see in (14.6). But then the energy density peaked at a certain frequency (the 'red' of a red-hot body) and fell off rapidly at higher frequencies. Planck found the solution, empirically at first, in terms of the quantization of the radiation modes, and his discovery heralded quantum theory, which has since been applied so successfully to a description of matter through atomic scales and down, at least, to the size of the nucleus.

† Negative integers do not give us new states, but just interchange terms in (14.2).

Figure 14.1 Black-body spectrum, compared to the Rayleigh–Jeans approximation.

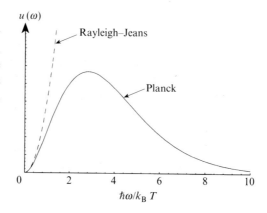

$u(\omega)$

Rayleigh–Jeans

Planck

0   2   4   6   8   10

$\hbar\omega/k_B T$

This chapter will deal with what is essentially the outcome of Planck's quantization of the modes of an electromagnetic cavity. This approach has led to the invention of the laser and to intriguing forms of non-thermal light which are at present on the frontiers of research.

14.1.2 Quantization of the electromagnetic modes in a cavity

The Rayleigh–Jeans argument, modified by Planck, continues as follows. The numbers (l, m, n) can be represented as integer points in a phase space in which l is counted in the x-direction, m in y and n in z (Fig. 14.2). From (14.4), the frequency ω corresponding to (l, m, n) is $\pi c / L$ times the distance of (l, m, n) from the origin. So the number of states with frequencies between ω and $\omega + \Delta\omega$ is the number within the positive quadrant of an onion-layer of radius $\omega L / c\pi$, thickness $\Delta\omega L / c\pi$, which has volume

$$\frac{1}{8} \cdot 4\pi \left(\frac{\omega L}{c\pi}\right)^2 \cdot \frac{\Delta\omega L}{c\pi} \tag{14.5}$$

and contains on average the same number of states, since there is one integer point per unit volume. This number, times two for the two independent polarizations, gives the *density of states* per unit interval $\Delta\omega$:

$$D(\omega) = \frac{L^3 \omega^2}{c^3 \pi^2}. \tag{14.6}$$

Planck's idea was that the electromagnetic energy was *quantized* in units of $\hbar\omega$. Each mode could then have any whole number of quanta of energy. The average number of such quanta would then be given by Boltzmann statistics, and he showed this average number to be[†]

$$\langle n \rangle = \left[\exp\left(\frac{\hbar\omega}{k_B T}\right) - 1\right]^{-1} \tag{14.7}$$

at temperature T. If the quantum $\hbar\omega$ is small compared with the classical average energy of a mode, $k_B T$, a large number of quanta is probable, and the classical result $\langle n \rangle \approx \hbar\omega / k_B T$ holds. But if the quantum is large compared with $k_B T$, there is little probability of there being even *one* quantum per mode in a cavity in thermal equilibrium. This is how we understand $\langle n \rangle$ in (14.7). For example, when we are in the region of the maximum of the black-body spectrum, where $\hbar\omega \approx k_B T$, the probable number of photons per mode is $(e-1)^{-1} \approx 0.6$. Only at frequencies much lower than $k_B T / \hbar$ is there a reasonable

---

† See any text on statistical mechanics.

probability of finding more than one photon in a mode. From (14.7), the total energy in the cavity between frequencies ω and $\omega + d\omega$ is $u(\omega)d\omega$ where

$$u(\omega) = \langle n \rangle \hbar \omega D(\omega) = \frac{\hbar L^3}{c^3 \pi^2} \cdot \frac{\omega^3}{e^{\hbar \omega / k_B T} - 1}. \tag{14.8}$$

This fits the observed black-body spectrum very well (Fig. 14.1), and can be integrated to find the total black-body radiation energy density in the cavity at temperature T (Stefan's law):

$$U(T) = \int_0^\infty u(\omega)d\omega = \frac{L^3 \pi^2 k_B^4 T^4}{15 \hbar^3 c^3}. \tag{14.9}$$

The quanta of radiation have been given the name *photons*, and can be considered in many ways like particles. This is because the distribution (14.7) is the same as that obtained for identical particles, which have integer spin and zero chemical potential, so it is tempting to consider photons as having these properties. But it is also dangerous, because photons cannot be localized in the way that massive particles can. We shall discuss some of the consequences in §14.1.3.

Another oscillator which has the same statistics is the quantum-mechanical simple harmonic oscillator, and it turns out to be very fruitful to establish the analogy between this and the electromagnetic wave, because we can then lift the solutions directly from quantum mechanics. In particular, it is usual nowadays to express the concepts of quantum electromagnetic fields in the language of second quantization, i.e. in terms of operators which create and anihilate photons and change the wave functions appropriately. We shall not develop this approach to the mathematical formulation of the theory (see, for example, Goldin, 1982; Loudon, 1983; Meystre and Sargent, 1991);

Figure 14.2
Distribution of
modes in the (l, m)
plane for a cubical
cavity.

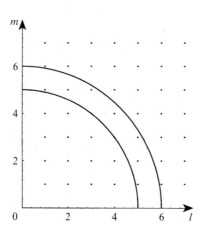

we shall only describe enough of it to see the physical basis of some of the newest ideas which lead to experimental results that can not be explained on the basis of classical electromagnetic theory. But first we shall go back to one of the oldest, and still puzzling, phenomena in photon optics.

14.1.3 Interference in the limit of very weak light

Can we observe interference in the limit of very weak intensity when, statistically, there may occasionally be a single photon within an interferometer, but very rarely more than one? Experiments done by G. I. Taylor in 1909 showed that an interference pattern could be recorded under such circumstances, given a long enough photographic exposure. Naively, we might expect that two photons must travel simultaneously through the system, one along each of the alternative paths in order to interfere when they are recombined. But from the experiment it is clear that one photon is sufficient. In fact the photon is not a localized particle, and any attempt to discover along which of the routes the photon travelled *destroys the interference pattern*. This apparently paradoxical situation has many implications in basic quantum theory, and has been discussed exhaustively, without any generally accepted understanding having emerged. [Discussions which cover several different approaches are given in the books by Rae (1986) and Peres (1993).] Because of the controversiality of the subject it is difficult to summarise it here without ending up with more questions than answers!

In general, the electromagnetic wave approach which characterizes this book gives the right average light intensity distribution in any given situation when large numbers of photons are involved. When the numbers are small, the average expectation is still correct, but the result in any particular experiment is modified by the statistics of arrival of the photons, whether the detector is a single unit or an array such as a photographic film. The statistics may be Poisson if the photons are uncorrelated, but may be modified by various techniques to be discussed below (§14.7). In many cases the statistics can be adequately described by analysis of the detector itself, for example in §14.3.1. It is as if the classical electromagnetic field *guides* the individual photons, in the same way as Schrödinger's wave field gives the probability density for matter particles, without telling us exactly what happens to each one.

To show what happens if we try to trace a photon (as if it were a localized particle) through an interferometer, we shall consider the following thought-experiment as an example on which many variations are possible. The apparatus is a Michelson interferometer (Fig. 14.3), and we shall show that an attempt to find out *which* of the two mirrors reflected a single photon traversing the instrument must result in destruction of the interference pattern. Imagine that all the components of the interferometer are infinitely massive, apart from the mirror M_2, which has finite mass. According to de Broglie's hypothesis, a photon with wavenumber k_0 has momentum $p = \hbar k_0$. When the photon is reflected, the mirror will recoil with momentum $2\hbar k_0$, which can be measured *after* the reflexion has occurred, and the measurement can therefore not affect the interference pattern. But in order to detect the recoil, we must know the initial momentum of M_2 to an accuracy δp better than $\hbar k_0$. So, *before* the reflexion, $\delta p < \hbar k_0$. The Heisenberg uncertainty principle relates the uncertainties of momentum and position in the form $\delta p \, \delta x \geq h$ (see §12.5); this means that the positional uncertainty δx of M_2 is at least $2\pi/k_0 = \lambda$. This much uncertainty in the mirror position makes the fringes unobservable!

Although this is only an example, *any* attempt to determine a photon's route through an interferometer is doomed to destroy the interference pattern that could be observed. We reach the inevitable conclusion that the photon must travel both routes at once, and interferes with itself. Some of the consequences of this line of thought are summarized by Greenberger, Horne and Zeikinger (1993). In particular, as food for thought, we suggest consideration of two experiments among the many which have been instrumental in focusing the conceptual problems. One involves interference between photons from independent lasers (Pfleegor and Mandel, 1968), and the second

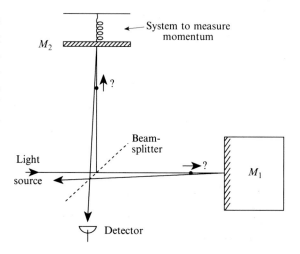

Figure 14.3 Thought-experiment to determine which mirror in a Michelson interferometer reflected the photon.

interference between photons emitted by down-converting crystals, in which a single input photon causes ejection of two photons which are coherently related to the input (Zou, Wang and Mandel, 1991).

14.2 Plane-wave modes in a linear cavity

We shall now return to the analogy between the photon and a simple harmonic oscillator (§14.1.2). It will be sufficient to consider a one-dimensional cavity of length L, in which the plane-wave mode has an electric field

$$E = E_0 \cos(kx - \omega t - \phi) \tag{14.10}$$

where the values of k and therefore ω are defined by L. The magnetic field is not independent, and is always related to E by the impedance. Note that k and E will now be written as scalars, since the choice of a particular mode (including its polarization) allows us to consider a single component of the field only. It is the *magnitude* of E_0 that will be shown to be quantized.

First we define two new quantities

$$q(t) \;=\; \frac{E_0}{\omega} \cos(\omega t + \phi) \tag{14.11}$$

$$p(t) \;=\; -E_0 \sin(\omega t + \phi), \tag{14.12}$$

noting that $dq(t)/dt = p(t)$, and write the field

$$
\begin{aligned}
E \;&=\; E_0[\cos(\omega t + \phi)\cos kx + \sin(\omega t + \phi)\sin kx] \\
&=\; \omega q(t) \cos kx - p(t) \sin kx. \tag{14.13}
\end{aligned}
$$

The total energy per unit cross-sectional area of the cavity (including both E and B fields) is then

$$
\begin{aligned}
U \;&=\; \int_0^L \epsilon_0 E^2 dx \\
&=\; \int_0^L \epsilon_0 (\omega q \cos kx - p \sin kx)^2 dx \\
&=\; \tfrac{1}{2}\epsilon_0 L(\omega^2 q^2 + p^2). \tag{14.14}
\end{aligned}
$$

Using new variables

$$
\begin{aligned}
Q(t) \;&=\; (\epsilon_0 L)^{\frac{1}{2}} q \tag{14.15} \\
P(t) \;&=\; (\epsilon_0 L)^{\frac{1}{2}} p, \tag{14.16} \\
U \;&=\; \tfrac{1}{2}(\omega^2 Q^2 + P^2). \tag{14.17}
\end{aligned}
$$

This has the same form as the energy of a mechanical simple harmonic oscillator, for which

$$U = \tfrac{1}{2}(Kx^2 + mv^2), \tag{14.18}$$

where K is the force constant and m the mass. This can also be written in the same way as (14.17), in which $\omega = (K/m)^{\frac{1}{2}}$ is the classical vibration frequency, $m^{\frac{1}{2}}x = Q$ and $m^{\frac{1}{2}}v = \mathrm{d}Q/\mathrm{d}t = P$.

14.2.1 Energy quantization and zero-point energy

The energy of a simple harmonic oscillator is $U_n = \hbar\omega(n + \tfrac{1}{2})$ where n can be any non-negative integer. *We thus deduce that the energy of a given mode of the electromagnetic field is quantized in the same way.* An important non-classical feature is the existence of zero-point energy

$$U_0 = \tfrac{1}{2}\hbar\omega \tag{14.19}$$

which is the lowest allowed energy level for that mode; it is not possible to eliminate field oscillations in any mode completely. Even the *vacuum field* (lowest energy in every mode of a cavity) contains this much energy in every mode.† The actual electric field resulting from the zero-point contributions of all the modes is their superposition. Since their phase relations are unspecified, they can be assumed for the moment to be random and give rise to an *inevitable fluctuating background field which adds noise to any physical measurement*, which we shall presently study in more detail. However, the past decade has seen development of methods to order the phases of these zero-point fluctuations, with the consequent possiblity of noise reduction. This is called *squeezed light* and will be discussed in more detail in §14.7.

14.2.2 Uncertainty relation

The uncertainty principle is one feature of quantum mechanics that can directly be applied to the electromagnetic field through the analogy with the harmonic oscillator. As we saw in §14.1.3, it can be written $\delta p\,\delta x \geq \hbar$. Now the conjugate variables we used in the harmonic oscillator above are $Q = m^{\frac{1}{2}}x$ and $P = m^{\frac{1}{2}}v = p/m^{\frac{1}{2}}$, so

$$\delta P\,\delta Q \geq \hbar \tag{14.20}$$

---

† When all modes are taken into account this adds up to an infinite total amount of energy since the cavity has an infinite number of possible modes; however, we have learnt to live with such infinite inaccessible background energies which occur frequently in quantum mechanics.

which relates, by analogy, the degree of accuracy with which we can specify the amplitudes of the $\cos kx$ and $\sin kx$ parts of the electromagnetic field.

It is illustrative to express the uncertainties on a $(P, \omega Q)$ diagram.† We plot P horizontally and ωQ vertically, as in Fig. 14.4. The energy (14.17) is then proportional to the square of the radius vector from the origin to $(P, \omega Q)$. Moreover, from (14.11) and (14.15), $(\epsilon_0 L)^{-\frac{1}{2}} \omega Q$ is the instantaneous amplitude of $\cos kx$ term, and likewise $(\epsilon_0 L)^{-\frac{1}{2}} P$ is the amplitude of $\sin kx$. Thus the phase $(\omega t + \phi)$ of the field (14.13) is given by the angle θ, and the amplitude by the radius vector. However, we know that the point $(P, \omega Q)$ can not be defined exactly because of the uncertainty principle. All we know is the average position of the point, and the product of the uncertainty $\delta P \, \delta Q$. From here on we shall ignore the ωt in the phase, so that the $(P, \omega Q)$ diagrams will be drawn as if in in a frame of reference rotating at angular velocity $-\omega$. Then the angle θ represents ϕ directly.

Since P and ωQ appear symmetrically in (14.17), we expect the values of δP and $\omega \delta Q$ to be equal, so that the defined region in Fig. 14.4 is a circle. This is the situation that would normally be found, and to which any other situation will naturally revert; light in a single mode with this property is called *chaotic light*, to be discussed further in §14.2.3. But all that quantum theory limits is the *area* of the region of uncertainty, and any experiment we propose which distorts its shape while retaining the area is allowed from the theoretical point of view. Let us look at some examples.

† In quantum mechanics, this is called a *Wigner diagram*.

Figure 14.4 $(P, \omega Q)$ diagram for light with minimum uncertainty (chaotic light).

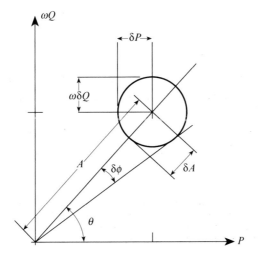

When we define an uncertainty region in the $(P, \omega Q)$ plane, the construction of a wave is quite elementary. We choose, randomly, a number of points within the uncertainty region and draw, one on top of the other, the waves that they represent. Each wave has amplitude and phase (A, ϕ), which are the polar coordinates of $(P, \omega Q)$ as in Fig. 14.4, in which $A^2 = P^2 + \omega^2 Q^2 = 2U$. The width covered by the resulting lines represents the uncertainty in the wave field. Fig. 14.5 shows what we get for the equilibrium form $\delta P = \omega \delta Q = \sqrt{\hbar \omega}$.

Various techniques have been designed to manipulate the shape of the uncertainty region (§14.7). For example, it has been shown to be possible to control the amplitude of a wave emitted by a diode laser by very careful stabilization of the excitation current (§14.7.1). Then δA is very small, and the uncertainty region is distorted as in Fig. 14.6. This gives rise to a wave whose phase is very unstable, which means that it has a large frequency spread. In fact, one can see an alternative form of the uncertainty principle here; the uncertainty area in Fig. 14.6 is:

$$\omega \delta P \, \delta Q = A \, \delta A \, \delta \phi = \tfrac{1}{2} \delta(A^2) \, \delta \phi \geq \hbar \omega \ . \qquad (14.21)$$

But $\tfrac{1}{2}\delta(A^2)$ is the uncertainty $\delta U = \delta n \, \hbar \omega$ in the energy per unit area (intensity $\times$ time), where n is the number of photons observed in an experiment. Thus

$$\delta n \, \delta \phi \geq 1 \ . \qquad (14.22)$$

A second example is just the opposite to the above. We stabilize the phase of the light by making $\delta \phi$ very small, in which case the amplitude fluctuates wildly (Fig. 14.7). Finally there is a form of light which has been achieved experimentally in which one of δP or $\omega \delta Q$ is

Figure 14.5 The wave represented by the previous figure. The uncertainty is indicated by the range of the superimposed waveforms.

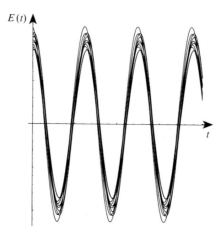

reduced at the expense of the other (Fig. 14.8). This has applications to accurate interferometry and will be discussed further in §14.7.3.

14.2.3 Fluctuations in chaotic light

Chaotic light has the equilibrium form $\delta P = \omega \delta Q = \sqrt{\hbar \omega}$. Then from (14.17), putting $\phi = 0$ to simplify things,

$$
\begin{aligned}
\delta U &= \omega^2 Q \delta Q + P \delta P \\
&= (\epsilon_0 L)^{\frac{1}{2}} \left[\omega^2 \frac{E_0}{\omega} \cos \omega t (\hbar/\omega)^{\frac{1}{2}} + E_0 \sin \omega t (\hbar \omega)^{\frac{1}{2}} \right] \\
&= (\epsilon_0 L \hbar \omega)^{\frac{1}{2}} (E_0 \cos \omega t + E_0 \sin \omega t).
\end{aligned}
\tag{14.23}
$$

Figure 14.6
Amplitude stabilized light, with consequent phase destabilization; (a) the $(P, \omega Q)$ diagram, (b) waveform.

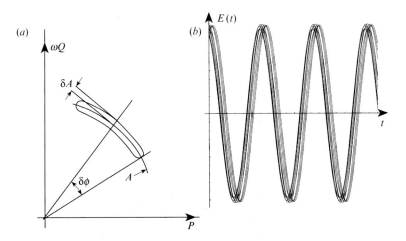

Figure 14.7 Phase stabilized light, with consequent amplitude destabilization; (a) the $(P, \omega Q)$ diagram, (b) waveform.

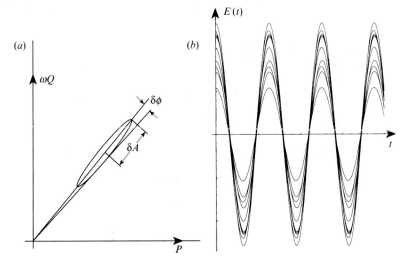

This shows that the contributions to δU from the $\cos \omega t$ and $\sin \omega t$ phases are equal. The root-mean-square fluctuation ΔU in each phase is

$$\Delta U = \langle \delta U^2 \rangle^{\frac{1}{2}} = (\epsilon_0 L \hbar \omega)^{\frac{1}{2}} E_0 \langle \cos^2 \omega t \rangle^{\frac{1}{2}} = E_0 \left(\frac{\epsilon_0 L \hbar \omega}{2} \right)^{\frac{1}{2}}. \quad (14.24)$$

It is most illustrative to compare this fluctuation with the mean intensity, when both are measured as numbers n of photons per unit area in a given time. Then we have

$$\delta n = \frac{\Delta U}{\hbar \omega} = E_0 \left(\frac{\epsilon_0 L}{2 \hbar \omega} \right)^{\frac{1}{2}}; \quad (14.25)$$

$$\langle n \rangle = \frac{U}{\hbar \omega} = \frac{1}{2} \langle \omega^2 Q^2 + P^2 \rangle \frac{1}{\hbar \omega}$$

$$= \frac{1}{2} \langle \omega^2 \frac{E_0^2}{\omega^2} \cos^2 \omega t + E_0^2 \sin^2 \omega t \rangle \frac{\epsilon_0 L}{\hbar \omega}$$

$$= E_0^2 \frac{\epsilon_0 L}{2 \hbar \omega}. \quad (14.26)$$

So for each phase $\quad (\delta n)^2 = \langle n \rangle. \quad (14.27)$

In §11.8 we used the same result for the detection probability of photo-electrons when observing a source of constant intensity, as a consequence of the Poisson statistics of uncorrelated events; the present result therefore indicates that *Poisson statistics apply to chaotic light.* In that discussion, we continued by adding the resultant classical intensity fluctuations arising from the partially coherent nature of a thermal source, and showed that the result was photon bunching, which can be

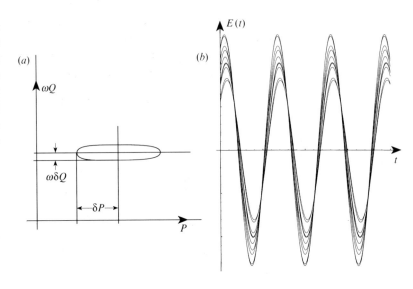

Figure 14.8 Light in which the fluctuations have been squeezed into the $\sin \omega t$ term; (a) the $(P, \omega Q)$ diagram, (b) waveform.

described loosely as *super-Poisson* in that there is a positive correlation between the times of detection of photons. The fluctuations result eventually in a limitation of the accuracy with which measurements can be made optically. On the other hand, any means of distributing them unequally between the $\sin \omega t$ and $\cos \omega t$ terms, and using the quieter one for measurement, acquire practical implications. What is needed is called *sub-Poisson light* (§14.7).

The 'darkest' state, which has the minimum number of photons, has $\langle n \rangle = 0$ in each mode, and is called the *vacuum field*. It still has energy $\frac{1}{2}\hbar\omega$, which is evident as a fluctuating wave field. Then the picture looks like Fig. 14.9, in which phase is completely indeterminate. The vacuum field is important in understanding spontaneous emission.

14.3 Interaction of light with matter

A detailed discussion of the interaction of light with matter is quite outside the scope of this book, but we must understand some of the essentials in order to appreciate the principle of the laser, which is a necessity for every student of optics. For a much deeper discussion of this subject see, for example, Loudon (1983).

We shall restrict our discussion to a pictorial description of the effect of an oscillating electromagnetic field on a single isolated one-electron atom with just two levels L_1 and L_2.† This is about the simplest relevant problem that we can imagine. The atom in state $j (= 1, 2)$ is described by an electronic wave function

$$\psi_j(\mathbf{r}, t) = f_j(\mathbf{r})e^{-i\omega_j t} \tag{14.28}$$

in which the spatial wave function $f(\mathbf{r})$ (assumed to be real) is separated from the temporal oscillations. The eigenvalues of the two wave functions are $\hbar\omega_1$ and $\hbar\omega_2$ ($\omega_2 > \omega_1$), and ψ is in each case a solution of Schrödinger's equation for the atomic potential $V(\mathbf{r})$. Each wave

† It is quite easy to see how to extend this to many levels.

Figure 14.9
Representation of the
vacuum field.

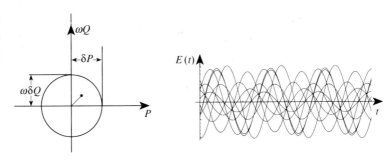

function corresponds to an *exact* solution of the Schrödinger equation and therefore an electron in either of the states will *stay there for ever*. All the time-dependence is in the exp($-i\omega t$). Any other possible electron distribution can always be written as a superposition of the wave functions $\psi_j(\mathbf{r}, t)$ since these form a complete set (like the sine and cosine functions in Fourier theory).

Now suppose an oscillating electric field is applied to the atom. The potential field is modified from $V(\mathbf{r})$ to $V(\mathbf{r}) + e\Phi(\mathbf{r}, t)$ where $\Phi(\mathbf{r}, t)$ is the electric potential of the oscillating field. The stationary-state wave functions $\psi(\mathbf{r}, t)$, corresponding to the new potential, are no longer the same solutions $\psi_j(\mathbf{r}, t)$ of Schrödinger's equation. But we can express $\psi(\mathbf{r}, t)$ as a linear superposition of the $\psi_j(\mathbf{r}, t)$. What does the resulting electron probability distribution look like? We write the superposition as

$$\psi(\mathbf{r}, t) = a\psi_1(\mathbf{r}, t) + b\psi_2(\mathbf{r}, t). \tag{14.29}$$

where $a^2 + b^2 = 1$. Remember that $\psi_j(\mathbf{r}, t)$ contains the factor $e^{-i\omega_j t}$. When we calculate the electron density $|\psi(\mathbf{r}, t)|^2$, we find a cross term

Figure 14.10 (*a*) and (*b*) Schematic representation of the functions $f_1(\mathbf{r})$ and $f_2(\mathbf{r})$ for electron density in an atom; (*c*) $f_1(\mathbf{r}) + f_2(\mathbf{r})$; (*d*) $f_1(\mathbf{r}) - f_2(\mathbf{r})$. In the latter two, (*e*) and (*f*), the region of maximum charge density is at an off-centre point Q, showing that the atom has a dipole moment.

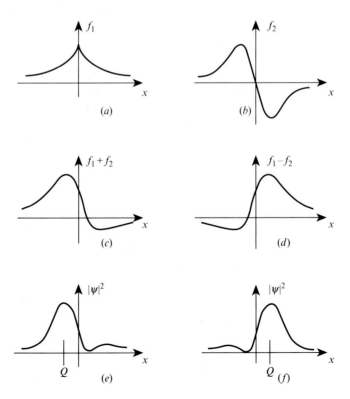

(underlined) which oscillates with frequency $(\omega_2 - \omega_1)$:

$$|\psi(\mathbf{r}, t)|^2 = |af_1(\mathbf{r})e^{-i\omega_1 t} + bf_2(\mathbf{r})e^{-i\omega_2 t}|^2 \qquad (14.30)$$
$$= a^2 f_1^2(\mathbf{r}) + \underline{2ab f_1(\mathbf{r})f_2(\mathbf{r})\cos[(\omega_2 - \omega_1)t]} + b^2 f_2^2(\mathbf{r}).$$

Pictorially the situation looks like Fig. 14.10. At time $t = 0$ or $2m\pi/(\omega_2 - \omega_1)$, where m is an integer, $|\psi|^2 = (af_1 + bf_2)^2$ as shown in Fig. 14.10(e) for $a = b$. On the left of this diagram, where f_1 and f_2 have the same sign, the charge density $|\psi|^2$ is larger than on the right, where f_1 and f_2 have opposite signs. At time $t = \pi(2m+1)/(\omega_2 - \omega_1)$ we have $|\psi|^2 = (af_1 - bf_2)^2$, Fig. 14.10($f$), and the charge density is larger on the right. In other words, the charge alternates between the two halves: we have an oscillating dipole. We know, from §5.3.2, that an oscillating dipole is a good radiator or absorber, so the atom absorbs or radiates at frequency $\omega = (\omega_2 - \omega_1)$; in general the atom couples to a radiation field tuned to the frequency difference between energy levels. The strength of the oscillating dipole represented by (e) and (f) corresponds to the underlined term in (14.30) which leads to a dipole moment of amplitude

$$2ab \int\!\!\int\!\!\int_{\text{all space}} e\mathbf{r} f_1(\mathbf{r})f_2(\mathbf{r})\mathrm{d}^3r \equiv 2ab\, eM_{12}. \qquad (14.31)$$

M_{12} is called the interaction *matrix element*, or the *oscillator strength* which was introduced empirically in §13.3.2. Because of the antisymmetric factor $\mathbf{r}$ in the integrand, the functions shown in (a) and (b) must have opposite symmetry for M_{12} to be large; this corresponds to a selection rule $\Delta l = \pm 1$ in quantum mechanics (where l is defined in the same way as in §10.3).

The above description might suggest that emission and absorption only occurs when $\omega = (\omega_2 - \omega_1)$ exactly. But this is not quite true. The result of energy transfer from the field is that a and b change with time, and so the interaction continues only for a time T during which both a and b are non-zero. So ω only needs to lie within $\omega_2 - \omega_1 \pm \pi/T$; $2\pi/T$ is the natural linewidth (§11.3.1). The larger the values of the matrix element and Φ, the faster a and b change and the wider the frequency range.

Although we have not carried out any detailed mathematics and have used an over-simplified model, the physics should be clear. Now we can make several very important observations.

(1) The situation is quite symmetric between the two levels. If the atom starts in level 1, initially $a = 1$ and $b = 0$, and the electromagnetic field causes a transition from the lower level to the upper. If the

atom were initially in the upper level, with $a = 0$ and $b = 1$, the same electromagnetic field would cause a transition to the lower level.

(2) Because the interaction is essentially that between an oscillating dipole and an electromagnetic field at the same frequency, the direction of energy transfer is determined by the phase relation between the two. In the first case in (1), energy $\hbar(\omega_2 - \omega_1)$ is absorbed from the field by the atom. In the second case, the dipole radiates the same energy coherently into the field.

(3) We have seen in §14.2 that the electromagnetic field is never zero. There are always the vacuum fluctuations at least. So an atom can not stay in the upper level for ever. One might be tempted to say the same about the lower level, but the atom has to absorb energy to ascend to the upper level. The field can not provide this energy because it is already in its lowest energy state, so there is no available source for an upward transition.

14.3.1 The photoelectric effect

Quantization of energy in a light wave was demonstrated by Einstein in his interpretation of the photoelectric effect, which applies to almost any sensitive photodetector. The argument in terms of quantized photons should be familiar to the reader, but here we shall describe it in terms of the above interaction picture, in which the electromagnetic field is classical.

Detection of light requires an interaction between the incident wave field and the electrons in a sensitive element, the photocathode, in which the electrons have many states. The lowest-lying ones refer to electrons bound within the cathode, but above a certain energy ε_w, called the *work function*, the states refer to free electrons having some amount of kinetic energy. The light wave with frequency ω causes a mixing, like (14.29), between the ground state ω_1 and a certain upper state ω_2 which satisfies $\omega = \omega_2 - \omega_1$. If $\hbar\omega_2 < \varepsilon_w$, the final state is bound and no free electrons are observed. When $\hbar\omega_2 > \varepsilon_w$, the final state is a free electron with kinetic energy $\hbar\omega_2 - \varepsilon_w$. The rate of transition to unbound states, and thus the rate of creation of free electrons, is proportional to the size of the perturbation, the intensity of the light wave. This, in a nutshell, is a description of the photoelectric effect; notice that the quantization has been introduced through the electron states in the photocathode, and not through the wave field. In fact, the photoelectric effect does not really prove that light energy is quantized!

14.3.2 Spontaneous and stimulated emission

The description in §14.3 leads us directly to the most important concepts involved in the laser. We have seen that in the presence of an electromagnetic wave, no atomic–electron wave function is completely stationary, except for the ground state in the presence of the vacuum field only. Otherwise, transitions occur in which energy is transferred backwards and forwards between the atom and the electromagnetic field. We emphasize that the atom behaves like an oscillating dipole antenna during a transition and the phase relation between this dipole and the electromagnetic field determines whether the atom absorbs or emits.

Spontaneous emission occurs when an atom is in the upper state L_2 and is influenced by the vacuum field. As we saw in §14.2.3, this generally has random phase and therefore the emitted waves have random phase. However, in principle the random vacuum field fluctuations can be ordered, and this possibility will be discussed further in §14.7. The dependence of spontaneous emission on the presence of a vacuum field has been beautifully demonstrated by experiments on radiation from atoms in microcavities. If the cavity dimensions are reduced until its first mode has frequency above that of the transition from level L_2 to the ground state, there are no vacuum fluctuations at the right frequency to stimulate that transition, and the lifetime of L_2 becomes infinite. The experiments are described in more detail by Jhe *et al.* (1987), Haroche and Kleppner (1989) and Haroche and Raimond (1993).

Stimulated emission occurs when an atom is in the same state L_2, but is influenced by an electromagnetic field larger than the vacuum field. The atom, perturbed at frequency ω, transits to state L_1 and the phase of the emitted wave is that of the oscillating dipole, which itself is that of the perturbing wave. Thus a second wave, coherent with the first, is emitted.

Stimulated absorption occurs when the atom is initially in state L_1. Then, the same description as the previous paragraph applies, but the phase is reversed and the atom absorbs the radiation.

The relationship between the stimulated and spontaneous emission rates can be compared by a simple argument due to Einstein. He considered the equilibrium of a large ensemble of atoms in the presence of equilibrium isotropic black-body radiation $u(\omega)$ at temperature T (14.8). From Boltzmann statistics we know the equilibrium ratio between the numbers of atoms n_1 in L_1 and n_2 in L_2 to be†

$$\frac{n_2}{n_1} = \exp\left(-\frac{\hbar(\omega_2 - \omega_1)}{k_{\mathrm{B}}T}\right). \tag{14.32}$$

The spontaneous transitions from L_2 to L_1 are dependent on the vacuum field, which is not included in $u(\omega)$. The stimulated transition rate is proportional to $u(\omega)$. Thus the rate of transition from L_2 to L_1 is

$$r_{21} = An_2 + Bu(\omega)n_2. \tag{14.33}$$

where A and B are constants. For transitions from L_1 to L_2, the spontaneous contribution is absent

$$r_{12} = Bu(\omega)n_1. \tag{14.34}$$

Putting (14.33) and (14.34) equal at equilibrium, and substituting (14.8) for $u(\omega)$, we find

$$\frac{A}{B} = \frac{\hbar\omega^3 L^3}{c^3 \pi^2}. \tag{14.35}$$

For the spontaneous component to be negligible, we require the energy density $u(\omega)$ to satisfy

$$Bu(\omega) \gg A \tag{14.36}$$

which, on substituting (14.35) for B/A, gives

$$u(\omega) \gg \hbar\omega^3 L^3/\pi^2 c^3. \tag{14.37}$$

On referring to (14.8), this implies that in the mode of frequency ω, the mean number of photons $\langle n \rangle \gg 1$.

It is of interest to see the order of magnitude of the threshold energy density (14.37), which is related to the intensity by $I = cu(\omega)$. At a microwave frequency, $\omega = 10^{11}\,\mathrm{s}^{-1}$ ($\lambda = 2$ cm) the threshold is $3 \times 10^{-20}\,\mathrm{J \cdot m^{-3}}$, corresponding to about $10^{-11}\,\mathrm{W \cdot m^{-2}}$, an extremely small intensity. At microwave frequencies, it therefore appears that spontaneous emission is quite negligible. At an optical frequency, $\omega = 3 \times 10^{15}\,\mathrm{s}^{-1}$ ($\lambda = 0.5\,\mu\mathrm{m}$), the threshold is $7 \times 10^{-7}\,\mathrm{J \cdot m^{-3}}$, corresponding to $20\,\mathrm{W \cdot m^{-2}}$. This is very intense, and led to considerable problems in constructing the first optical lasers (§14.4.3). Those facing the designer of an X-ray laser are even more formidable.

---

† All levels are assumed to be non-degenerate throughout this discussion. Introduction of a degeneracy factor does not introduce new physics at this level.

14.4 Lasers

The acronym 'LASER' means 'Light Amplifier by the Stimulated Emission of Radiation'. Today it understood to refer to a light source from which the stimulated emission is dominant, although the initial stimulus that triggers the emission is usually spontaneous.

The important difference between stimulated and spontaneously emitted waves is in their phase coherence. Each stimulated photon is exactly in phase with the photon which provided the stimulation, and so the wave grows as a continuous wave with complete temporal coherence; if we know the phase at one time we can, in principle, predict the phase of the wave at any later time because all its components are exactly in phase. This idyll is spoilt by the spontaneous emission, which is caused by the randomly-phased vacuum fluctuations. These provide a noisy background which results in a degradation of the complete phase coherence. It will be convenient in what follows to ignore spontaneous emission; to create the large energy density $u(\omega)$ that this demands, one must usually put the lasing material in a cavity which is resonant at the frequency ω (§§9.5.2 and 9.5.3).

Returning to equations (14.33) and (14.34), without the spontaneous term, we have the rate of stimulated emission of light from (14.33)

$$I_e = \hbar\omega r_{21} = Bu(\omega)n_2 \tag{14.38}$$

and that for absorption of the same frequency (14.34),

$$I_a = \hbar\omega r_{12} = Bu(\omega)n_1 . \tag{14.39}$$

For (14.38) to be larger than (14.39) it is necessary for n_2 to be larger than n_1, which from (14.32) is clearly impossible in an assembly of atoms in equilibrium at any (positive) temperature. The laser therefore requires that the atoms be excited to a non-equilibrium distribution, in which there are more atoms in the upper level L_2 than in the lower one L_1. This is called *population inversion*. As long as this situation is maintained, stimulated emission dominates over absorption.

14.4.1 Population inversion in a chemical laser

Conceptually, the simplest process to achieve population inversion is probably the chemical laser. A reaction takes place which generates large amounts of energy and the resultant molecules are formed in an excited state (signified by a star after the molecular formula). Then at the time of formation, there are no molecules in the ground state, only the new ones in the excited state, and so population inversion is achieved. The reaction takes place within a cavity which resonates at the frequency of the transition from the excited state to the ground state. For example, fluorine and hydrogen react in the required manner:

$$H_2 + F_2 \rightarrow 2HF^{\star}. \tag{14.40}$$

Stimulated emission occurs when a photon of frequency ω in the cavity excites the transition from $HF^{\star}$ to HF, with the emission of a second photon of the same frequency and phase as the first one:

$$\hbar\omega + HF^{\star} \rightarrow 2\hbar\omega + HF. \tag{14.41}$$

Laser action continues as long as H_2 and F_2 are burnt to provide the excited molecules and the ground-state HF is swept out of the cavity. However, this type of laser is not convenient or safe for everyday use!

14.4.2 Atomic fluoresence

Suppose that we flash a short burst of light onto an atom with a number of levels, and the light is absorbed. This is called *optical pumping* and induces transitions from the ground state to an excited state. If the atom subsequently re-radiates radiation at a longer wavelength, it is clear that there must exist a radiative route back to the ground state via at least one intermediate level. Such a *fluorescent system* provides us with a means of achieving population inversion. Suppose that just one intermediate level is involved. We shall call the ground state L_0, the uppermost state L_2 (to which we excite the atoms by the flash), and the intermediate one L_1, as in Fig. 14.11(a). Denote the lifetime of the atom in level L_i by T_i; this is the average time for which it stays excited before spontaneously emitting (T_0 is of course infinite). If the system is fluorescent as described above, a fraction of the atoms in L_2 decay to L_0 via L_1.

First suppose that $T_2 < T_1$. The short lifetime T_2 indicates that the matrix element M_{02} is large and means that the pump radiation can be absorbed efficiently by the atom. Then L_0 can be substantially emptied by atoms being excited to L_2, from where they rapidly decay to L_1. In L_1 they remain for the longer T_1, and population inversion between L_1 and L_0 arises, provided that the occupation of L_0 is at least half depleted by the pumping, as in Fig. 14.11(b). The ruby and erbium-doped fibre lasers (§14.4.3) operate essentially with this scheme.

Of course real life is rarely quite so simple. Usually, more levels are involved, but some cases are close to the ideal. For example, using a fourth level L_3 as in Fig. 14.11(c) makes it much easier to maintain the inverted population since the ground state does not have to be substantially depopulated; the neodymium-YAG laser is an important example. In addition, the lifetime of an atom in the upper state of the lasing pair is shortened once stimulated emission begins, and the balance may be upset. This can lead to pulsed behaviour.

Another possible situation is $T_2 > T_1$, as in Fig. 14.11(d). In this case, level L_1 empties faster than L_2 can fill it, so that there are always more atoms in L_2 than L_1 and population inversion occurs between them. Laser action between L_2 and L_1 is then possible while 'pumping' atoms from L_0 to L_2. Because T_2 is long in this scheme, M_{02} is small and optical pumping is inefficient; the argon ion, CO_2 and He–Ne lasers (§14.4.4) use this idea, but are pumped by electrical discharge.

14.4.3 Optically-pumped ruby and erbium lasers

Ruby is an Al_2O_3 crystal with a small amount of Cr^{3+} impurity, which gives it its red colour. The ruby laser constructed by Maiman in 1960 was the first laser working at an optical frequency and employed the energy levels of dilute Cr^{3+} shown schematically in Fig. 14.12(a). This is similar to the three-level laser in Fig. 14.11(b). Because the three-level scheme requires depopulation of the ground state by at least one-half, this laser is relatively inefficient and needs a very intense pump, provided by a xenon flash tube whose output is focused onto the ruby crystal.

The erbium-doped silica laser and amplifier which are now widely used in optical communication systems at 1.5 µm are also examples of optically-pumped three-level lasers, Fig. 14.11(b). They are constructed

Figure 14.11 (a) Energy levels in a fluorescing atom; (b) and (d) show three-level lasers, and (c) a four-level laser based on the same atom, optical pumping being indicated by the upward arrow on the left. The lengths of the lines represent the relative populations of the levels; their thicknesses indicate their decay rates, $1/T_i$. Fast transitions are indicated by broad arrows. The thermal equilibrium populations are indicated in (a).

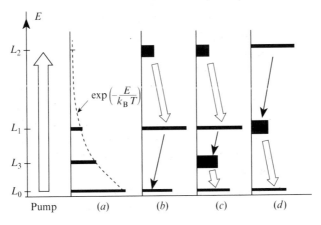

from silica glass fibres containing about 35 p.p.m. of Er^{3+} ions and are pumped by light from diode laser sources (§14.4.5) at either 1.48 µm or 0.98 µm. The level scheme of Er^{3+} is shown in Fig. 14.12(*b*), which includes the two pumping possibilities. Because the wavelengths involved are relatively long and the fibre construction concentrates both the pumping light and the emitted radiation in the core region, a high degree of population inversion can easily be achieved and the emission is stimulated very efficiently. As a result, in the absence of a resonator (§14.5.1) the system behaves as an optical amplifier, while the addition of a resonator makes it into a laser.

14.4.4 Discharge pumped gas lasers

These lasers employ mixed gases to create the population inversion. In the He–Ne laser, He is electrically excited by a discharge to an excited state, $He^\star$. During collision between He and Ne atoms, the excitation energy can be transferred to the Ne, some of whose energy levels are shown in Fig. 14.13. Thus population inversion is achieved between L_2 and L_1. The figure shows the levels involved in only one of the many possible transitions of the Ne, that at 6328 Å.

The CO_2 laser has a generally similar scheme, with N_2 as the excitation gas instead of He. Several wavelengths between 9.6 and 10.6 µm can be radiated, depending on the resonator tuning.

14.4.5 Population inversion in semiconductor *p–n* junctions

A lasing system of considerable importance, particularly in optical communication, is the reverse-biased *p–n* junction. This is different

Figure 14.12 Ruby (*a*) and erbium (*b*) laser level schemes.

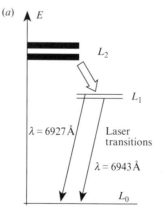

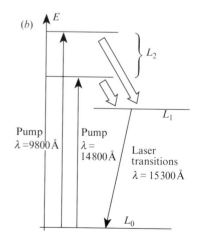

from the previous examples in that the energy levels involved are not those of individual atoms, but those of free carriers in a doped semiconductor crystal: electrons at the bottom of the conduction band and holes at the top of the valence band.† Their exact energy values are a function of the position in the junction because of its structure (*p*-type on one side and *n*-type on the other). As a result of heavy doping, there are free electrons in the conduction band of the *n*-type side, and free holes in the valence band of the *p*-type side. In thermal equilibrium, the energies of the bands are as in Fig. 14.14(*a*). There is no point in space where there are *both* free electrons and holes in more than negligible densities. When a reverse bias ΔV is applied, the bands are moved energy-wise as shown in Fig. 14.14(*b*). Now it is energetically favourable for the electrons to drift towards the positive side and the holes towards the negative side, and in doing so, both move into the junction region. When they reach the same place, we have a population inversion, in that there are substantial densities of free electrons and holes in the same place, which is a higher-energy situation than the recombined one in which the electron has filled the hole and both are annihilated.‡ Provided the recombination process results in the emission of a photon alone, which occurs in a class called *direct gap semiconductors* including many III–V materials such as GaAs, InP and InSb, but not Si or Ge, we have the makings of a laser. The corresponding fluorescent device is

† The physics of the semiconductor junction is not material for this book; see, for example, Kittel (1986), Myers, (1990).
‡ The lifetime T_2 of the uncombined state can be very short, of order $10^{-9} - 10^{-10}$ s.

Figure 14.13 Level scheme in a helium–neon laser.

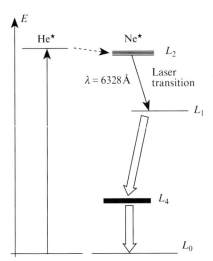

the LED (light-emitting diode) which works on the same principle but radiates spontaneous and not stimulated emission. The wavelength of the radiation emitted corresponds closely to the band gap of the semiconductor; for GaAs this is about 8700 Å; for a member of the quaternary system InGaAsP it can be designed to be about 1.5 µm which is most attractive for optical communication, because this is about the wavelength of minimum attenuation that has been attained in glass fibres (§10.3.3).

14.5 Laser hardware

Lasers are discussed in detail in many books, such as Svelto (1989), Yariv (1991), Wilson and Hawkes (1989), and in the limited space available to us it is impossible to do justice to the many facets of laser technology which have developed since the 1960s. All we shall do in this section is to point out how some of the physical ideas which have been discussed in this and other chapters of the book have been used in the design of lasers of various types.

14.5.1 The optical resonator

The gain provided by one of the mechanisms of population inversion exemplified in the previous section must now be harnessed to provide a source of coherent radiation. This can be done by incorporating it in a positive-feedback amplifier system; a familiar acoustic example is a public-address system which starts to whistle when the microphone 'hears' the loudspeaker's output. Starting from random noise, this creates a coherent sound wave, whose frequency is near that of the amplifier's maximum gain, but is determined exactly by the acoustic delay (distance/sound velocity) between loudspeaker and microphone. The delay must be such that the phase of the input to the amplifier is $2n\pi$ different from the output, so that the two reinforce exactly. When the amplifer gain is sufficiently large to overcome

Figure 14.14 Semiconductor diode laser (a) with no voltage difference, (b) with applied voltage ΔV.

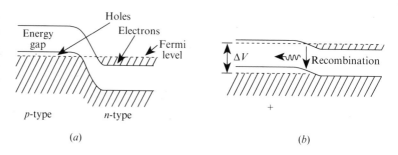

the losses in a single round trip of the wave, a sustained oscillation occurs.

It is easy to see the analogy with the laser (Fig. 14.15). The laser medium is the amplifer, in which an incident photon creates new photons with the same phase by stimulated emission. Its bandwidth is determined by the linewidth of the emission, which involves the lifetimes of the levels and processes such as Doppler broadening (§11.3.2). Feedback is provided by an optical resonator, usually of the type described in §3.10. The actual selection of the frequencies that can be radiated is determined by the optical length of the resonator, and there are sometimes several such frequencies within the linewidth of the transition. These are called the *modes* of the laser.

Quantitatively, the laser gain is determined by the pump power and atomic parameters. It has to overcome the losses occurring in the resonator due to imperfect reflexion as well as providing the useful output of the laser (which, from the point of view of the laser itself, is also a loss). Spontaneous emission is also undesirable, as it uses the inverted population to create waves with the wrong phases, although it was necessary as the original stimulus which started the oscillations.

The longitudinal lasing modes correspond to the condition that the optical length $\bar{L}$ of a complete trip back and forth through the resonator in an integral number n of wavelengths: $\bar{L} = 2\bar{\mu}l = n\lambda$ where $\bar{\mu}$ is an average refractive index (which may change with intensity). The frequencies of the modes are separated by $2\pi c/\bar{L}$. The length l also depends on the direction the ray takes along the resonator, and sometimes several transverse modes are possible with the same n, but with the rays at different angles to the optical axis. This aspect is best treated as a diffraction problem (§9.5.3).

A frequency analysis of the output from a typical laser is shown in Fig. 14.16, when several longitudinal modes are excited.

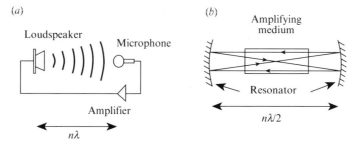

Figure 14.15 Positive feedback in (a) a public-address system and (b) a laser.

14.5.2 Continuous-wave versus pulsed lasers

The lifetime T_j of a level depends on the way it can decay. Many lasers can be pumped sufficiently strongly that population inversion is maintained in the presence of stimulated emission and *continuous-wave* emission occurs. On the other hand, an excited level will generally have a shorter lifetime when its emission is stimulated than it had naturally. Then, when laser action starts, it is possible that the condition for a population inversion, ($T_2 > T_3$ in Fig. 14.11(c), for example) is destroyed, and so lasing stops. As a result, we have a *pulsed laser*. Some lasers can be operated in either way.

There are various ways of controlling and ordering pulses, by changing factors coming into the gain. An important one, which allows regular giant pulses to be created, is called *mode-locking*. When a laser operates in several longitudinal modes, as we saw in §14.5.1, the waveform obtained is the superposition of the waves corresponding to the individual modes. If these have random phases, the result is similar to the waves we constructed in §11.2.1, except that because the modes are equally spaced in frequency by $2\pi c/\bar{L}$, the waveform repeats itself at intervals of $\bar{L}/c$. But if the modes have the same phase, their combined waveform is a series of well-defined wave-groups (Problem 14.2); if more modes are involved, the shorter and more intense are the individual groups. This situation can be forced on the laser by including within the resonator a variable attenuator, which is transparent once in every cycle time of $\bar{L}/c$.

14.5.3 Structure of the He–Ne laser

The helium–neon laser is the commonest laser to be found in elementary laboratories, and its structure will be familiar to many students. In §14.4.4 we described the type of level scheme it uses. It is constructed from a sealed discharge tube containing a mixture of He and

Figure 14.16
Frequency spectrum
of longitudinal
modes in a laser.

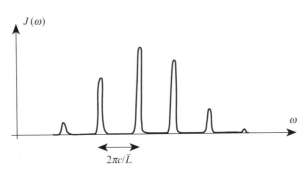

Ne with about 10:1 ratio in pressure that is situated within a confocal resonator (§9.5.3), one of whose mirrors transmits a few percent of the radiation to produce the output beam. The laser transition of choice is encouraged by using multilayer dielectric mirrors, with peak reflectivity at the required wavelength (§10.4.4). The windows used to seal the discharge tube must have the smallest possible reflexion losses and may be anti-reflexion coated (§10.4.3), uncoated but mounted at the Brewster angle (§5.4.3) as in Fig. 14.17, or may be the confocal resonator reflectors themselves. Since the laser amplification in this system is weak, it is important to reduce the losses to a minimum by these means; if Brewster-angle windows are used, one polarization will have less reflexion losses than the other, so the output beam is polarized.

14.5.4 Structure of a semiconductor laser

The design of semiconductor lasers is still developing rapidly because of their applications in communications, and we shall only discuss briefly some of the most basic aspects. A typical laser diode, shown in Fig. 14.18(a), consists of a 0.5 mm cube of GaAs, containing a p–n junction. The junction region, where the population inversion occurs and light is emitted, is $d = 1$–$3\,\mu m$ thick. Because this dimension is so small, $\lambda/d \approx 0.3$ and the emitted light is quite divergent; much of it is therefore reabsorbed in the inactive n and p regions on each side. The resonator is formed by cleaving the crystal normal to the junction, providing two plane and accurately parallel surfaces, a resonator which is marginally stable (§3.10). The refractive index of GaAs is high (about 3.6) and so these surfaces are quite highly reflecting without any coating.

Because of the absorption of the fairly divergent beam of radiation in the inactive regions, such a laser is relatively inefficient, and can usually be operated only in a pulsed mode. Considerable improvement in efficiency can be obtained by employing the principles of the planar waveguide (§10.2.2). By using different materials, particularly

Figure 14.17 Diagram of a He–Ne laser: E, electrodes to excite discharge in the gas; B, Brewster-angle windows; M confocal resonator mirrors.

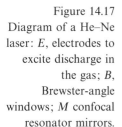
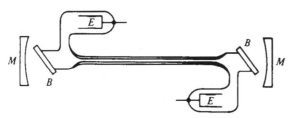

$Ga_xAl_{1-x}As$, it can be arranged that the emitting region of thickness
$<1\,\mu m$ has a higher refractive index than its surrounds, and so a
large part of the radiation is trapped in waveguide modes which have
maximum intensity in that region and therefore contribute efficiently
to the stimulated emission. Since higher refractive index usually goes
together with lower electronic band-gap, it is possible to employ the
same structure to confine the overlapping electron and hole densities
to the emitting region. As a result there is sufficiently high gain to allow continuous-wave operation. An example of such a *heterostructure
laser* is shown in Fig. 14.18.

14.6 Laser light

Consider a laser in which only one longitudinal mode of the cavity
is excited. Stimulated emission results in a very large number $\langle n \rangle$ of
photons in this one mode (14.37). This is what distinguishes laser light
from thermal light for which, as we pointed out in §14.1.2, there is on
average much less than one photon per mode. The fluctuation δn is
given by (14.27) because even laser light has to fulfil the uncertainty
relation. Thus $\delta n = \langle n \rangle^{\frac{1}{2}}$ and so, according to (14.22), $\delta\phi = \langle n \rangle^{-\frac{1}{2}}$,
which is very small.

    Laser light is therefore characterized by having very well-defined

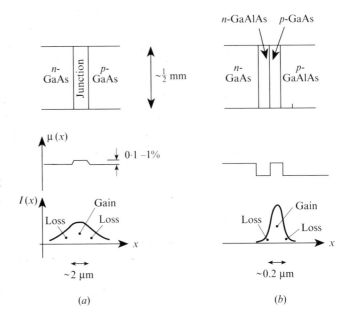

Figure 14.18
Semiconductor diode
lasers: (*a*) basic
structure;
(*b*) heterostructure
using GaAs and
GaAlAs. Each
diagram shows the
geometry, the
refractive index $\mu(x)$
and the resulting
intensity distribution
$I(x)$ of the emitted
light.

phase.† In addition, when only a few longitudinal modes are excited the light is very well-defined in its direction, the angular spread of the beam being determined by diffraction as if it were restricted by an aperture corresponding to its actual physical extent (§8.2.7). These three properties – phase coherence, high intensity and directionality – are the most characteristic properties of laser light.

14.6.1 Coherence function

As described in §11.5.2, the temporal coherence function is the normalized Fourier transform of the spectral intensity. If a laser were to operate in a single longitudinal mode, ideally the spectrum would be a single spike. It is not quite a delta-function because there are phase fluctuations; the time-scale of these fluctuations must be at least the lifetime of the lasing transition, and so the coherence time τ_c is at least equal to this lifetime (for example, 10^{-7} s for T_2 in Fig. 14.13). The corresponding coherence length is $c\tau_c$ (30 m). In practice, the coherence time may be shortened by mechanical fluctuations (due to temperature etc.) in the optical round-trip length $\bar{L}$ of the cavity.

Many lasers operate in more than one longitudinal mode simultaneously (Fig. 14.16). If the modes have random phases, the coherence function, the Fourier transform of a few spectral lines separated by $\delta\omega = 2\pi c/\bar{L}$, has the form of Fig. 14.19, in which the coherence disappears and reappears at intervals of $\bar{L}/c$. There is no simply-defined coherence length when the coherence function behaves in this manner, but for many practical purposes, since coherence disappears first after time $\bar{L}/2c$, the effective coherence length is $\bar{L}/2$, the optical length of the cavity. If the laser is pulsed or mode-locked, as in §14.5.2, the coherence time equals the duration of an individual pulse.

The coherence area of a single or multi-longitudinal mode laser is just the beam area, since the light distribution comes from a single coherent mode or a superposition of such modes.

14.7 Advanced topic: squeezed light and its applications

In §14.2.2 we introduced the idea that in the $(P, \omega Q)$ diagram the end of the amplitude and phase vector representing a monochromatic

---

† This is not necessarily true of *any* bright light. The large number of photons has to be concentrated in a single mode.

wave can only be located to a limited accuracy, represented by a region of minimum area $\hbar$. However, it is possible to change the *shape* of this region (as shown in Figs. 14.4–14.8) and trade one uncertainty for another. We shall expand these ideas in the present section; for a more substantial discussion, see Teich and Saleh (1990) and Reynauld *et al.* (1992).

14.7.1 Sub-Poisson light

We saw in §14.2.2 that chaotic light, which is light with many photons in a single mode having $\delta P = \omega \delta Q = \hbar^{\frac{1}{2}}$, fluctuates in a way corresponding to Poisson statistics, which also represent the smallest fluctuations possible in the classical model. Poisson statistics indicate that the arrival of each photon at a detector is uncorrelated with that of the previous one (§11.8), which necessarily limits the accuracy of optical measurements. If we can force them to be correlated, the stream of arrivals will be steadier and an improvement in accuracy is to be expected. For example, were the photons to arrive at exactly equal intervals T, the arrival of no photon within a period marginally greater than T would indicate zero field with no shadow of doubt. Such non-classical light is called *sub-Poisson*, and corresponds to Fig. 14.6. One should notice that the intensity has been stabilized at the expense of phase fluctuations, in accordance with (14.22).

Several ways of doing this have been invented.

(1) The operation of a semiconductor laser diode (§§14.4.5, 14.5.4) from a stabilized constant current (pumping) source (Machida, Yamamoto and Itaya, 1987; Tapster, Rarity and Satchell, 1987). This controls the flow of electrons which, because of the very short lifetime T_2 for electron-hole pairs in the junction, regulates the photon output.

Figure 14.19 Coherence function for a laser emitting a few longitudinal modes.

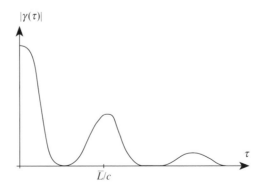

(2) The emission of resonance fluorescence from a single atom, which has to be re-excited after every emission. The re-excitation creates a 'dead-time' after each emission which smooths the flow of photons, by creating a dependence of each on the previous one (Kimble, Dagenais and Mandel, 1977; Teich and Saleh, 1985).

(3) Emission of pairs of photons in a cascade from three levels of ^{40}Ca. One photon creates the beam. The other triggers a gate by which an artificial electronic dead time is achieved. This is called *conditionally anti-bunched* light (see Teich and Saleh, 1990).

Although methods (2) and (3) seem rather artificial, the essential point is that they only exist because light is emitted as individual quanta, and therefore can not be described classically. Fig. 14.20 shows a *simulation* of Poisson and sub-Poisson light. In (a) we see a series of uncorrelated photon events at rate r. In (b) we have taken the sequence (a), doubled the rate to $2r$ and then introduced after every registered event a detection dead-time with average value equal to $1/r$, during which any event occurring is erased, so that the mean rate is once again r. It is easy to see that a steadier stream of photons has been achieved, and is equivalent to method (2) above.

14.7.2 Sub-Poisson light and digital optical communication

Suppose that a digital communication channel signals '0' by being off (0 photons) and '1' by being on for a time T (average number of photons rT, where r is the mean rate). What is the error probability for the cases illustrated by Fig. 14.20(a) and (b)?

In the Poisson case, when the beam is on, the mean number of photons in T is rT and its fluctuation $(rT)^{\frac{1}{2}}$. Approximating the Poisson distribution by a Gaussian (the central limit theorem, valid for large rT), with variance $\sigma = (rT)^{\frac{1}{2}}$, the probability of registering n photons is

$$p(n) = (2\pi r T)^{-\frac{1}{2}} \exp[-(n - rT)^2/2rT]. \tag{14.42}$$

Figure 14.20 Simulation of (a) Poisson, (b) sub-Poisson and (c) super-Poisson events; (c) corresponds to $\gamma_2(\tau)$ shown in Fig. 11.13.

(a)

(b)

(c)

Thus the probability of getting zero *by mistake* (i.e. as an estimate for rT) is

$$p(0) = (2\pi rT)^{-\frac{1}{2}} \exp[-(rT)^2/2rT] = (2\pi rT)^{-\frac{1}{2}} \exp(-rT/2). \quad (14.43)$$

If we require an error rate $p(0) < 10^{-9}$, this gives $rT > 34$; each '1' pulse must contain at least 34 photons. In the simulated sub-Poisson case, we have a basic rate of pulses $2r$, with dead time r^{-1} after each pulse which is received. The probability of getting zero pulses in time T can then be calculated in the same way using a basic rate $2r$, for a time which is at least $T - r^{-1}$ (the case when there was a pulse immediately preceding the start of T). The equivalent Gaussian now has $\sigma = (2rT - 2)^{\frac{1}{2}}$ and

$$p(0) = [4\pi(rT - 1)]^{-\frac{1}{2}} \exp\{-(2rT)^2/4(rT - 1)\} . \quad (14.44)$$

If $p(0) < 10^{-9}$, this gives a mean number of photons in time T, $rT > 18$, so that the '1' pulse now need contain no more than 18 photons.†

The price of this sub-Poisson light is that we lose control over its phase. But that is of no importance in this particular application.

14.7.3 Squeezed light and interferometry

In interferometry the phase of the light wave is important and we should like to use squeezed light to improve the accuracy of measurement (Xiao, Wu and Kimble, 1987). However, since any observation is eventually made by measuring an intensity, we can not use phase-squeezed light in which the fluctuations have been transferred to the intensity as in (Fig. 14.7). The most suitable squeezing is that of Fig. 14.8 in which the ωQ component ($\cos \omega t$) alone is used (P would do just as well).

Let us first consider an ideal Michelson interferometer (Fig. 14.21) as an example of a system, in which the movement of the mirror M_2 is to be measured as accurately as possible. We use a very intense laser source, so that its fluctuations can be neglected, and set the mirrors M_1 and M_2 so as to create destructive interference, with zero output from the detector. Then any movement of M_2 will give rise to a signal.

---

† Of course, if the sub-Poisson distribution were really produced by gating, such as method (3) above, this would be no real gain, since half the photons would be thrown away in order to get it. But if a method like (1) were used, there would be a real advantage.

The smallest signal which can be confidently detected is equal to the minimum† noise in the system.

There are two entrances for radiation into the ideal system. One is where the laser light enters, L, and the other is where the detector is placed, N. Light entering each way can be seen by the detector after reflexions in M_1 and M_2 with phase difference Φ for the light from L (where $\Phi = \pi$ in the nominal, destructive interference, case). At the same setting, the noise from N is seen with phase difference $\Phi - \pi$ (see §§9.3.2 and 5.6.2). If the incident amplitudes are L for the laser and N for the noise, then the detector sees an intensity signal

$$S = L^2 \sin^2(\Phi/2) + N^2 \cos^2(\Phi/2) \qquad (14.45)$$

where we have assumed an ideal 50% beam-splitter. The smallest value of Φ that can be distinguished from zero is that which makes the first term in (14.45) equal to the second (i.e. signal = noise):

$$|\tan(\Phi_{min}/2)| = N/L. \qquad (14.46)$$

It is instructive to represent these two contributions on the $(P, \omega Q)$ diagram, Fig. 14.22(a). The tip of the laser amplitude vector $\mathbf{L}$ is within a circle C_L which represents its fluctuations; the almost-destructive interference signal $\mathbf{L'} = \mathbf{L} \sin \Phi/2$ thus lies within a much smaller

---

† We assume that all noise from avoidable practical effects has been eliminated, and only quantum noise remains.

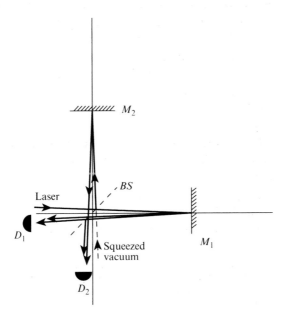

Figure 14.21 Michelson interferometer to be used with squeezed light.

circle C'_L nearer the origin, because $\sin \Phi/2 \ll 1$. The vacuum noise **N** is represented by the region C_N which is centred on the origin because the mean value of a noise signal is zero, and $N' = N \cos \Phi/2 \approx N$. If the laser is very intense, the diameter of C'_L is very small (because C'_L is C_L scaled down by the ratio L'/L) and can be neglected with respect to that of C_N. The signal L' will be measurable in the presence of the noise provided that C'_L lies outside C_N, so that vectors from a point in one of them to a point in the other have a non-zero mean. The situation with signal equal to noise (C'_L on the edge of C_N) is shown in Fig. 14.22(a). Now if the vacuum fluctuations are squeezed, C_N will not be a circle but more of an ellipse (Fig. 14.8), and we have the situation in Fig. 14.22(b). Clearly, the marginal situation is improved if the Q component is measured; the accuracy is now limited by the Q-component of the new C_N, N_{SQ} which is reduced to the extent of the squeezing.

14.7.4 Production of squeezed light

A possible way of squeezing one phase component of light (e.g. the Q component at the expense of P) is to use a phase-conjugate mirror to replace one of the usual mirrors in the cavity of an optical amplifier (Yurke, 1984). Although other, maybe more practical, methods have been described (see, for example, Wu *et al.*, 1986, and other papers

Figure 14.22 $(P, \omega Q)$ diagram for the Michelson interferometer (a) with normal vacuum fluctuations; (b) with squeezed vacuum.

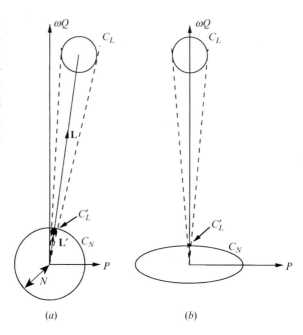

(a)   (b)

in Meystre and Walls, 1991) we shall use this as an example because the background has already been discussed in §13.5.3. Let us compare the conventional cavity, Fig. 14.23(*a*), with mirrors M_1 and M_2 at $x = 0$ and $x = h$ with one in which M_2 is replaced by a phase-conjugate mirror, Fig. 14.23(*b*). The wave leaving M_1 has the form of a plane-wave

$$E_3 = e_3 \cos(kx - \omega t + \phi_3) \tag{14.47}$$

where $\phi_3 = 0$ corresponds to the P phase, and $\phi_3 = \pi/2$ the Q phase. After reflexion in the conventional mirror M_2 the wave has the form

$$E_4 = e_4 \cos(-kx - \omega t + \phi_4). \tag{14.48}$$

For perfect reflectors, the total field at $x = h$ must be zero at all t, and so $e_3 = -e_4$ and $\phi_4 = \phi_3 + 2kh$. After reflexion of E_4 again at M_1, the wave becomes

$$E_3' = e_3' \cos(kx - \omega t + \phi_4), \tag{14.49}$$

and for a resonant mode this must be identical to (14.47), which means that $e_3' = e_3$ and $2kh = 2\pi m$, where m is an integer. No particular value of ϕ_3 is dictated by the resonance condition, so that the P and Q modes have equal amplitudes.

Now replace M_2 by the phase-conjugate mirror. From §13.5.3 we have $\phi_4 = -\phi_3$, so that after reflexion at M_1 the second time,

$$E_3' = e_3' \cos(kx - \omega t + \phi_4) = e_3' \cos(kx - \omega t - \phi_3). \tag{14.50}$$

For constructive interference with the original wave (14.47) we now

Figure 14.23
Comparison of
resonators (*a*) with
two conventional
mirrors and (*b*) one
conventional and one
phase-conjugate
mirror; both are
shown with resonant
modes.

(*a*)

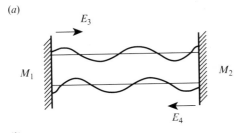

(*b*)

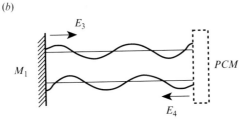

require $\phi_3 = -\phi_3$, i.e. $\phi_3 = 0$. This means that the P component is resonant, and the Q component is not. It is interesting to note that this cavity has no resonant condition for k.

Where has the symmetry between P and Q been broken? The relationship $\phi_4 = -\phi_3$ was obtained in §13.5.3 because we used pump beams in the phase-conjugate mirror with the form $\cos(\pm\mathbf{k}\cdot\mathbf{r} - \omega t)$. If we had shifted the phase of one of them by $\pi/2$ we should have obtained $\phi_4 = -\phi_3 + \pi$, leading us to a resonant Q phase rather than P. The squeezed phase is therefore dictated by the phases of the pump waves.

Other methods of squeezing light have also been invented, in particular using parametric amplification, but this subject is too close to the frontiers of research to be discussed adequately in a textbook! The interested reader will find the compendium by Meystre and Walls (1991) an excellent beginning for a deeper study of modern quantum optics.

CHAPTER FIFTEEN

Problems

General remarks

$\lambda = 0.5\,\mu m$ unless otherwise stated.

The problems often spill over from one chapter to another. It is the student's job to find the neccessary information from other chapters to understand the question. On the whole, these are not examination-type problems, but are intended to encourage the student to think. For this reason we are not including a section on 'answers to the problems', but intend to produce a booklet containing answers and further discussion of the problems, which will be available from the authors.

Chapter 2

2.1 A chain of masses m are situated on the x-axis at equally-spaced points $x = na$. They are connected by springs having spring-constant K and are restrained to move only along the x-axis. Show that longitudinal waves can propagate along the chain, and that their dispersion relation is

$$\omega = 2(K/m)^{\frac{1}{2}}|\sin\tfrac{1}{2}ka|. \tag{15.1}$$

Explain in physical terms why the dispersion relation is periodic. Calculate the phase and group velocities when $\omega \to 0$ and $\omega = 2(K/m)^{\frac{1}{2}}$. [For answers, consult Kittel, 1986; Myers, 1990.]

2.2 Flexural waves on a bar have a wave equation of the form

$$\frac{\partial^2 y}{\partial t^2} = -B^2 \frac{\partial^4 y}{\partial x^4}, \tag{15.2}$$

where B is a constant. Find the dispersion relation. Under what conditions are the waves evanescent?

2.3   X-rays in a medium have refractive index $\mu = (1 - \Omega^2/\omega^2)^{\frac{1}{2}}$ (§13.3.3) Show that the product of group and phase velocities is c^2.

2.4   A gradient index ('GRIN') lens is made from a plate with thickness d in the z-direction which has refractive index $\mu(x, y) = \mu_2 - \alpha(x^2 + y^2)$ where μ_2 and α are constants. Use Fermat's principle to show that this acts as a converging lens with focal length $(2d\alpha)^{-1}$ in the paraxial approximation (see Marchand, 1978).

2.5   Use Fermat's principle and the properties of conic sections to prove that a point source at one focus of an ellipsoidal or hyperbolic mirror is imaged at the other focus (the most common example is the paraboloid, which has one focus at infinity). The Cassegrain and Gregorian reflecting telescopes use both concave and convex mirrors to obtain high magnification (like a telephoto lens). What profiles of mirror should ideally be used?

2.6   A lens, corrected for spherical aberration, is used to image a distant axial point source. The lens has diameter 100 mm and focal length 500 mm. How close to the focal point, on the axis, will it be possible to detect that the image is out of focus (i.e. what is the depth of field)? Use Fermat's principle.

2.7   A spherical wave emanates from a point source. Show that energy is conserved if the amplitude decays like r^{-1}. What is the corresponding result for a cylindrical wave emanating from a line source?

Chapter 3

3.1   The foci of a mirror in the form of an ellipsoid of revolution are conjugate points (Problem 2.5). What is the magnification produced, in terms of the eccentricity of the ellipsoid? (Tricky.)

3.2   In order to use a microscope to observe an inaccessible specimen, one can introduce a relay lens between the specimen and the objective, so that the microscope looks at a real image of the specimen. Draw a ray diagram of the system, and find the influence of the relay lens on the exit pupil and the field of view.

3.3   Design a periscope having a length of 2 m and a tube diameter of 0.1 m. The field of view must be a cone of semi-angle 30°. The periscope needs several relay and field lenses. Use paraxial optics only.

3.4   A compound lens consists of two positive thin lenses L_1 and L_2, with focal lengths 90 mm and 30 mm and apertures 60 mm and 20 mm respectively. $L_1L_2 = 50$ mm. Between the lenses, in the plane 30 mm from L_1 there is an axial aperture with diameter 10 mm. Where is the aperture stop, for a given axial object 120 mm in front of L_1? Find also the positions of the entrance and exit pupils.

3.5   The following is a useful method of finding the refractive index of a transparent material in the form of a parallel-sided plate with thickness d. A microscope is focused on an object. The plate is inserted between the object and the microscope objective, and the microscope is re-focused. The distance that the microscope moves in refocusing is measured. Find the relationship between this distance, the refractive index and d. Estimate the accuracy of the method (Problem 2.6 may help you).

3.6   Find the matrix representing transmission of light along a GRIN rod of length l in the paraxial approximation, using the parameters of Problem 2.4. Show that, in the limit $l \to 0$ it behaves as a thin lens.

3.7   Within the limitations of Gaussian optics, is it possible to replace a glass sphere of any refractive index by a single lens ?

3.8   A zoom lens consists of two lenses, with focal lengths 100 mm and -20 mm respectively. Plot a graph showing the effective focal length and f-number of the combination, as a function of the distance between the two lenses.

3.9   A glass shell with refractive index 1.5 has equal radii of cuvature on both sides (one is convex, the other concave). The radii are both 100 mm and the thickness is 1.5 mm.

   (*a*) Without carrying out any calculation, decide whether the shell acts as a lens with positive or negative optical power.
   (*b*) Find its focal length and principal planes.

3.10 Write a computer program based on the Gaussian matrices to find the cardinal points of any paraxial optical system defined by coaxial spherical interfaces between regions of given refractive indices, and/or thin lenses. Use it to check the results of Problems 3.7–3.9.

3.11 Show that in a symmetrical imaging system with unit paraxial magnification the distortion must be zero.

3.12 Design a lens of the type shown in Fig. 3.17(*b*) with $\mu = 2$ and $f = \infty$. What is m when O is at the aplanatic point? Explain

physically why the lens magnifies, even though its effective focal
length is infinite.

Chapter 4

4.1   Given that the transform of $f(x)$ is $F(k)$, find general expressions
       for the transforms of $\int_0^x f(x')\,dx'$ and df/dx.

4.2   The 'Hartley transform' of a real function $v(x)$ is defined as

$$H(k) = \int_{-\infty}^{\infty} v(x)[\cos kx + \sin kx]\,dx. \qquad (15.3)$$

Show that it is related to the Fourier transform by

$$H(k) = \mathrm{Re}[V(k)] + \mathrm{Im}[V(k)]. \qquad (15.4)$$

Devise a method, using Fourier optics, to record the Hartley
transform optically. Note that it contains all the information
of the Fourier transform in a single real function, but only
works for real functions. See Bracewell (1986) for a thorough
discussion; an experimental method is given by Villasenor and
Bracewell (1987).

4.3   Find the Fourier transform of a decaying series of δ-functions:

$$f(t) = \sum_{n=0}^{\infty} \delta(t - nt_0)e^{-\alpha n}. \qquad (15.5)$$

How can this result be used to understand the Fabry–Pérot
interferometer?

4.4   Derive the Fourier transform of a periodic triangular wave,
       defined in one period as $y = |x|\ (-\pi < x \le \pi)$. How can the
       result be related to the auto-correlation of a square pulse?

4.5   The period of a square wave is b, the value being 1 for a period
       c within the cycle and 0 for $b - c$ (this is called 'duty cycle'
       $= c/b$). Use the convolution theorem to study how its Fourier
       transform changes as a function of c/b. What happens when
       $c/b \to 1$?

4.6   Show that the self-convolution of $\mathrm{sinc}(ax/2)$ is the same function
       multiplied by a constant.

4.7   Compare the functions $(f_1 \otimes f_2) \times f_3$ and $f_1 \otimes (f_2 \times f_3)$ and
       their transforms, when $f_1 = \sum_{n=-\infty}^{\infty} \delta(x - nb)$, $f_2 = \mathrm{rect}(x/b)$
       and $f_3 = \exp(i\alpha x)$.

4.8   The convolution has some odd properties. For example, a
       group of three δ-functions at intervals b can be represented as
       the product of the infinite periodic array $\sum \delta(x - nb)$ multiplied

by rect(x/c) where c has any value between $2b + \epsilon$ and $4b - \epsilon$ (ϵ arbitrarily small). Show that the transform, which can be expessed as a convolution, is indeed independent of the argument of rect between these limits. (This problem is not easy!) The solution is discussed in detail by Collin (1991).

4.9   A periodic array of δ-function has every fifth member missing. What is its Fourier transform?

4.10   The image of a thick black straight line at an angle to the x- and y-axes on a white background is digitized on a grid of $N \times N$ squares, so that a square is white or black depending on whether it is more or less than half covered by the line (Fig. 15.1). The digital Fourier transform of the line is then calculated. When N is small, the transform is predominantly along the k_x and k_y axes (transforms of elementary squares). As N increases, the transform approaches a limit which is predominantly along the axis at right-angles to the line. How does the transition take place?

4.11   A 'wavelet transform' is the Fourier transform of a function whose spectrum changes with time, and consists of a representation of the Fourier transform measured during an interval δt as a function of time (for examples see Combes, Grossman and Tchamitchian 1990). It is often used in speech and music analysis. With the help of the convolution theorem, show that δt and the frequency resolution $\delta\omega$ of the wavelet transform are related by $\delta t \cdot \delta\omega \approx 2\pi$.

4.12   A long one-dimensional quasi-periodic array of δ-functions is created as follows. It has a basic period b, and within each cell there is a δ-function at either $x = 0$ or $x = h$ where $h < b/2$.

Figure 15.1 Digital representation of a diagonal line.

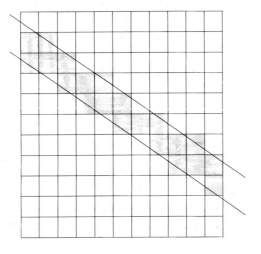

The probability of each is 50%. Use the concept of the auto-correlation function to calculate the power spectrum of this array.

Chapter 5

5.1   Two equal and opposite dipoles $\pm\mathbf{p}$ separated by vector $\mathbf{l}$ constitute a *quadrupole*. There are two generic types, with $\mathbf{l} \parallel \mathbf{p}$ and $\mathbf{l} \perp \mathbf{p}$. When $l \ll \lambda$, show that $\mathbf{E}_q(\mathbf{r})$, the field of an oscillating quadrupole, can be related to that of the dipole $\mathbf{E}_p$ by

$$\mathbf{E}_q = -ik_0\mathbf{E}_p(\mathbf{l} \cdot \mathbf{r})/r. \qquad (15.6)$$

and find the frequency dependence and radiation polar diagram for the power radiated by each of the generic types.

5.2   Light tunnels between two prisms as in Fig. 5.8. What is the wave velocity in the tunnelling region? Now consider a Gaussian wave-group tunnelling through; can it propagate a signal faster than c? [For a discussion of this topic, see Chiao, Kwait and Steinberg (1993).]

5.3   The *Fresnel rhomb* is a device which converts linearly-polarized light into circularly-polarized, employing two stages of total internal reflexion (Fig. 15.2). Calculate the angles of a rhomb made from glass of refractive index 1.5.

5.4   Suppose that the reflecting surface in Fig. 5.11 has completely symmetrical construction (e.g. a quarter-wavelength thick free-standing plate, which has non-zero reflexion coefficient). How can $\mathcal{R}$ and $\bar{\mathcal{R}}$ have opposite signs?

Figure 15.2 The Fresnel rhomb. It consists of a glass prism with parallelogram cross-section as shown.

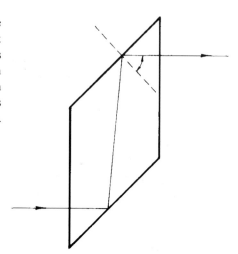

5.5 Show that at the Brewster angle the reflected and transmitted rays are orthogonal. When incidence is in the air, consider the reflected ray as originating in dipole radiation (Huygens-like) from surface dipoles on the interface, and show that the reflexion coefficient for the ∥ mode must indeed be zero at this angle. Can you extend this argument to Brewster angle reflexion when the incidence is *within* the medium, at an interface with air? (We can't!)

5.6 A pile of glass plates (μ=1.5) is used to polarize light by Brewster-angle reflexion. If the incident light is incoherent and unpolarized, derive an expression for the degree of polarization (ratio of ∥ to ⊥) of the transmitted light as a function of the number of plates.

5.7 MgF_2 has refractive index 1.38. A 45°,45°,90° prism is constructed from this material and is used to turn a light beam by 90° by internal reflexion at its hypotenuse. Relate the polarization vector after the reflexion to that before, both for linearly- and circularly-polarized light (cf. §9.6).

5.8 A simple (free-electron) metal has conductivity σ. What thickness of it is necessary as a coating to make a beam-splitter which reflects and transmits equal intensities of a wave with frequency ω, and how much of the light is absorbed in the process? Ignore the substrate, to make things simpler.

Chapter 6

6.1 A beam of light is known to be partly unpolarized and partly circularly polarized. How would you find the proportions?

6.2 Devise a method to find *absolutely* the sense of rotation of an elliptically-polarized wave.

6.3 A quartz plate has thickness d and its optic axis makes an angle 45° to its faces. A ray of unpolarized light enters normal to the plate and leaves as two separate polarized rays. Given that $\mu_o = 1.544$ and $\mu_e = 1.533$, find the separation between the two exiting rays.

6.4 Mica (biaxial) has refractive indices 1.5998 and 1.5948 for propagation normal to the cleavage plane. A sheet of mica, observed between crossed polarizers, is seen to have a purple colour, i.e. it transmits red and blue light, but not green. Estimate the thickness of the sheet. How does the colour change (*a*) as the mica is turned in its own plane, (*b*) as one of the polarizers is turned in its own plane?

6.5   A parallel beam of sodium light (spectral doublet with $\lambda = 5890\,\text{Å}$ and $5896\,\text{Å}$) passes through a pair of parallel polarizers separated by a calcite plate whose optic axis lies in the plane of its faces, at $45°$ to the axes of the polarizers. One line of the doublet is transmitted and one is absorbed. Calculate the thickness of the plate. Given that in the above spectral region:
$\mu_e = 1.486$ and $d\mu_e/d\lambda = -3.53 \times 10^{-6}\text{Å}^{-1}$,
$\mu_o = 1.658$ and $d\mu_o/d\lambda = -5.88 \times 10^{-6}\text{Å}^{-1}$.

6.6   How is Pöverlein's construction used to describe *reflexion* at the plane surface of a crystal? An unpolarized ray enters obliquely into an parallel-sided crystal plate, whose optic axis is at an arbitrary angle. The ray is reflected to and fro between the surfaces many times. Into how many distinct rays does it separate after n reflexions?

6.7   In the photo-elastic effect in an isotropic material the degree of birefringence $\mu_o - \mu_e$ is proportional to the difference between the principal stresses, $p_x - p_y$. Describe the pattern of fringes that is observed in a plastic model of a cantilever beam of uniform cross-section, rigidly-supported horizontally at one end and with a weight at the other. What orientation of polarizers is necessary to make the effect clearest?

6.8   A linearly-polarized light beam is incident normally on a parallel-sided transparent plate. After transmission, it is reflected back to its starting point by a plane metal-coated mirror. Compare the final state of polarization to the initial state for the following types of plate:

(*a*) a birefringent plate with its optic axes in arbitrary directions;
(*b*) a plate showing the Pockels effect, with the applied electric field parallel to the light beam;
(*c*) an optically-active plate, with optic axis parallel to the light beam;
(*d*) a magneto-optic plate with applied magnetic field parallel to the light beam.

6.9   Consider possible practical ways of creating a completely polarized light beam from an unpolarized source. For example, a polarizing prism is used to create two orthogonally-polarized beams, one of which is then rotated in polarization before the two are recombined. Show that the brightness of the output beam can never exceed that of the input (brightness is defined as power per unit area, per unit wavelength interval, per unit solid angle for a given polarization) and therefore the second law of thermodynamics is obeyed.

Chapter 7

7.1   A plane-wave is incident normally on a mask containing a 1 mm hole. What is the furthest distance from the mask at which one can observe a diffraction pattern with zero intensity at its centre?

7.2   A 5 mm-diameter disc is used to demonstrate the classic experiment showing the bright spot at the centre of its Fresnel diffraction pattern. The screen is at 1 m distance. What irregularity in the edges of the disc can be tolerated? Estimate the diameter of the bright spot on the screen.

7.3   Calculate the distances of the bright and dark fringes from the edge of the geometrical shadow in the diffraction pattern of a straight edge, observed in parallel light on a screen at 1 m distance from the edge.

7.4   What is the dispersive power (§3.8.2) of a Fresnel zone-plate, considered as a lens?

7.5   Find the variation of intensity along the axis of an annular aperture with inner and outer radii R_1 and R_2 illuminated by parallel light.

7.6   Find the Fresnel diffraction pattern of the axially-symmetric light distribution represented by

$$f(r) - \exp(-r^2/2\sigma^2 - ir^2/2\beta^2).$$

Show that this function represents the basic stable mode in a laser cavity, when σ and β are suitably chosen. See §§3.10 and 9.5.3.

7.7   A pinhole camera forms an image of a distant object on a screen at distance d from the pinhole. What diameter of pinhole gives the sharpest image? (Take into account both diffraction and convolution of the image with the aperture of the pinhole.)

7.8   Use the Cornu spiral to calculate the Fresnel diffraction pattern of a slit of width 1 mm on a screen at 1 m distance, when illuminated by parallel light. Compare this with the Fraunhofer pattern, obtained by inserting a lens of focal length 1 m immediately after the slit.

7.9   Devise a method of using amplitude–phase diagrams for Fraunhofer diffraction patterns, and apply it to finding the pattern of a periodic array of six thin slits. (This method is used in many elementary texts on optics.)

Chapter 8

Note: many variants on the problems on diffraction patterns can be generated using the *Atlas of Optical Transforms* (Harburn, Taylor and Welberry, 1975).

8.1 Deduce the Fraunhofer diffraction pattern of a set of three equally-spaced slits by considering them as the sum of a single slit and a pair of slits.

8.2 Deduce the Fraunhofer diffraction pattern of four equally-spaced slits as

(*a*) a pair of pairs of slits (use the convolution theorem);
(*b*) a double slit in the centre flanked by a double slit of three times the spacing.

8.3 Deduce the Fraunhofer diffraction pattern of a square frame, by subtracting the pattern of a smaller square from that of a larger one.

8.4 Deduce the diffraction pattern of four pinholes at $(x, y) = (\pm a, 0)$ and $(0, \pm a)$. Compare your result with Fig. 12.4(f).

8.5 Fig. 15.3 shows 20 apertures and Fig. 15.4 shows 20 Fraunhofer diffraction patterns. All the patterns were photographed to the same scale, but may have had different exposures. Some of them are not correctly oriented. Match each pattern to its aperture, and find its correct orientation.

8.6 Fig. 15.5 shows five diffraction patterns and four pairs of triangular holes. Which diffraction pattern corresponds to each pair of holes? The fifth pattern corresponds to a pair of holes not shown; what is it? (There are two answers, related by symmetry.)

8.7 What is the Fraunhofer diffraction pattern of a mask in the form of a chess-board of opaque and transparent squares?

8.8 A square aperture is half covered by a sheet of mica which changes the phase of the transmitted light by $\pi/2$. What is its Fraunhofer diffraction pattern?

8.9 Calculate (either analytically or numerically) the intensity of the Fraunhofer diffraction pattern of a set of six pinholes at the corners of a regular hexagon. Now repeat the calculation with an extra pinhole at the centre of the hexagon. Show how comparison of the two patterns allows you to determine the phase at each point of the first pattern. Compare this with the heavy-atom method in crystallography.

8.10 A one-dimensional object consists of several equal positive δ-functions arranged symmetrically about the origin at integer values of x within the range $-32 \leq x \leq 32$. This object is repeated periodically along the x-axis with period 64. The

Figure 15.3
Apertures.

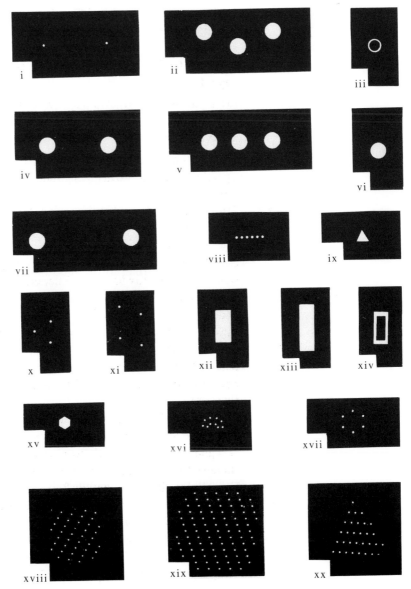

diffraction spots observed at $u = n\pi/32$ have amplitudes $a(n)$ as follows:

n	1	2	3	4	5	6	7	8
$a(n)$	0	2.11	0	0.23	0	0.96	0	0.41

n	9	10	11	12	13	14	15	16
$a(n)$	0	1.39	0	2.85	0	2.66	0	1.00

(15.7)

n	17	18	19	20	21	22	23	24
$a(n)$	0	0.66	0	0.85	0	0.61	0	2.41

n	25	26	27	28	29	30	31	32
$a(n)$	0	2.96	0	1.77	0	0.11	0	1.00

(a) What are the implications of the fact that all the odd values of n have zero amplitude?

(b) Find the phases of all the spots with amplitudes greater than unity, using Sayre's equation (8.92).

(c) Reconstruct the object using the amplitudes and phases that you have found.

(d) Find the phases of the remaining spots.

8.11 A plane-wave of wavelength λ_0 is incident normally on a diffraction grating moving at a constant velocity $v \ll c$ in its own plane in a direction normal to the slits. Find the wavelengths of the various diffracted waves.

Chapter 9

9.1 An amplitude diffraction grating (i.e. one which does not affect the phase) has a line profile $f(x)$ ($\in 0 < x < d$) which is real and positive. What profile maximizes the diffraction efficiency in the first order?

9.2 An echelon grating has construction like a staircase, with highly-reflecting treads of width b and height h. It has N steps (Fig. 15.6). Deduce its Fraunhofer diffraction pattern, and the resolving power attainable for high resolution spectroscopy.

9.3 Find the resolving power of a diffraction grating according to the Sparrow criterion.

9.4 A reflexion grating is blazed for $\lambda = 7000\,\text{Å}$ in the first order. The zero order is found to have intensity 0.09 compared to the first order at that wavelength. Assuming the grating to be constructed of flat mirrors, find the relative intensities of the other orders. Find also the relative intensities at $\lambda = 5000$ Å.

9.5   Why does a soap film appear black in reflected light as its thickness approaches zero? Light reflected from a certain film has a spectrum which peaks at 6660 Å, 5450 Å and 4620 Å. What is its thickness. Take the refractive index as 1.4 and assume normal incidence.

9.6   A Michelson interferometer is used in exactly parallel monochromatic light, and is adjusted so that the two optical paths SM_1 and SM_2 differ by exactly $\lambda/2$. The output intensity is therefore zero. Where has the energy gone?

Figure 15.4
Diffraction patterns.

9.7 Where are fringes from a broad source localized in (*a*) interference from a thin film, (*b*) a Mach–Zehnder interferometer? Can multiple-reflexion fringes be localized, and if so, under what conditions?

9.8 A Fabry–Perot interferometer is constructed with plates which are not quite parallel. Assuming the reflexion coefficient of each plate to be close to unity and the mean separation to be d, work out approximately how the resolving power is affected by the angle θ between the plates.

9.9 A *Lummer–Gherke* plate is constructed as in Fig. 15.7. It uses multiple reflexion at internal angles just less than critical in a parallel-sided plate of thickness d, length L and refractive index μ to create a large number of parallel output beams. Find the phase difference between them, as a function of the output angle θ. What is the resolving power?

Figure 15.5 Pairs of triangular holes and their Fraunhofer diffraction patterns.

9.10 The interferometer constructed by Chiao *et al.* (1988) to investigate the Berry phase is shown in Fig. 15.8. Find the phase difference between the interference patterns observe for right- and left-handed circularly-polarized light traversing the interferometer. (Answer in their paper.)

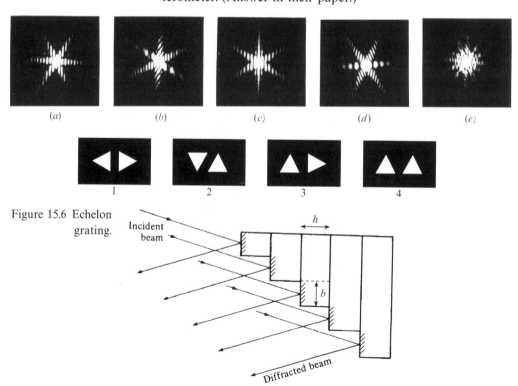

(*a*)   (*b*)   (*c*)   (*d*)   (*e*)

1   2   3   4

Figure 15.6 Echelon grating.

Incident beam

h

b

Diffracted beam

Chapter 10

10.1 Incoherent light is to be focused onto the plane end of a multi-mode fibre, with core refractive index μ_2 and cladding μ_1. What is the largest useful numerical aperture (NA) of the focusing optics? Can this be increased by making the end of the fibre non-planar?

10.2 Show that the brightness (Problem 6.9) of light entering an optical fibre can not be increased by tapering the input end, so as to collect light over a larger area and concentrate it (optical funnel).

10.3 Show that the numbers of $\perp$ and $\parallel$ modes in a slab waveguide are equal. If the cladding has higher losses than the core, which type of mode travels further?

Figure 15.7
Lummer–Gherke
plate.

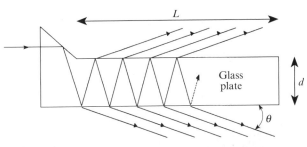

Figure 15.8
Berry-phase
interferometer.

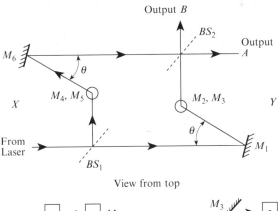

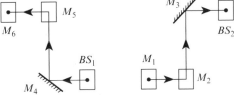

10.4 An asymmetrical slab waveguide of thickness a and refractive index μ_2 is made on a substrate of index μ_0 and is covered with cladding of index μ_1. What is the smallest value of a/λ for which a single mode propagates, assuming that $\mu_2 - \mu_1 \ll \mu_2 - \mu_0$?

10.5 Fig. 10.8 shows the light distribution in several modes of round fibres. Some of the modes are mixed. Determine which ones are pure modes, and designate them by mode numbers (l, m).

10.6 An interference filter has transmission wavelength λ and band-width $\delta\lambda$ when light is incident normally on it. How do λ and $\delta\lambda$ change as the filter is tilted with respect to the incident light? Take into account the effect of polarization.

10.7 Design a multilayer beam-splitter at oblique incidence, which transmits one polarization completely and reflects 99% of the other.

10.8 Using physical ideas only, explain why the multilayer mirror system 'glass $(HL)^{q+1}$ air' has a lower reflexion coefficient than 'glass $(HL)^q H$ air', despite the latter's having one layer less. The symmetrical arrangement $(HL)^q H$ is called a *quarter-wave stack* and is widely used in filter design.

10.9 Write a computer program to study multilayer systems built from two materials H and L on a substrate. Use it to calculate $\mathcal{R}(g)$ and $\mathcal{T}(g)$, $\in 0 < g < 2\pi$. Investigate the following ideas with it:

(a) broad-band, sharp-edged filters using the idea of coupled potential wells – 'glass $(HL)^q H(HL)^p H(HL)^q$ air', where p is small;

(b) high- and low-pass filters based on a multilayer mirror in which the layer thickness changes monotonically through the stack.

Chapter 11

11.1 Estimate the Doppler and collision line widths of emission from H_2O molecules at $\lambda = 0.5\,\mu m$, at $300\,K$ and atmospheric pressure. Assume the collision cross-section to be the same as the geometrical size of the molecule.

11.2 Monochromatic light is scattered at $90°$ from a cell containing 10^{-16} g particles in suspension at $300\,K$. Estimate the coherence time and linewidth of the scattered light.

11.3 A laser beam is spatially filtered by focusing it through a pinhole, to give it a smooth intensity profile across the wavefront. The laser beam has a Gaussian profile (with the addition of noise) with $\sigma = 1$ mm, and is focussed by a $\times 50$ microscope objective having focal length 5 mm. What size of pinhole would be suitable to transmit a homogeneous beam?

11.4 A Fourier transform spectrometer gives an interferogram, the positive half of which is shown in Fig. 15.9. What qualitative deduction can you make about the source spectrum?

11.5 The spectrum of a light source is $J(\omega) = \delta(\omega - \omega_1) + \delta(\omega - \omega_2)$. Discusss the output spectra obtained from a Fourier spectrometer as a function of d_{max}, and the improvement obtained when the observed interferogram $I - I_M(d/c)$ is multiplied by an 'apodizing function' of the form $\cos(d\pi/2d_{max})$ before being transformed. How is the resolution limit affected by the apodization?

11.6 Calculate the spatial coherence function in the plane of a table illuminated by a standard fluorescent tube at a height of 10 m. Assume monochromatic light.

11.7 Two incoherently illuminated points sources A and B of the same wavelength are situated on a vertical line at heights h_1 and h_2 above the plane surface of a metal mirror M (Fig. 15.10). The interference pattern is observed on a screen at distance L. What is seen? Use this as a model to explain the idea of the coherence area resulting from an extended source AB, which has an angular size α. (Assume all angles to be small.)

11.8 A strange star consists of a laser with a long coherence time of the order of seconds. Why would it not be possible to measure its diameter with a Brown–Twiss interferometer, but only with a Michelson stellar interferometer?

Figure 15.9 Fourier interferogram.

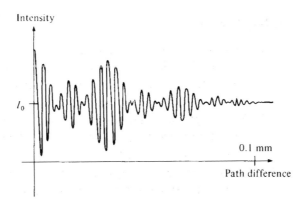

Intensity

I_0

0.1 mm

Path difference

11.9   Prove (11.84), that the intensity correlation coefficient $c(\tau) = 1 + |\gamma(\tau)|^2$, by evaluting $\gamma(\tau)$ in terms of the model of §11.2.1. (Take care how you calculate averages!)

11.10 Find the equivalent point spread function for a pair of antennae observing the region of the pole star for one day at 10 cm wavelength, using very-long-baseline interferometry. The distance between the antennae is 100 km projected on the plane normal to the Earth's axis. Qualitatively, how would your answer be modified for other directions of observation?

11.11 Two radio-receiving antennae in an aperture synthesis observatory are situated at $\mathbf{r} = (\pm x_0/2, 0, 0)$ on horizontal ground, and observe two equally bright quasi-monochromatic radio-stars, one at the zenith (z-axis) and one at angle θ to the z-axis in the $z-x$ plane.

(a) Calculate the field $E^R(\mathbf{r}, t)$ seen by the antennae, and the real coherence function $\gamma^R(x_0)$. [To simulate incoherence between the signals from the two stars, assume them to have frequencies ω_1 and ω_2 which are close but not equal, and integrate over a time long compared with $|\omega_1 - \omega_2|^{-1}$.]

(b) Calculate the associated complex fields, and the complex coherence function $\gamma(\mathbf{r}_0)$. Show that its Fourier transform retrieves the angular positions of the sources.

Chapter 12

12.1   An astronomical telescope is used in conjunction with a camera to produce a highly-magnified image. The photographic plate used has resolution 0.05 mm. If the primary mirror has diameter 1 m and focal length 12 m, what extra magnification can be provided by the camera optics?

Figure 15.10 Lloyd's mirror with two incoherent sources.

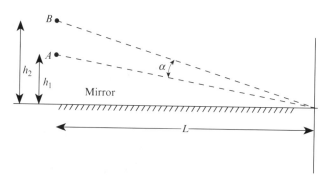

12.2   An object consists of two white points on a dark background. Their separation is 3λ. Calculate the image which is obtained when the object is viewed in a microscope under the following illumination conditions:

(*a*) axial coherent illumination, objective with NA=0.5;

(*b*) axial coherent illumination, objective with NA=0.2;

(*c*) incoherent illumination, objective with NA=0.2.

Treat the problem as one dimensional.

12.3   A photographic transparency shows a monkey behind a fence consisting of equally-spaced narrow vertical bars. How would you use a spatial filtering technique to remove the fence, hurting the monkey as little as possible?

12.4   For certain types of grey-scale object (with no phase structure) the dark-ground image is the negative of the normal image, in the photographic sense. What condition is necessary for this to be true?

12.5   Two point objects are illuminated coherently in antiphase, so that they are resolved by a microscope however close they may be. What is the apparent separation between them, as a function of the NA of the microscope, when the separation is less than the Abbe limit?

12.6   A phase object consists of many identical small transparent discs on a uniformly illuminated field. The discs are randomly arranged without overlapping, and together they cover half the field of view. The discs change the phase of the transmitted light by angle ϕ. What spatial filter will give maximum contrast beween the discs and their surroundings?

12.7   Calulate the dimensions of a Wollaston prism from calcite for use in a Nomarski DIC microscope, with objective of focal length 5 mm, NA=0.6. It should be designed so that doubling of the image is not observable.

12.8   A popular form of phase-contrast imaging consists of simply defocussing the microscope a little. Express this in terms of a complex spatial filter, and apply it to the phase slit of §12.4.5.

12.9   A telescope lens is apodized in order to reduce the prominence of the diffraction rings in the point spread function. If the radius of the objective is R and the amplitude transmission is reduced by a mask having Gaussian transmission function with parameter σ, find the value of σ which reduces the intensity of the first diffraction ring to 10% of its original value. How is its resolving power affected, according to Rayleigh and Sparrow?

12.10 Calculate the resolution limit of a confocal microscope whose two lenses are masked by annular apertures, of radii equal to those of the lenses.

12.11 A hologram of a certain object is made using light of wavelength λ_1. The reconstruction is made using a similar reference beam having wavelength λ_2. How is the reconstruction distorted, and where is it observed? (Assume all angles involved in the problem to be small.)

12.12 Calculate the longitudinal and transverse resolution of a holographic reconstruction, in terms of the wavelength, the dimensions of the illuminated part of the hologram and the image position. (Use Fermat's principle.)

12.13 What is the relationship between the reconstructions produced by an amplitude hologram and its negative?

Chapter 13

13.1 In what direction, relative to the Sun, should a photograph be taken so that a polarizing filter will be most effective in reducing scattering by dust in the atmosphere?

13.2 In the anomalous dispersion region, $d\mu/d\lambda > 0$. In principle, this could lead to a group velocity $v_g > c$. Why can this not be used to transmit information at superluminal velocities? Try to find a physical answer; a mathematical one is given by Brillouin (1960).

13.3 A material has a spectral absorption line at wavelength λ_0, which can be represented as a δ-function of strength a_0. Use the Kramers–Krönig relations to deduce $\mu(\lambda)$.

13.4 A uniaxial non-linear crystal has $\mu_o = 1.40$ and $\mu_e = 1.45$. Its dispersion in both polarizations is $\lambda d\mu/d\lambda = -2.5 \times 10^{-2}$. At what angle to the optic axis would phase matching be observed for (a) second harmonic generation, (b) third harmonic generation?

13.5 Explain why a polycrystalline non-linear material can be used for second harmonic generation without any particular care in orientation. When would this allow you to see second and third harmonic generation simultaneously?

13.6 Derive the phase-matching condition required for mixing two frequencies ω_1 and ω_2 to obtain their sum $\omega_1 + \omega_2$.

13.7 What interference pattern would you expect to see when one mirror in a Michelson interferometer is replaced by a phase-conjugate mirror? (Answer in Fischer and Sternklar, 1985.)

13.8 You look at yourself in a phase-conjugate mirror. What do you see? (Answer in Pepper, 1986.)

Chapter 14

14.1 A cube beam-splitter which reflects and transmits 50% of the radiation incident on its faces is used to mix a reference wave $E_0 \cos \omega t$ with a signal wave of frequency ω and unknown amplitude and phase (Fig. 15.11). Show how the phase and amplitude of the signal can be determined by measurements of the long-term averages of the two outputs at detectors D_1 and D_2. This is called an *optical phase-sensitive detector*.

14.2 Several output modes of a laser, indicated by the small integer n which lies between, say, $+5$ and -5, are represented by the waves

$$E_n = a \exp\{-i[(\omega_0 + n\omega_1)t + \phi_n]\}. \qquad (15.8)$$

where ω_1 is the mode-spacing frequency. To illustrate mode-locking, calculate the wave resulting from superposition of these modes when (a) ϕ_n is a random variable and (b) all $\phi_n = 0$. (It is convenient to do this by computer.)

14.3 An atomic nucleus contains an approximately uniform charge distribution throughout of sphere of radius of order 10^{-14} m. It undergoes a transition in which it emits a γ-ray of energy about 1000 keV. Explain why the selection rule $\Delta l = \pm 1$ may not be obeyed in this transition.

14.4 An atom has a transition from its first excited state to the ground state with wavelength λ. It is situated in a cubical metal

Figure 15.11 Optical phase-sensitive detector.

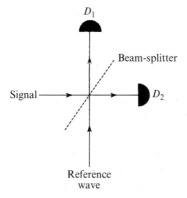

cavity with side $l \approx \lambda/2$. How would you expect the lifetime of the excited state to depend on the *exact* value of l within the range $0 < l < 3\lambda/2$?

14.5 A material has six energy levels A to F at 2, 1.9, 1.7, 1.6, 1.1 and 0.4 eV above the ground state, G. The time-constants for the various possible transitions in nanoseconds are shown in Fig. 15.12. Suggest possible lasers working with this material, and give the pump and output wavelengths of each one.

14.6 A weak source emits N photons per second. The light goes to an ideal beam-splitter so that half goes to each of two fast, ideal ($\eta = 1$) detectors. The correlation between the outputs from the detectors is recorded, a positive correlation meaning that both detectors emitted an electron within a given period $T \ll N^{-1}$. Analyse this experiment classically (i.e. each detector sees a wave of half the incident intensity) and from a quantum point of view (the incident photon goes to one detector *or* the other), taking into account the Poisson statistics of the source. Show that the results in the two cases are identical; but not so if the source is not Poisson.

14.7 What are the problems involved in building an X-ray laser working at $\lambda = 500$ Å? Consider in particular the threshhold intensity required and design of the resonator.

14.8 What is the coherence function corresponding to the spectrum of the laser shown in Fig. 14.16? (Compare your answer with Fig. 14.19.)

Figure 15.12 Energy scheme for lasing medium. The arrows show transition times in nanoseconds.

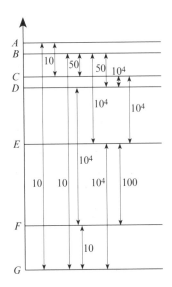

Appendix I

Bessel functions in wave optics

AI.I Bessel functions

Bessel functions come into wave optics because most optical elements – lenses, apertures, mirrors – are circular. We have met them in several places (§§8.2.7, 12.3, 12.5.2 for example), although since most students are not very familiar with them (and probably becoming less so with the ubiquity of computers) we have restricted our use of them as far as possible. The one unavoidable meeting is the Fraunhofer diffraction pattern of a circular aperture, the Airy pattern, which is the diffraction-limited point-spread-function of an aberration-free optical system (§12.3.1).

In this Appendix we simply intend to acquaint the reader with the results which are necessary for elementary wave optics. The proofs can be found in Watson's (1958) treatise and other places.

It is most convenient to start with Bessel's integral formulation of the function $J_n(x)$:

$$J_n(x) = \int_0^{2\pi} \exp[\mathrm{i}(x\cos\phi + n\phi)]\mathrm{d}\phi. \tag{A1.1}$$

The functions have the forms shown in Fig. A1.1. Typically, $J_n(x)$ starts from $x = 0$ like x^n, but when $x > n\pi/2$ it develops damped oscillations $\sim x^{-\frac{1}{2}}\cos[x-(n-\frac{1}{2})\pi/2]$. Thus alternate functions behave roughly as cosine and sine at large x, with a $\pi/4$ shift.

For proving differential and integral properties of the functions, it is often convenient to express them as power series:

$$J_n(x) = \left(\frac{x}{2}\right)^n \sum_{j=0}^{\infty} \frac{(-1)^j}{j!(j+n)!} \left(\frac{x}{2}\right)^{2j} \tag{A1.2}$$

from which it is easy to see the $\sim x^n$ behaviour at $x \ll 1$.

A1.1.1 Fraunhofer diffraction by an annular aperture

The zero-order Bessel function arises as the diffraction pattern of an annular aperture. It has radius a and width $\delta a \ll a$. The Fourier transform is, following (8.33),

$$F_0(\rho, \phi) = 2\pi a \delta a \int_0^{2\pi} \int_0^\infty \delta(r - a) \exp[-i\rho r \cos(\phi - \theta)] \, dr \, d\theta \quad (A1.3)$$

By symmetry, this is not a function of ϕ and so, putting $\phi = 0$,

$$F_0(\rho) = 2\pi a \, \delta a \int_0^{2\pi} \exp[i\rho a \cos \theta] d\theta$$

$$= 2\pi a \, \delta a \, J_0(\rho a). \quad (A1.4)$$

This is the diffraction pattern you will find for (iii) in Problem 8.5 and can see in the central region of Fig. A1.2(b). It is also beautifully illustrated in Harburn et al. (1975).

A1.1.2 A circular aperture

The diffraction pattern of a circular aperture is obtained by integrating (A1.4) from $a = 0$ to $a = R$:

$$F_0(\rho) = 2\pi \int_0^R J_0(\rho a) a \, da = \frac{2\pi}{\rho^2} \int_0^{R\rho} J_0(\rho a) \rho a \, d(\rho a). \quad (A1.5)$$

From (A1.2) one can prove easily that

$$\int_0^\rho x^{n+1} J_n(x) dx = \rho^{n+1} J_{n+1}(\rho) \quad (A1.6)$$

from which, putting $n = 0$,

$$F_1(\rho) = 2\pi \frac{R}{\rho} J_1(\rho R) = 2\dot{\pi} R^2 \frac{J_1(\rho R)}{\rho R}. \quad (A1.7)$$

Let us stress the similarity between the transforms of equivalent linear and circular systems

Figure A1.1 $J_0(x)$, $J_1(x)$ and $J_2(x)$.

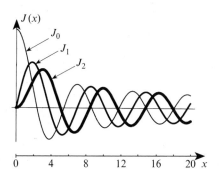

– A slit of width $2R$ has transform $2R\sin(uR)/uR$.
– A hole of radius R and area $A = \pi R^2$ has transform $2A\,J_1(\rho R)/\rho R$.
– Two narrow slits of width w at $x = \pm R$ have transform $2w\cos(uR)$.
– A narrow annular ring of width w and radius R has transform $2\pi Rw\,J_0(\rho a)$.

One can see that, roughly, $J_1(x)$ replaces $\sin x$, $J_0(x)$ replaces $\cos x$.

A1.1.3 A ring of equally-spaced holes

An illustration of the use of higher Bessel functions is worth presenting because of the beauty of the diffraction patterns. A ring of m pinholes can be represented roughly by the function

$$f(r,\theta) = [1 + \cos(m\theta/2\pi)]\delta(r-a) \tag{A1.8}$$

which has m peaks equally spaced around a circle of radius a. The transform is

$$F_m(\rho,\phi) =$$

$$\int_0^{2\pi} [1 + \tfrac{1}{2}\exp(im\,\theta/2\pi) + \tfrac{1}{2}\exp(-im\,\theta/2\pi)]\exp[i\rho a\cos(\phi-\theta)]\mathrm{d}\theta$$

$$= \; J_0(\rho a) + \tfrac{1}{2}\int \exp\{i[-\rho a\cos(\phi-\theta) + m\theta/2\pi]\}\mathrm{d}\theta$$

$$+\tfrac{1}{2}\int \exp\{i[-\rho a\cos(\phi-\theta) - m\theta/2\pi]\}\mathrm{d}\theta$$

$$= \; J_0(\rho a) + \tfrac{1}{2}[e^{im\phi/2\pi} + e^{-im\phi/2\pi}]J_m(\rho a)$$

$$= \; J_0(\rho a) + \cos(m\phi/2\pi)J_m(\rho a) \;. \tag{A1.9}$$

The intensity $|F_m(\rho,\phi)|^2$ has two contributions which overlap very little when m is large. At the centre ($\rho a \sim 1$) there is the usual $J_0(\rho a)$

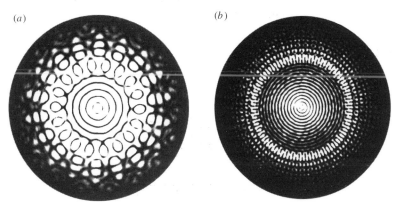

Figure A1.2 Fraunhofer diffraction patterns of (a) 18 and (b) 47 pinholes equally-spaced around a circle. In (b), the contributions of $J_0(\rho a)$ and $J_{47}(\rho a)$ are well-separated radially.

(a)   (b)

pattern from the ring at $r = a$; the individual holes are not resolved at small ρ. On the other hand, the function $J_m^2(\rho a)$ is very weak when $\rho a < m\pi/2$ but develops decaying oscillations at larger ρ. This function is modulated by $\cos^2(m\phi/2\pi)$ which has $2m$ peaks around the full circle. Fig. A1.2 shows the diffraction patterns for $m = 18$ and $m = 47$.

Lecture demonstrations in Fourier optics

A2.1 Introduction

Optics is the ideal subject for lecture demonstrations. Not only is the output of an optical experiment usually visible (and today, with the aid of closed circuit television, can be made visible to large audiences), but often the type of idea which is being put across can be made clear pictorially, without measurement and analysis being required. Recently, several institutes have cashed in on this, and offer for sale video films of optical experiments carried out under ideal conditions, done with equipment considerably better than that available to the average lecturer. Although such films have some place in the lecture room, we firmly believe that the student learns far more from seeing real experiments carried out by a live lecturer, with whom he can interact personally, and from whom he can sense the difficulty and limitations of what may otherwise seem to be trivial experiments. Even the lecturer's failure in a demonstration, followed by advice and help from his audience which result in ultimate success, is bound to imprint on the student's memory far more than any video film can do.

The purpose of this appendix is to transmit a few ideas which we have, ourselves, found particularly valuable in demonstrating the material covered in this book, and can be prepared with relatively cheap and easily-available equipment. Many other ideas are given by Taylor (1988). Need we say that we also *enjoyed* developing and performing these experiments?

A2.2 Correlation and convolution by the pinhole camera

The projection apparatus shown schematically in Fig. A2.1 produces an image which is the correlation between two real positive two-dimensional functions. The description below will complete some of the details which were omitted in §4.6.1. A projector incoherently illuminates a semi-transparent mask $f(x, y)$ in contact with a translucent screen in plane A. In plane B there is a second screen with cutouts representing the function $g(x, y)$. A third semi-transparent screen is placed in plane C positioned so that the distances AB and BC are equal. We shall show that the illumination $h(x, y)$ on plane C is the correlation function of $f(x, y)$ and $g(x/2, y/2)$, or the convolution of $f(x, y)$ and $g(-x/2, -y/2)$. It is viewed from its reverse side and can conveniently be projected with the TV camera (Fig. 4.11).

In one dimension, it can easily be seen from the figure that if $f(x) = \delta(x - a)$ and $g(x) = \delta(x - b)$ then $h(x) = \delta(x + a - 2b)$, i.e. a point on C at $x = 2b - a$ is illuminated. The correlation function of $f(x) = \delta(x - a)$ and $g(x/2) = \delta(x/2 - b)$ is indeed $h(x) = \delta(x + a - 2b)$. Moreover, the intensity of the point on C is proportional to the product of the source intensity f and the transmission of the point g. Since a general function can be described as a sum of δ functions and since the operation of correlation is associative the following equation is then generally valid:

$$h(x) = \int f(x' + x)g(x'/2)\mathrm{d}x' . \qquad (A2.1)$$

The function $h(x)$ therefore describes the correlation function of $f(x)$ and $g_1(x)$ where $g_1(x) = g(x/2)$.

Some particular cases can be emphasized. If the masks are similar but $g(x)$ is half the size of $f(x)$, i.e. $g(x/2) = f(x)$ then $h(x)$ describes the autocorrelation function of $f(x)$. This can be shown for sets of holes – note the very strong central peak in Fig. 4.11(f) – and for continuous functions such as the square hole – Fig. 4.11(g). Showing the gradation of intensity in the image in the latter case (its section

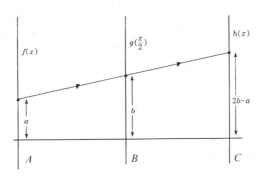

Figure A2.1
Geometry of the
correlation
apparatus.

has a roof-top profile) needs careful adjustment of the TV camera. By rotating the second function 180° about its origin, so that the axes of $g(x, y)$ are now down and to the left, we get the *convolution* of the two functions. An important demonstration shows the periodic function to be the convolution of the unit with a periodic lattice of small holes (δ-functions) either in one dimension (diffraction grating) or two (crystal).

In constructing this apparatus it is important to use translucent screens which diffuse uniformly in all directions, otherwise there are angular effects which complicate the analysis, because a convenient-sized apparatus involves quite large angles (Fig. A2.2). Drafting film works well, although one can see from Fig. 4.11 that the edges of the image are weaker than the centre.

A2.3 Fraunhofer diffraction

The demonstration of Fraunhofer diffraction was developed initially for crystallographic analysis (Taylor and Lipson, 1964) using a mercury arc source. Since the advent of inexpensive He–Ne lasers ($\lambda = 0.633\,\mu$m) the demonstration of diffraction and spatial filtering effects in a classroom has become very easy. In addition, closed circuit TV can make the output clearly visible.

Figure A2.2 The correlation apparatus.

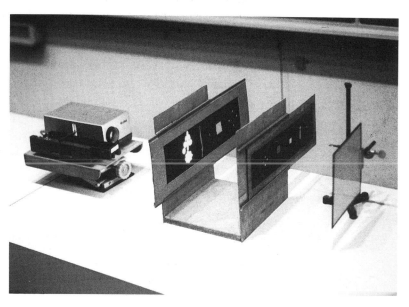

A2.3.1 Optical bench

We describe here an optical bench which was developed for this purpose (Lipson, 1972) shown in Fig. A2.3.

A low-power (0.5 mW) laser is used. This is safe for classroom use; there is almost no danger of damage to the eye at this power, except if one looks for a long time directly into the beam. The laser beam is expanded with the help of lenses L_1 ($F \approx -50$ mm) and L_2 ($F \approx 250$ mm) to about 8 mm diameter; one can then use diffraction masks of a reasonable size. (We found it unneccessary to use a pinhole spatial filter to 'clean up' the laser beam; this makes the apparatus very delicate, and means frequent readjustment.) With a typical mask being constructed with round 0.5 mm holes, at a mask–screen distance of 5 m the radius of the first dark ring of the Airy disk of the envelope is only about 7 mm, which is rather small. Two lenses L_3 and L_4 acting as a telephoto combination are therefore used to magnify the image by a factor of three or more, depending on the mask and the details one wants to see. This way, the effective focal length is multiplied without making the apparatus any longer.

With modern video technology one can achieve much more dramatic images visible without turning off the room lights. One can use a CCD video camera (which is not be damaged by overexposure to the zero order) without its lens, and project the diffracted image directly onto the light-sensitive element, which has a typical dimension of 1 cm. This image is then magnified by a video projector to the full screen size. This allows much more detail to be shown, without a powerful laser being necessary. The glare of the zero order is often a problem (it can saturate the camera); a way round this is to use the camera with its own lens to look at a real diffraction pattern projected on to a semi-transparent screen.

Often one wants to see an image of the mask on the same screen as the diffraction pattern. This can be done by inserting a beam-splitter after the mask; a simple lens L_5 is then used to generate the required image. The beam-splitter has to be of reasonable quality, otherwise it produces unwanted interference patterns. A cube beam-splitter with $\mathcal{R} \approx 0.5$ works quite well.

Figure A2.3
Diffractometer.

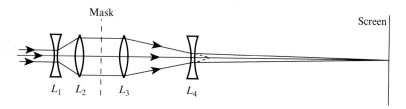

A variation of the same idea can be used to demonstrate the Abbe theory and spatial filtering in the Fourier plane (Fig. A2.4). The expanded beam is now made slightly convergent to a point P, around which point the diffraction pattern appears. The beam-splitter S is inserted directly after P and the lens L_5 in one of the outputs is used to image the filtered optical transform onto the screen. Meanwhile, the other output from the beam-splitter is used to create the filtered image. This has to be formed with a telephoto combination, otherwise it is too small to be visible. (There is actually a conflict of requirements here, which only a telephoto combination can solve. On the one hand, a large object would be required to give a large image. But then the diffraction pattern is small, and spatial filtering become a delicate operation. On the other hand, a small object with fine detail would solve this problem, but the reimaging lens can not be put close enough to it in order to get a highly-magnified image because it can only be placed after P. The telephoto combination is the answer.)

A2.3.2 Objects

Diffracting masks with outside dimensions up to 8 mm must be constructed. Photography is not a good way because of the phase changes introduced by the emulsion. The best masks are made from thin cardboard, unexposed photographic film (X-ray film is ideal – it is somewhat thicker than optical photographic film) or thin phosphor-bronze foil (about 0.1 mm). Taylor and Lipson (1964) described a pantograph for producing the masks which is easily modified to a simplified form. The rectangular holes in punched computer cards are useful for showing single and multiple apertures. Other shapes can be drilled and filed in foil. The patterns of multiple apertures are very beautiful, particularly if the latter are symmetrical – for example rings

Figure A2.4 Diffractometer used to illustrate spatial filtering.

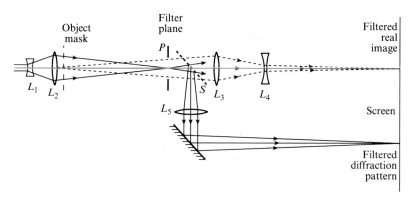

of pinholes (Appendix 1). Note in particular the symmetry relations when the numbers of holes in the ring are odd and even.

A demonstration of Babinet's principle can be done simply by comparing the patterns produced by a thick wire (say 1.5 mm diameter) and a slit of the same width.

Dynamic demonstrations are particularly impressive. The effect of changing the separation of a pair of apertures can be shown with the aid of a double slit having variable separation. This can be constructed from a pair of long narrow slits that are not quite parallel, a region of their length being selected for use by a sliding cursor. Likewise, a pair of parallel wedge slits allows one to show the effect of changing the individual aperture dimension with constant separation. The sequence of diffraction patterns produced by $1, 2 \ldots n$ parallel apertures can be shown by superposition of a coarse periodic square-wave grating (Ronchi ruling) and a slit of variable width, which selects the required number of periods. This demonstration is particularly valuable because the mask is obviously a product, and the two diffraction patterns which are convoluted are clearly visible (particularly when the variable slit is accidentally not quite parallel to the lines of the grating).

Phase masks can be produced in various ways. Of course, photography can be used, together with bleaching. More simply, cleaved mica and thin glass cover slips can be used to construct phase masks, for which the actual magnitude of the phase differences can be changed by tilting the plane of the mask with respect to light beam. Phase objects with much detail (to show phase contrast, for example) can be made by smearing transparent glue on a microscope slide. Fingerprints are also good phase objects. We should also mention blazed gratings, which are quite easily available. The hot air field around a candle flame gives a dynamical phase object, particularly useful for showing schlieren filtering.

A2.3.3 Spatial filters

The spatial filters that can be inserted into the Fourier plane include first of all an iris diaphagm (to demonstrate Porter's experiments, Fig. 12.4) to which wires and other obstacles mounted on microscope slides can be added to select almost any combination of diffraction spots. It is convenient to use a microscope specimen stage to get fine x–y control of the position of the spatial filters, because they need to be positioned quite accurately with respect to the diffraction pattern.

For phase objects, the filters in Fig. 12.18 – a black dot at the centre of a microscope slide, a razor blade and a small hole in a transparent celluloid sheet – can be used to show dark ground, schlieren and phase-contrast filtering (the last is very tricky and, demonstrated this way, in our opinion not very convincing!)

A2.4 Fresnel diffraction

Fresnel diffraction patterns can easily be shown by defocusing the Fraunhofer apparatus, although pedagogically it is better to take the lenses out completely. To avoid scattering at the edges of the apertures, it is best to make them from film and not from metal foil.

Particularly impressive is the bright spot at the centre of the shadow of the disc, which is easily shown with the aid of a ball-bearing about 5 mm in diameter. It can be glued to a microscope slide or supported on a wire (which affects the spot very little, but distorts the rest of the diffraction pattern noticeably). The use of a TV camera is important here, both because the spot is very small and because its non-linear response can be used to boost the intensity of the spot (i.e. by letting the outer parts of the diffraction pattern saturate the camera).

Bibliography

Agrawal, G. W. (1989). *Nonlinear Fiber Optics*. Academic Press, Boston.

Azzam, R. M. A. and Bashra, N. M. (1989). *Ellipsometry and Polarized Light*, 2nd edn. North Holland, Amsterdam.

Altman, C. and Suchy, K. (1991). *Reciprocity, Spatial Mapping and Time Reversal in Electromagnetics*. Kluwer Academic Press, Dordrecht.

Beckers, J. M. and Merckle, F. (eds) (1992). *High-Resolution Imaging by Interferometry II*, ESO Conference and Workshop Proceedings no. 39, European Southern Observatory, Garching-bei-München.

Bell, R. J. (1972). *Introductory Fourier Transform Spectroscopy*. Academic Press, New York.

Berry, M. V. (1984). Quantal Phase Factors Accompanying Adiabatic Changes, *Proc. R. Soc. London* A**392**, 45–57.

Berry, M. V. (1987). Interpreting the Anholonomy of Coiled Light, *Nature* **326**, 277.

Betzig, E., Trautman, J. K., Harris, T. D., Weiner, J. S. and Kostelak, R. L. (1991). Breaking the Diffraction Barrier: Optical Microscopy on a Nanometric Scale, *Science* **251**, 1468–70.

Betzig, E. and Trautman, J. K. (1992). Near-field Optics: Microscopy, Spectroscopy and Surface Modification beyond the Diffraction Limit, *Science* **257**, 189–95.

Binder, R. C. (1973). *Fluid Mechanics*, 5th edn. Prentice Hall.

Binnig, G. and Rohrer, H. (1985). The Scanning Tunnel Microscope, *Scient. Am.* Aug. 1985, p. 40.

Bloembergen, N. (1982). *Non-linear Optics*, 2nd edn. Benjamin, New York.

Blundell, H. and Johnson, L. N. (1976). *Protein Crystallography*. Academic Press, New York.

Born, M. and Wolf, E. (1980). *Principles of Optics*, 6th edn. Pergamon, Oxford.

Bracewell, R. N. (1986). *The Hartley Transform*. Oxford University Press, Oxford.

Brigham, E. O. (1988). *The Fast Fourier Transform and Applications*. Prentice Hall, New York.

476

Brillouin, L. (1960). *Wave Propagation and Group Velocity*. Academic Press, New York.

Brown, R. Hanbury (1974). *The Intensity Interferometer*. Taylor and Francis, London.

Brown, R. Hanbury and Twiss, R. Q. (1956). Correlation between photons in two beams of light, *Nature* **177**, 27–9.

Budden, K. G. (1966). *Radio Waves in the Ionosphere*. Cambridge University Press, Cambridge.

Chiao, R. Y., Antaramian, A., Ganga, K. M., Jaio, H. and Wilkinson, S. R. (1988). Observation of a Topological Phase by means of a Non-planar Mach–Zehnder Interferometer, *Phys. Rev. Letts* **60**, 1214 .

Chiao, R. Y., Kwiat, P. G. and Steinberg, A. M. (1993). Faster Than Light?, *Scient. Am.* p. 38.

Chow, W. W., Gea-Banacloche, J., Pedrotti, L. M., Sanders, V. E., Schleich, W. and Sculley, M. O. (1985). The Ring Laser Gyro, *Rev. Mod. Phys.* **57**, 61–104.

Clarke, D. and Grainger, J. F. (1971). *Polarized Light and Optical Measurement*. Pergamon Press, Oxford.

Cohen-Tannoudji, C., Diu, B. and Laloe, F. (1977). *Quantum Mechanics*. Wiley-Interscience, New York.

Collier, R. J., Burckhardt, C. B. and Lin, L. H. (1971). *Optical Holography*. Academic Press, New York.

Collin, R. E., (1991). *Field Theory of Guided Waves*, 2nd edn. IEEE, New York.

Combes, J. M., Grossman, A. and Tchamitchian, Ph. (1990) (eds). *Time–Frequency Methods and Phase Space*, 2nd edn. Springer, Berlin.

Cowley, J. M. (1984). *Diffraction Physics*, 2nd edn. North Holland, Amsterdam.

Desurvire, E. (1992). Lightwave Communications: the 5th Generation, *Scient. Am.* Jan. 1992, p. 114.

Ewald, P. P. (1962). *Fifty Years of X-ray Diffraction*. Oosthoek, Utrecht.

Fienup, J. R. (1982). Phase Retrieval Algorithms: A Comparison, *Appl. Opt.* **21**, 2758.

Fischer, B. and Sternklar, S. (1985). Image Transmission and Interferometry with Multimode Fibers Using Self-pumped Phase Conjugation. *Appl. Phys. Lett.* **46**, 113.

Foot, C. J. (1991). Laser Cooling and Trapping of Atoms, *Contemp. Phys.* **32**, 369.

Gasiorowicz, S. (1974). *Quantum Physics*. Wiley, New York.

Ghatak, A. and Thyagarajan, K. (1980). Graded-Index Optical Waveguides: A Review, in *Progress in Optics*, XVIII, p. 1. North Holland, Amsterdam.

Gloge, D. (1979). The Optical Fibre as a Transmission Medium, *Rep. Prog. Phys.* **42**, 1777.

Goldin, E. (1982). *Waves and Photons – an Introduction to Quantum Optics*. Wiley, New York.

Goodman, J. W. (1985). *Statistical Optics*. Wiley, New York.

Grant, I. S. and Phillips, W. R. (1975). *Electromagnetism*. Wiley, London.

Greenberger, D. M., Horne, M. A. and Zeikinger, A. (1993). Multiparticle Interferometry and the Superposition Principle, *Physics Today*, p. 22.

Guenther, R. (1990). *Modern Optics*. Wiley, New York.

Harburn, G. (1972). Optical Fourier Synthesis, p. 189 in *Optical Transforms*, ed. H. Lipson. Academic Press, London.

Harburn, G., Taylor, C. A. and Welberry, T. R. (1975). *Atlas of Optical Transforms*. G. Bell and Sons, London.

Hariharan, P. (1985). *Optical Interferometry*. Academic Press, Sydney.

Hariharan, P. (1989). *Optical Holography*. Cambridge University Press, Cambridge.

Haroche, S. and Kleppner, D. (1989). Cavity Quantum Electrodynamics, *Physics Today* **42**, 24–30.

Haroche, S. and Raimond, J. M. (1993). Cavity Quantum Electrodynamics, *Scient. Am.*, April 1993, p. 26.

Hecht, E. (1987). *Optics*, 2nd edn. Addison-Wesley, Reading MA.

Hauptman, H. A. (1991). The Phase Problem of X-ray Crystallography, *Rep. Prog. Phys.* **54**, 1427.

Hooke, R. (1665). *Micrographia*.

Howells, M. R., Kirz, J. and Sayre, D. (1991). X-ray Microscopes, *Scient. Am.* Feb. 1991, p. 42.

van de Hulst, H. C. (1984). *Scattering of Light by Small Particles*. Dover, New York.

Hutley, M. C. (1982). *Diffraction Gratings*. Academic Press, London.

Jacquinot, P. and Roizen-Dossier, B. (1964). Apodization, in *Progress in Optics*, ed. E. Wolf, vol. **III**, p. 29. North-Holland, Amsterdam.

Jackson, J. D. (1975). *Classical Electrodynamics*. Wiley, New York.

Jeans, J. (1982). *An Introduction to the Kinetic Theory of Gases*. Cambridge University Press, Cambridge.

Jhe, W., Anderson, A., Hinds, E. A., Meschede, D., Moi, L. and Haroche, S. (1987). Suppression of Spontaneous Decay at Optical Frequencies, *Phys. Rev. Lett.* **58**, 666–9.

Jones, R. and Wykes, C. (1989). *Holographic and Speckle Interferometry*. Cambridge University Press, Cambridge.

Keiser, G. (1991). *Optical Fiber Communications*, 2nd edn. McGraw Hill, New York.

Kellerman, K. I. and Thompson, A. R. (1988). Very Long Baseline Array, *Scient. Am.* Jan. 1988, p. 44.

Kim, B. Y. and Shaw, H. J. (1986). Fiber-Optic Gyroscopes, *IEEE Spectrum*, March 1986, p. 54.

Kimble, H. J., Dagenais, M. and Mandel, L. (1977). Photon Antibunching in Resonance Fluorescence, *Phys. Rev. Lett.* **39**, 691–5.

Kingslake, R. (1978). *Lens Design Fundamentals*. Academic Press, New York.

Kingslake, R. (1983). *Optical System Design*. Academic Press, Orlando.

Kittel, C. (1986). *Introduction to Solid-State Physics*, 6th edn. Wiley, New York.

Knox, K. T. and Thompson, B. J. (1974). Recovery of Images from Atmospherically-degraded Short-exposure Photographs, *The Astrophysical Journal* **193**, L45–8.

Korpel, A. (1988). *Acousto-Optics*. Marcel Dekker Inc., New York.

Krug, W., Rienitz, J. and Schulz, G. (1968). *Contributions to Interference Microscopy*, translated by J. H. Dickson. Hilger and Watts, London.

Labeyrie, A. (1976). High Resolution Techniques in Optical Astronomy, in *Progress in Optics* **XIV** p. 47, ed. E. Wolf. North Holland, Amsterdam.

Landau, L. D. and Lifshitz, E. M. (1980). *Statistical Physics*. Pergamon, Oxford.

Lefevre, H. (1993). *The Fiber-Optic Gyroscope*, Artech House, Boston.

Levenson, M. D. (1993). Wavefront Engineering for Microlithography, *Physics Today*, July, p. 26.

Lewis, A., Isaacson, M., Harootunian, A. and Murray, A. (1984). Development of a 500 Å Resolution Light Microscope: I – Light is Efficiently Transmitted through a $\lambda/6$ Diameter Aperture. *Ultramicroscopy* **13**, 227–30.

Liddell, H. M. (1981). *Computer-aided Techniques for the Design of Multilayer Filters*. Hilger, Bristol.

Lipson, H. (1968). *The Great Experiments of Physics*. Oliver and Boyd, Edinburgh.

Lipson, S. G. (1972). Optical Transforms in Teaching, p. 349 in *Optical Transforms*, ed. H. Lipson. Academic Press, London.

Lipson, S. G. (1990). Berry's Phase in Optical Interferometry – a Simple Interpretation, *Optics Letters* **15**, 154.

Loewen, E. G. (1983). Diffraction Gratings, Ruled and Holographic, in *Applied Optics and Optical Engineering*, Vol. **IX**, ed. R. R. Shannon and J. C. Wyant. Academic Press, New York.

Loudon, R. (1983), *The Quantum Theory of Light*, 2nd edn. Clarendon Press, Oxford.

Machida, S., Yamamoto, Y. and Itaya, Y. (1987). Observation of Amplitude Squeezing in a Constant-Current-Driven Semiconductor Laser, *Phys. Rev. Lett.* **58** 1000–3.

Macleod, H. A. (1989). *Thin-film Optical Filters*. Hilger, Bristol.

Marchand, E. W. (1978). *Gradient Index Optics*. Academic Press, London.

Marcuse, D. (1989) *Light Transmission Optics*, 2nd edn. van Nostrand-Reinhold, New York.

Marcuse, D. (1991a) *Theory of Dielectric Optical Waveguides*, 2nd edn. Academic Press, Boston.

Marcuse, D. (1991b). *Dielectric Optical Waveguides*, 2nd edn. Academic Press, San Diego CA.

Meystre, P. and Sargent, M. (1991). *Elements of Quantum Optics*, 2nd edn. Springer, Berlin.

Meystre, P. and Walls, D. F. eds (1991). *Non-classical Effects in Quantum Optics* (Key papers in physics, no. 4). American Institute of Physics, New York.

Michelson, A. A. (1927). *Studies in Optics*, reprinted by University of Chicago Press, 1962.

Michette, A. G. (1988). X-ray microscopy, *Rep. Prog. Phys.* **51**, 1525.

Morgan, B. L. and Mandel, L. (1966). Measurement of Photon Bunching in a Thermal Light Beam, *Phys. Rev. Lett.* **16**, 1012–5.

Moyen, P. and Paesler, M. (1995). *Near-field Optics.* Wiley Interscience, New York. (In press.)

Myers, H. P. (1990). *Introductory Solid-State Physics.* Taylor and Francis, London.

Paczyński, B. and Wambsganss, J. (1993). Gravitational Microlensing, *Physics World* **6**, no. 5, 26.

Parker, T. J. (1990). Dispersive Fourier Transform Spectroscopy, *Contemp. Phys.* **31**, 335.

Pearson, J. E., Freeman, R. H. and Reynolds, H. C. (1979). Adaptive Optical Techniques for Wavefront Correction, in *Applied Optics and Optical Engineering*, Vol. **VII**, ed. R. R. Shannon and J. C. Wyant. Academic Press, New York.

Pepper, D. M. (1986). Applications of Optical Phase Conjugation, *Scient. Am.* Jan. 1986, p. 56.

Pepper, D. M., Feinberg, J. and Kukhtarev, N. V. (1990). The Photorefractive Effect, *Scient. Am.* Oct. 1990, p. 34.

Peres, A. (1993). *Quantum Theory – Concepts and Methods.* Kluwer, Dordrecht.

Perley, R. A., Schwab, F. R. and Bridle, A. H. (1986). Synthesis Imaging (NRAO Summer School). NRAO, Green Bank WV, USA.

Pfleegor, R. L. and Mandel, L. (1968). Further Experiments on Interference of Independent Photon Beams at Low Light Levels, *J. Opt. Soc. Am.* **58**, 946.

Pohl, D. W., Denk, W. and Lanz, M. (1984). Optical Stethoscopy: Image Recording with Resolution $\lambda/20$, *Appl. Phys. Lett.* **44**, 651.

Rae, A. I. M. (1986). *Quantum Physics: Illusion or Reality?* Cambridge University Press, Cambridge.

Reynauld, S., Heidmann, A., Giacobino, E. and Fabre, C. (1992). Quantum Fluctuations in Optical Systems, in *Progress in Optics*, vol. **XXX**, ed. E. Wolf. North-Holland, Amsterdam.

Rohlfs, K. (1986). *Tools of Radio Astronomy.* Springer-Verlag, Berlin.

Segré, E. (1984). *From Falling Bodies to Radio Waves.* Freeman and Co., New York.

Schneider, P., Ehlers, J. and Falco, E. E. (1992). *Gravitational Lenses.* Springer, Berlin.

Smith, F. G. (1962). *Radioastronomy.* Penguin Books, Harmondsworth.

Squires, G. L. (1978). *Introduction to the Theory of Thermal Neutron Scattering.* Cambridge University Press, Cambridge.

Steel, W. H. (1983). *Interferometry.* Cambridge University Press, Cambridge.

Svelto, O. (1989). *Principles of Lasers*, 3rd edn. Plenum, New York.

Tapster, P. R., Rarity, J. G. and Satchell, J. S. (1987). Generation of Sub-Poissonian Light by High-Efficiency Light-Emitting Diodes, *Europhys. Lett.* **4**, 293.

Taylor, C. A. (1988). *The Art and Science of Lecture Demonstration.* Hilger, Bristol.

Taylor, C. A. and Lipson, H. (1964). *Optical Transforms.* G. Bell and Sons, London.

Teich, M. C. and Saleh, B. E. A. (1985). Observation of sub-Poisson Franck–Hertz Light, *J. Opt. Soc. Am.* **B2**, 275–82.

Teich, M. C. and Saleh, B. E. A. (1990). Squeezed and anti-bunched light, *Physics Today*, June 1990, p. 26.

Tolansky, S. (1973). An Introduction to Interferometry, 2nd edn. Wiley, New York.

Tomita, A. and Chiao, R. Y. (1986). Observation of Berry's Topological Phase by Use of an Optical Fiber, *Phys. Rev. Letts.* **57**, 937.

Tonomura, A. (1986). Electron Holography, in *Progress in Optics*, vol. **XXIII**, p. 183, ed. E. Wolf. North Holland, Amsterdam.

Toraldo di Francia, G. (1952). Super-gain Antennas and Optical Resolving Power, *Suppl. Nuovo Cimento* **9**, 426.

Tyson, R. K. (1991). *Principles of Adaptive Optics.* Academic Press, Boston.

Villasenor, J. and Bracewell, R. N. (1987). Optical phase obtained by analogue Hartley transformation, *Nature* **330**, 735.

Walker, J. S. (1988). *Fourier Analysis.* Oxford University Press, New York.

Watson, G. N. (1958). *A Treatise on the Theory of Bessel Functions*, 2nd edn. Cambridge University Press, Cambridge.

Weigelt, G. (1991). Triple-Correlation Imaging in Optical Astronomy, in *Progress in Optics*, vol. **XXIX**, ed. E. Wolf. North Holland, Amsterdam.

Welford, W. T. (1986). *Aberrations of Optical Systems.* Adam Hilger, Bristol.

Welford, W. T. and Winston, R. (1989). *High Collection Non-imaging Optics.* Academic Press, San Diego.

Wille, K. (1991). Synchrotron Radiation Sources, *Rep. Prog. Phys.* **54**, 1005.

Wilson, J. and Hawkes, J. F. B. (1989). *Optoelectronics – an Introduction*, 2nd edn. Prentice-Hall, Englewood Cliffs, New Jersey.

Wilson, T. and Sheppard, C. J. R. (1984). *Theory and Practice of Scanning Optical Microscopy.* Academic Press, London.

Wood, R. W. (1934). *Physical Optics*, MacMillan; 3rd edn. published by the Optical Society of America, 1989.

Woolfson, M. M. (1970). *X-ray Crystallography.* Cambridge University Press, Cambridge.

Woolfson, M. M. (1971). Direct Methods in Crystallography, *Rep. Prog. Phys.*, p. 369.

Wu, L., Kimble, H. J., Hall, J. L. and Wu, H. (1986). Generation of Squeezed States by Parametric Down-Conversion, *Phys. Rev. Lett.* **57** 2520–3.

Xiao, M., Wu, L. and Kimble, H. J. (1987). Precision Measurement beyond the Shot-Noise Limit, *Phys. Rev. Lett.* **59** 278–81.

Yariv, A. (1989). *Quantum Electronics*, 3rd edn. Wiley, New York.

Yariv, A. (1991). *Optical Electronics*, 4th edn. Holt, Reinhart and Winston, Philadelphia.

Yariv, A. and Yeh, P. (1984). *Optical Waves in Crystals*. Wiley-Interscience, New York.

Yablonovitch, E. (1993). Photonic Band Structures, *J. Opt. Soc. Am.* **B10**, 283–95.

Yurke, B. (1984). Use of Cavities in Squeezed-State Generation, *Phys. Rev.* A**29**, 408–10.

Zou, X. Y., Wang, L. J. and Mandel, L. (1991). Induced Coherence and Indistinguishability in Optical Interference, *Phys. Rev. Lett.* **67**, 318.

Index